T0279768

STRUCTURAL PROOF

Structural proof theory is a branch of logic that studies the general structure and properties of logical and mathematical proofs. This book is both a concise introduction to the central results and methods of structural proof theory and a work of research that will be of interest to specialists. The book is designed to be used by students of philosophy, mathematics, and computer science.

The book contains a wealth of new results on proof-theoretical systems, including extensions of such systems from logic to mathematics and on the connection between the two main forms of structural proof theory – natural deduction and sequent calculus. The authors emphasize the computational content of logical results.

A special feature of the volume is a computerized system for developing proofs interactively, downloadable from the web and regularly updated.

Sara Negri is Docent of Logic at the University of Helsinki.

Jan von Plato is Professor of Philosophy at the University of Helsinki and author of the successful *Creating Modern Probability* (Cambridge, 1994).

STRUCTURAL PROOF THEORY

SARA NEGRI

JAN VON PLATO

With an Appendix by Aarne Ranta

CAMBRIDGE UNIVERSITY PRESS
Cambridge, New York, Melbourne, Madrid, Cape Town, Singapore, São Paulo

Cambridge University Press
The Edinburgh Building, Cambridge CB2 8RU, UK

Published in the United States of America by Cambridge University Press, New York

www.cambridge.org
Information on this title: www.cambridge.org/9780521793070

© Sara Negri & Jan von Plato 2001

This publication is in copyright. Subject to statutory exception
and to the provisions of relevant collective licensing agreements,
no reproduction of any part may take place without the written
permission of the copyright holder.

First published 2001
This digitally printed version 2008

A catalogue record for this publication is available from the British Library

Library of Congress Cataloguing in Publication data
Negri, Sara, 1967–
Structural proof theory / Sara Negri, Jan von Plato.
p.   cm.
Includes bibliographical references.
ISBN 0-521-79307-6
1. Proof theory.   I. von Plato, Jan.   II. Title.
QA9.54. N44   2001
511.3 – dc21   00-040327

ISBN 978-0-521-79307-0 hardback
ISBN 978-0-521-06842-0 paperback

Contents

Preface

This book grew out of our fascination with the contraction-free sequent calculi. The first part, Chapters 1 to 4, is an introduction to intuitionistic and classical predicate logic as based on such calculi. The second part, Chapters 5 to 8, mainly presents work of our own that exploits the control over proofs made possible by the contraction-free calculi.

The first of the authors got her initial training in structural proof theory in a course given by Prof. Anne Troelstra in Amsterdam in 1992. The second author studied logic in the seventies, when by surprise Dag Prawitz mailed a copy of his book *Natural Deduction*. We thank them both for these intellectual stimuli, brought to fruition by the second author with a considerable delay. Since 1997, collaboration of the first author with Roy Dyckhoff has led us to the forefront of research in sequent calculi.

Dirk van Dalen, Roy Dyckhoff, and Glenn Shafer read all or most of a first version of the text. Other colleagues have commented on the manuscript or papers and talks on which part of the book builds, including Felix Joachimski, Petri Mäenpää, Per Martin-Löf, Ralph Matthes, Grigori Mints, Enrico Moriconi, Dag Prawitz, Anne Troelstra, Sergei Tupailo, René Vestergaard, and students from our courses, Raul Hakli in particular; we thank them all.

Aarne Ranta joined our book project in the spring of 1999, implementing in a short time a proof editor for sequent calculus. He also wrote an appendix describing the proof editor.

Little Daniel was with us from the very first day when the writing began. To him this book is dedicated.

Sara Negri
Jan von Plato

Introduction

The idea of mathematical proof is very old, even if precise principles of proof have been laid down during only the past hundred years or so. Proof theory was first based on axiomatic systems with just one or two rules of inference. Such systems can be useful as formal representations of what is provable, but the actual finding of proofs in axiomatic systems is next to impossible. A proof begins with instances of the axioms, but there is no systematic way of finding out what these instances should be. Axiomatic proof theory was initiated by David Hilbert, whose aim was to use it in the study of the consistency, mutual independence, and completeness of axiomatic systems of mathematics.

Structural proof theory studies the general structure and properties of mathematical proofs. It was discovered by Gerhard Gentzen (1909–1945) in the first years of the 1930s and presented in his doctoral thesis *Untersuchungen über das logische Schliessen* in 1933. In his thesis, Gentzen gives the two main formulations of systems of logical rules, **natural deduction** and **sequent calculus**. The first aims at a close correspondence with the way theorems are proved in practice; the latter was the formulation through which Gentzen found his main result, often referred to as Gentzen's "Hauptsatz." It says that proofs can be transformed into a certain "cut-free" form, and from this form general conclusions about proofs can be made, such as the consistency of the system of rules.

The years when Gentzen began his researches were marked by one great but puzzling discovery, Gödel's incompleteness theorem for arithmetic in 1931: Known principles of proof are not sufficient for deriving all of arithmetic; moreover, no single system of axioms and rules can be sufficient. Gentzen's studies of the proof theory of arithmetic led to **ordinal proof theory**, the general task of which is to study the deductive strength of formal systems containing infinitistic principles of proof. This is a part of proof theory we shall not discuss.

Of the two forms of structural proof theory that Gentzen gave in his doctoral thesis, natural deduction has remained remarkably stable in its treatment of rules of proof. Sequent calculus, instead, has been developed in various directions. One line leads from Gentzen through Ketonen, Kleene, Dragalin, and Troelstra to what are known as **contraction-free** systems of sequent calculus. Each of these logicians added some essential discovery, until a gem emerged. What it is can be only intimated at this stage: There is a way of organizing the principles of proof so that one can start from the theorem to be proved, then analyze it into simpler parts in a guided way. The gem is this "guided way"; namely, if one lays down what the last rule of inference was, the premisses of that last step are uniquely determined. Next, one goes on analyzing these premisses, and so on. Gentzen's basic discovery is reformulated as stating that a proof can be so organized that the premisses of each step of inference are always simpler than its conclusion. (To be more accurate, it can also happen that the premisses are not more complicated than the conclusion.)

Given a purported theorem, the question is whether it is provable or unprovable. In the first case, the task is to find a proof. In the second case, the task is to show that no proof can exist. How can we, then, prove unprovability? The possibility of such proofs depends crucially on having the right kind of calculus, and these proofs can take various forms: In the simplest cases we go through all the rules and find that none of them has a conclusion of the form of the claimed theorem. For certain classes of theorems, we can show that it makes no difference in what order we analyze the theorem to be proved. Each step of analysis leads to simpler premisses and therefore the process stops. From the way it stops we can decide if the conclusion really is a theorem or not. In other cases, it can happen that the premisses are at least as complicated as the conclusion, and we could go on indefinitely trying to find a proof. Some ingenious discovery is usually needed to prove unprovability, say, some analyses stop without giving a proof, and we are able to show that all of the remaining alternatives never stop and thus never give a proof.

One line of division in proof theory concerns the methods used in the study of the structure of proofs. In his original proof-theoretical program, Hilbert aimed at an "absolutely reliable" proof of consistency for formalized mathematics. The methods he thought acceptable had to be finitary, but the goal was shown to be unattainable already for arithmetic by Gödel's results. Later, parts of proof theory remained **reductive**, using different constructive principles, whereas other parts have studied proofs by unrestricted means. Most of our methods can be classified as reductive, but the reasons for restricted methods do not depend on arguments such as reliability. It is rather that we want results about systems of formal proof to have a computational significance. Thus it would not be sufficient to show by unrestricted means the mere existence of proofs with some desirable property. Instead, a constructive method for actually transforming any given proof so that it has the property is sought. From this point of view, our treatment of structural proof theory belongs, with a few exceptions, to what can be described as **computational proof theory**.

Since 1970, a branch of proof theory known as **constructive type theory** has been developed. A theorem typically states that a certain claim holds under given assumptions. The basic idea of type theory is that proofs are **functions** that convert any proofs of the assumptions of a theorem into a proof of its claim. A connection to computer science is established: In the latter, formal languages have been developed for constructing functions (programs) that act in a specified way on their input, but there has been no formal language for expressing what this specified way, the **program specification**, is. Logical languages, in turn, are suitable for expressing such specifications, but they have totally lacked a formalism for constructing functions that effect the task expressed by the specification. Constructive type theory unites specification language and programming language in a unified formalism in which the task of verifying the correctness of a program is the **same** as the logical task of controlling the correctness of a formal proof. We do not cover constructive type theory in detail, as another book would be needed for that, but some of the basic ideas and their connection to natural deduction and normalization procedures are explained in Appendix B.

At present, there are many projects in the territory between logic, mathematics, and computer science that aim at fully formalizing mathematical proofs. These projects use computer implementations of **proof editors** for the interactive development of formal proofs, and it cannot be said what all the things are that could come out of such projects. It has been observed that even the most detailed informal proofs easily contain gaps and cannot be routinely completed into formal proofs. More importantly, one finds imprecision in the conceptual foundation. The most optimistic researchers find that formalized proofs will become the standard in mathematics some day, but experience has shown formalization beyond the obvious results to be time-consuming. At present, proof editors are still far from being practical tools for the mathematician. If they gain importance in mathematics, it will be due to a change in emphasis through the development of computer science and through the interest in the computational content of mathematical theories. On the other hand, proof editors have been used for program verification even with industrial applications for some years by now. Such applications are bound to increase through the critical importance of program correctness.

Gentzen's structural proof theory has achieved perfection in pure logic, the predicate calculus. Intuitionistic natural deduction and classical sequent calculus are by now mastered, but the extension of this mastery beyond pure logic has been limited. A new approach that we exploit is to formulate mathematical theories as systems of **nonlogical rules** added to a suitable sequent calculus. As examples of proof analyses, theories of order, lattice theory, and plane affine geometry are treated. These examples indicate a way to an emerging field of study that could be called **proof theory in mathematics**. It is interesting to note that a large part of abstract mathematical reasoning seems to be finitary, thus not requiring any strong transfinite methods in proof analysis.

In the rest of this Introduction we comment on use of this book in teaching proof theory and what is new in it.

USE OF THIS BOOK IN TEACHING

Chapters 1–4 are based on courses in proof theory we have given at the University of Helsinki. The main objective of these courses was to give to the students a concise introduction to contraction-free intuitionistic and classical sequent calculi. The first author has also given a more specialized course on natural deduction, based on Chapter 1, the first two sections of Chapter 5, Chapter 8, and Appendices A and B.

The presentation is self-contained and the book should be readable without any previous knowledge of logic. Some familiarity with the topic, as in Van Dalen's *Logic and Structure*, will make the task less demanding.

Chapter 1 starts with general observations about logical languages and rules of inference. In a first version, we had defined logical languages through categorial grammars, but this was judged too difficult by most colleagues who read the text. With some reluctance, the categorial grammar approach was moved to Appendix A. Some traces of the definition of logical languages through an abstract syntax remained in the first section of Chapter 1, though.

The introduction rules of natural deduction are explained through the computational semantics of intuitionistic logic. A generalization of the inversion principle, to the effect that "whatever follows from the direct grounds for deriving a proposition, must follow from that proposition," determines the corresponding elimination rules. By the inversion principle, three rules, those of conjunction elimination, implication elimination, and universal elimination, obtain a form more general than the standard natural deduction rules. Using these general elimination rules, we are able to introduce sequent calculus rules as formalizations of the derivability relation in natural deduction. Contraction-free intuitionistic and classical sequent calculi are treated in detail in Chapters 2 and 3. These chapters work as a concise introduction to the central methods and general results of structural proof theory. The basic parts of structural proof theory use combinatorial reasoning and elementary induction on formula length, height of derivation, and so on, therefore perhaps giving an impression of easiness on the newcomer. The main difficulty, witnessed by the long development of structural proof theory, is to find the right rules. The first part of our text shows in what order structural proof theory is built up once those rules have been found. The second part of the book, Chapters 5–8, gives ample further illustration of the methodology. There is usually a large number of details, and a delicate order is required for putting things together, and mistakes happen. For such reasons, our first cut elimination theorem, in Chapter 2, considers, to our knowledge, absolutely all cases, even at the expense of perhaps being a bit pedantic.

In Chapter 3, following a suggestion of Gentzen, multisuccedent sequents are presented as a natural generalization of single succedent sequents into sequents with several (classical) open cases in the succedent. We find it important for students to avoid the denotational reading of sequents in favor of one in terms of formal proofs.

Chapter 4 contains a systematic treatment of quantifier rules in sequent calculi, again introduced through natural deduction and the general inversion principle.

Connected to the book is an **interactive proof editor** for developing formal derivations in sequent calculi. The system has been implemented by Aarne Ranta in the functional programming language Haskell. A description of the system, with instructions on how to access and use it, is given in Appendix C written by Ranta.

The proof editor serves several purposes: First, it makes the development of formal derivations in sequent calculi less tedious, thereby helping the student. It also checks the formal correctness of derivations. The user can give axiomatic systems to the editor that converts them into systems of nonlogical rules of inference by which the logical sequent calculi are extended. Formal derivations are quite feasible to develop in those extensions we have so far studied. Even though the extensions need not permit a terminating proof search, the user will soon notice how neatly the search space can become limited, often into one or two applicable rules only, or no rules at all, which establishes underivability.

The proof editor produces provably correct LaTeX code, with the advantage that the rewriting of parts of sequents is not needed. The editor is in its early stages; more is hoped to be included in later releases, including translation algorithms between various calculi, cut elimination algorithms, a natural language interface, and so on.

Exercises, mostly to Chapters 1–4 and 8, can be found in the book's home page (see p. 243). We welcome suggestions for further exercises. Basic exercises are just derivations of formulas in the various calculi. Other exercises consist in filling in details of proofs of theorems. Another type of task, for those conversant with the Haskell language, is to formalize theorems about sequent calculi. Since these theorems are, almost without exception, proved constructively in the book, their formalizations give proof-theoretical algorithms for the transformation of proofs. An example is the proof of Glivenko's theorem in Section 5.4.

Through the use of contraction-free sequent calculi, it is possible for students to find proofs of results that were published as research results only a few decades ago, say, Harrop's theorem in Section 2.5. This should give some idea of what a powerful tool is being put into their hands.

Finally, a description of what is new in this book (for the expert, mostly).

WHAT IS NEW IN THIS BOOK?

Chapter 1 contains a generalization of the inversion principle of Gentzen and Prawitz, one that leads to elimination rules that are more general than the usual ones. Contrary to earlier inversion principles that only justify the elimination rules, our principle actually determines what the elimination rules corresponding to given introduction rules should be. The elimination rules are all of the form of disjunction elimination, with an arbitrary consequence.

Starting from natural deduction with general elimination rules, the rules of sequent calculus are presented in Section 1.3 as straightforward formalizations of the derivability relation of natural deduction.

Section 3.3 gives a proof of completeness of the contraction-free invertible sequent calculus for classical propositional logic known as **G3c**-calculus in the literature. The proof is an elaboration of Ketonen's original 1944 proof. It uses a novel notion of validity defined as a negative concept, the inexistence of a refuting valuation.

Chapter 4 contains proofs of height-preserving α-conversion and the substitution lemma that we have not found done elsewhere in such detail. Section 4.4 gives a proof of completeness of classical predicate calculus worked out for the definition of validity as a negative notion.

Chapter 5 studies various sequent calculi, most of which are new to the literature. Section 5.1 gives a sequent calculus with independent contexts in all two-premiss rules and explicit rules of weakening and contraction. The calculus is motivated by the independent treatment of assumptions in natural deduction. A classical multisuccedent version is also given. Proofs of cut elimination for these calculi are given that do not use Gentzen's mix rule, or rule of multicut. In Section 5.2, the calculi of Section 5.1 are modified so that weakening and contraction are treated implicitly as in natural deduction. Section 5.4 gives a single succedent calculus for classical propositional logic based on a formulation of the law of excluded middle as a sequent calculus rule. A proof of Glivenko's theorem is given through an explicit proof transformation. It is shown that in the derivation of a sequent $\Gamma \Rightarrow C$, the rule can be restricted to atoms of C, from which a full subformula property follows.

Chapter 6 studies extensions of logical sequent calculi by nonlogical rules corresponding to axioms. Contrary to widespread belief, it is possible to add axioms to sequent calculus as rules of a suitable form while maintaining the eliminability of cut. These extensions have no structural rules, which gives a strong control over the structure of possible derivations. As a first application, a formulation of predicate calculus with equality as a cut-free system of rules is given. It is proved through an explicit proof transformation that predicate logic with equality is conservative over predicate logic. It is essential for the proof that no cuts, even on atoms, be permitted. In Section 6.6, examples of structural proof analysis in mathematics are given. Topics covered include intuitionistic and classical theories of order, lattice theory, and affine geometry. The last one goes beyond the expressive means of first-order logic, but the methods of structural proof analysis of the previous chapters still apply. As an example of such analysis of derivations without structural rules, a proof of independence of Euclid's fifth postulate in plane affine geometry is given.

In Chapter 8 the structure of derivations in natural deduction with general elimination rules is studied. A uniform definition of normality of derivations is given: A derivation is normal when all major premisses of elimination rules are assumptions. This structure follows from the applicability of permutation conversions to all cases in which the major premiss of an elimination rule is concluded by another elimination rule. Translations are given that establish isomorphism of normal derivations and cut-free derivations in the sequent calculus with independent contexts of Chapter 5. It is shown what the role of the structural rules of sequent calculus is in terms of natural deduction: Weakening and contraction in sequent calculus correspond to the vacuous and multiple discharge of assumptions in natural deduction. Cuts in which the cut formula is principal in the right premiss correspond to such steps of elimination in which the major premiss has been derived. (No other cuts can be expressed in terms of natural deduction.) A translation from non-normal derivations to derivations with cuts is given, from which follows a normalization procedure consisting of said translation, followed by cut elimination and translation back to natural deduction.

In the Conclusion, a uniform logical calculus is given that encompasses both sequent calculus and natural deduction.

1

From Natural Deduction to Sequent Calculus

We first discuss logical languages and rules of inference in general. Then the rules of natural deduction are presented, with introduction rules motivated by meaning explanations and elimination rules determined by an inversion principle. A way is found from the rules of natural deduction to those of sequent calculus. In the last section, we discuss some of the main characteristics of structural proof analysis in sequent calculus.

1.1. LOGICAL SYSTEMS

A logical system consists of a formal language and a system of axioms and rules for making logical inferences.

(a) Logical languages: A logical language is usually defined through a set of inductive clauses for well-formed formulas. The idea is that expressions of a formal language are special sequences of symbols from a given alphabet, as generated by the inductive definition. An alternative way of defining formal languages is through **categorial grammars**. Such grammars are well-known for natural languages, and categorial grammars for formal languages are in use with programming languages, but not so often in logic.

Under the first approach, expressions of a logical language are formulas defined inductively by two clauses: *1*. A statement of what the **prime** formulas are. These are formulas that contain no other formulas as parts. *2*. A statement of what the **compound** formulas are. These are built from other simpler formulas by logical connectives, and their definition requires reference to **arbitrary** formulas and how these can be put together with the symbols for connectives to give new formulas. Given a formula, we can find out how it was put together from other formulas and a logical connective. Parentheses may be needed to indicate the composition uniquely. Then we can find out how the parts were obtained until we arrive at the prime formulas. Thus, in the end, all formulas consist of prime formulas, logical connectives, and parentheses.

1

We shall define the language of **propositional logic**:

1. The prime formulas are the **atomic** formulas denoted by
$P, Q, R, \ldots$, and **falsity** denoted by $\perp$.

2. If A and B are formulas, then the **conjunction** $A\&B$, **disjunction**
$A \vee B$, and **implication** $A \supset B$ are formulas.

For unique readability of formulas, the components should always be put in
parentheses but in practice these are left out if a conjunction or disjunction is a
component of an implication. Often $\perp$ is counted among the atomic formulas,
but this will not work in proof theory. It is best viewed as a zero-place connec-
tive. **Negation** $\sim A$ and **equivalence** $A \supset\subset B$ are defined as $\sim A = A \supset \perp$ and
$A \supset\subset B = (A \supset B)\&(B \supset A)$.

Expressions of a language should express something, not just be strings of
symbols from an alphabet put together correctly. In logic, the thing expressed
is called a **proposition**. Often, instead of saying "proposition expressed by formula A" one says simply "proposition A." There is a long-standing debate in
philosophy on what exactly propositions are. When emphasis is on logic, and not
on what logic in the end of a philosophical analysis is, one considers expressions
in the formal sense and talks about formulas.

In recent literature, the definition of expressions as sequences of symbols is referred to as **concrete syntax**. Often it is useful to look at expressions from another
point of view, that of **abstract syntax**, as in categorial grammar. The basic idea
of categorial grammar is that expressions of a language have a **functional structure**. For example, the English sentence *John walks* is obtained by representing
the intransitive verb *walk* as a function from the category of noun phrases *NP* to
the category of sentences S, in the usual notation for functions, $walk : NP \to S$.
John is an element of the category *NP* and **application** of the function *walk* gives
as value *walk(John)*, an element of the category of sentences S. One further step
of **linearization** is required for hiding the functional structure, to yield the original sentence *John walks*. In logic and mathematics, no consideration is given
to differences produced by this last stage, nor to differences in the grammatical
construction of sentences. Since Frege, one considers only the logical content of
the functional structure.

We shall briefly look at the definition of propositional logic through a categorial
grammar. There is a **basic category** of propositions, designated *Prop*. The atomic
propositions are introduced as parameters $P, Q, R, \ldots$, with no structure and with
the categorizations

$$P : Prop, \quad Q : Prop, \quad R : Prop, \quad \ldots$$

and similarly for falsity, $\perp : Prop$. The connectives are two-place functions for
forming new propositions out of given ones. Application of the function & to the

two arguments A and B gives the proposition $\&(A, B)$ as value, and similarly for $\vee$ and $\supset$. The functional structure is usually hidden by an **infix** notation and by the dropping of parentheses, $A \& B$ for $\&(A, B)$, and so on. This will create an ambiguity not present in the purely functional notation, such as $A \& B \supset C$ that could be both $\&(A, \supset (B, C))$ and $\supset (\&(A, B), C)$. As mentioned, we follow the convention of writing $A \& (B \supset C)$ for the former and $A \& B \supset C$ for the latter, and in general, having conjunction and disjunction bind more strongly than implication. Appendix A explains in more detail how logical languages are treated from the point of view of categorial grammar.

Neither approach, inductive definition of strings of symbols nor generation of expressions through functional application, reveals what is special about logical languages. Logical languages of the present day have arisen as an abstraction from the informal language of mathematics. The first work in this direction was by Frege, who invented the language of **predicate logic**. It was meant to be, wrote Frege, "a formula language for pure thought, modeled upon that of arithmetic." Later Peano and Russell developed the symbolism further, with the aim of formalizing the language of mathematics. These pioneers of logic tried to give definitions of what logic is, how it differs from mathematics, and whether the latter is reducible to the former, or if it is perhaps the other way around.

From a practical point of view there is a clear understanding of what logical languages are: The prime logical languages are those of propositional and predicate logic. Then there are lots of other logical languages more or less related to these. Logic itself is, from this point of view, what logicians study and develop. Any general definition of logic and logical languages should respect this situation.

An essential aspect of logical languages is that they are **formal** languages, or can easily be made into such, an aspect made all the more important by the development of computer science. There are many connections between logical languages and programming languages; in fact, logical and programming languages are brought together in one language in some recent developments, as explained in Appendix B.

(b) **Rules of inference:** Rules of inference are of the form: "If it is the case that A and B, then it is the case that C." Thus they do not act on propositions, but on **assertions**. We obtain an assertion from a proposition A by adding something to it, namely, an assertive mood such as "it is the case that A." Frege used the assertion sign $\vdash A$ to indicate this but usually the distinction between propositions and assertions is left implicit. Rules seemingly move from given propositions to new ones.

In **Hilbert-style** systems, also called **axiomatic systems**, we have a number of basic forms of assertion, such as $\vdash A \supset A \vee B$ or $\vdash A \supset (B \supset A)$. Each instance of these forms can be asserted, and in the case of propositional logic there is just one rule of inference, of the form

$$\frac{\vdash A \supset B \quad \vdash A}{\vdash B}$$

Derivations start with instances of axioms that are decomposed by the rule until the desired conclusion is found.

In **natural deduction** systems, there are only rules of inference, plus **assumptions** to get derivations started, exemplified by

$$\frac{\vdash A \quad \vdash B}{\vdash A\&B} \qquad \frac{\begin{array}{c}[\vdash A]\\ \vdots\\ \vdash B\end{array}}{\vdash A \supset B}$$

Instances of the first rule are single-step inferences, and if the premisses have been derived from some assumptions, the conclusion depends on the same assumptions. In the second rule instead, in which the vertical dots indicate a derivation of $\vdash B$ from $\vdash A$, the assumption $\vdash A$ is **discharged** at the inference line, as indicated by the square brackets, so that $\vdash B$ above the inference line depends on $\vdash A$ whereas $\vdash A \supset B$ below it does not.

In **sequent calculus** systems, there are no temporary assumptions that would be discharged, but an explicit listing of the assumptions on which the derived assertion depends. The derivability relation, to which reference was made in natural deduction by the four vertical dots, is an explicit part of the formal language, and sequent calculus can be seen as a formal theory of the derivability relation.

Of the three types of systems, the first, axiomatic, has some good properties that are due to the presence of only one rule of inference. However, it is next to impossible to actually use the axiomatic approach because of the difficulty of finding the instances of axioms to start with. Systems of the second type correspond to the usual way of making inferences in mathematics, with a good sense of structure. Systems of the third type are the ones that permit the most profound analysis of the structure of proofs, but their actual use requires some practice. Moreover, the following is possible in natural deduction and in sequent calculus:

> *Two systems of rules can be equivalent in the sense that the same assertions can be derived in them, but the addition of the same rule to each system can destroy the equivalence.*

This lack of modularity will not occur with the axiomatic Hilbert-style systems.

Once a system of rules of logical inference has been put up, it can be considered from the formal point of view. The assertion sign is left out, and rules of inference are just ways of writing a formula under any formula or formulas that have the form of the premisses of the rules. In a complete formalization of logic, also the formation of propositions is presented as the application of rules of proposition formation. For example, conjunction formation is application of the rule

$$\frac{A : Prop \quad B : Prop}{A\&B : Prop}$$

Rules of inference can be formalized in the same way as rules of proposition formation: They are represented as functions that take as arguments formal proofs of the premisses and give as value a formal proof of the conclusion. A hierarchy of functional categories is obtained such that all instances of rules of proposition formation and of inference come out through functional application. This will lead to **constructive type theory**, described in more detail in Appendix B.

The viewpoint of proof theory is that logic is the theory of correct demonstrative inference. Inferences are analyzed into the most basic steps, the formal correctness of which can be easily controlled. Moreover, the semantical justification of inferences can be made compositional through the justification of the individual steps of inference and how they are put together.

Compound inferences are synthesized by the composition of basic steps of inference. A system of rules of inference is used to give an inductive, formal definition of the notion of **derivation**. Derivability then means the existence of a derivation. The correctness of a given derivation can be mechanically controlled through its inductive definition, but the finding of derivations typically is a different matter.

1.2. NATURAL DEDUCTION

Natural deduction embodies the operational or computational meaning of the logical connectives and quantifiers. The meaning explanations are given in terms of the **immediate grounds** for asserting a proposition of corresponding form. There can be other, less direct grounds, but these should be reducible to the former for a coherent operational semantics to be possible. The "BHK-conditions" (for Brouwer–Heyting–Kolmogorov), which give the explanations of logical operations of propositional logic in terms of **direct provability** of propositions, can be put as follows:

1. A direct proof of the proposition $A\&B$ consists of proofs of the propositions A and B.

2. A direct proof of the proposition $A \vee B$ consists of a proof of the proposition A or a proof of the proposition B.

3. A direct proof of the proposition $A \supset B$ consists of a proof of the proposition B from the assumption that there is a proof of the proposition A.

4. A direct proof of the proposition $\perp$ is impossible.

In the third case it is only **assumed** that there is a proof of A, but the proof of the conclusion $A \supset B$ does not depend on this assumption temporarily made in order to reduce the proof of B into a proof of A. Proof here is an informal notion. We shall gradually replace it by the formal notion of derivability in a given system of rules.

We can now make more precise the idea that rules of inference act on assertions; namely, an assertion is warranted if there is a proof available, and therefore, on a formal level, rules of inference act on derivations of the premises, to yield as value a derivation of the conclusion. From the BHK-explanations, we arrive at the following **introduction rules**:

$$\frac{A \quad B}{A \& B} \, \&I \qquad \frac{A}{A \vee B} \, \vee I_1 \qquad \frac{B}{A \vee B} \, \vee I_2 \qquad \frac{\begin{array}{c}[A]\\ \vdots \\ B\end{array}}{A \supset B} \, \supset I$$

The assertion signs are left out. (There will be another use for the symbol soon.) In the last rule the auxiliary assumption A is discharged at the inference, which is indicated by the square brackets. We have as a special case of implication introduction, with $B = \bot$, an introduction rule for negation. There is no introduction rule for $\bot$.

There will be **elimination rules** corresponding to the introduction rules. They have a proposition of one of the three forms, conjunction, disjunction, or implication, as a **major premiss**. There is a general principle that helps to find the elimination rules: We ask what the conditions are, in addition to assuming the major premiss derived, that are needed to satisfy the following:

> **Inversion principle:** *Whatever follows from the direct grounds for deriving a proposition must follow from that proposition.*

For conjunction $A \& B$, the direct grounds are that we have a derivation of A and a derivation of B. Given that C follows when A and B are assumed, we thus find, through the inversion principle, the elimination rule

$$\frac{A \& B \qquad \begin{array}{c}[A, B]\\ \vdots \\ C\end{array}}{C} \, \&E$$

The assumptions A and B from which C was derived are discharged at the inference. If in a derivation the premises A and B of the introduction rule have been derived and C has been derived from A and B, the derivation

$$\frac{\begin{array}{cc} \vdots & \vdots \\ A & B \end{array}}{A \& B} \&I \qquad \frac{\begin{array}{c} [A, B] \\ \vdots \\ C \end{array}}{C} \&E$$

converts into a derivation of C without the introduction and elimination rules,

$$\begin{array}{cc} \vdots & \vdots \\ A & B \\ & \vdots \\ & C \end{array}$$

Therefore, if $\&I$ is followed by $\&E$, the derivation can be simplified.

For disjunction, we have two cases. Either $A \vee B$ has been derived from A and C is derivable from assumption A, or it has been derived from B and C is derivable from assumption B. Taking into account that both cases are possible, we find the elimination rule

$$\frac{A \vee B \qquad \begin{array}{c} [A] \\ \vdots \\ C \end{array} \qquad \begin{array}{c} [B] \\ \vdots \\ C \end{array}}{C} \vee E$$

Assume now that A or B has been derived. If it is the former and if C is derivable from A and C is derivable from B, the derivation

$$\frac{\frac{\begin{array}{c} \vdots \\ A \end{array}}{A \vee B} \vee I_1 \qquad \begin{array}{c} [A] \\ \vdots \\ C \end{array} \qquad \begin{array}{c} [B] \\ \vdots \\ C \end{array}}{C} \vee E$$

converts into a derivation of C without the introduction and elimination rules,

$$\begin{array}{c} \vdots \\ A \\ \vdots \\ C \end{array}$$

In the latter case of B having been derived, the conversion is into

$$\begin{array}{c} \vdots \\ B \\ \vdots \\ C \end{array}$$

Again, an introduction followed by the corresponding elimination can be removed from the derivation.

The elimination rule for implication is harder to find. The direct ground for deriving $A \supset B$ is the existence of a **hypothetical** derivation of B from the assumption A. The fact that C can be derived from the existence of such a derivation can be expressed by:

> *If C follows from B, then it already follows from A.*

This is achieved precisely by the elimination rule

$$\frac{A \supset B \quad A \quad \begin{matrix}[B]\\ \vdots \\ C\end{matrix}}{C} \supset E$$

In addition to the major premiss $A \supset B$, there is the **minor premiss** A in the $\supset E$ rule. If B has been derived from A and C from B, the derivation

$$\frac{\dfrac{\begin{matrix}[A]\\ \vdots \\ B\end{matrix}}{A \supset B}\supset I \quad A \quad \begin{matrix}[B]\\ \vdots \\ C\end{matrix}}{C} \supset E$$

converts into a derivation of C from A without the introduction and elimination rules,

$$\begin{matrix}\vdots \\ A \\ \vdots \\ B \\ \vdots \\ C\end{matrix}$$

Finally we have the zero-place connective $\perp$ that has no introduction rule. The immediate grounds for deriving $\perp$ are empty, and we obtain as a limiting case of the inversion principle the rule of **falsity elimination** ("ex falso quodlibet") that has only the major premiss $\perp$:

$$\frac{\perp}{C} \perp E$$

We have still to tell how to get derivations started. This is done by the **rule of assumption** that permits us to begin a derivation with any formula. In a given derivation tree, those formula occurrences are assumptions, or more precisely,

open assumptions, that are neither conclusions nor discharged by any rule. Discharged assumptions are also called **closed** assumptions.

Rules $\&E$ and $\supset E$ are usually written for only the special cases of $C = A$ and $C = B$ for $\&E$ and $C = B$ for $\supset E$, as follows:

$$\frac{A\&B}{A} \,_{\&E_1} \qquad \frac{A\&B}{B} \,_{\&E_2} \qquad \frac{A \supset B \quad A}{B} \,_{\supset E}$$

These "special elimination rules" correspond to a more limited inversion principle, one requiring that elimination rules conclude the immediate grounds for deriving a proposition instead of arbitrary consequences of these grounds. The first two rules just conclude the premises of conjunction introduction. The third gives a one-step derivation of B from A by a rule that is often referred to as "modus ponens." The more limited inversion principle suffices for justifying the special elimination rules but is not adequate for determining what the elimination rules should be. In particular, it says nothing about $\perp E$.

The special elimination rules have the property that their conclusions are **immediate subformulas** of their premises. With conjunction introduction, it is the other way around: The premises are immediate subformulas of the conclusion. Further, in implication introduction, the formula above the inference line is an immediate subformula of the conclusion. It can be shown that derivations with conjunction and implication introduction and the special elimination rules can be transformed into a **normal form**. The transformation is done by **detour conversions**, the removal of applications of introduction rules followed by corresponding elimination rules. In a derivation with special elimination rules in normal form, first, assumptions are made, then elimination rules are used, and finally, introduction rules are used. This simple picture of normal derivations, moving by elimination rules from assumptions to immediate subformulas and then by introductions the other way around, is lost with the disjunction elimination rule. However, when all elimination rules are formulated in the general form, a uniform subformula structure for natural deduction derivations in normal form is achieved. The normal form itself is characterized by the following property:

> **Normal form:** *A derivation in natural deduction with general elimination rules is in normal form if all major premises of elimination rules are assumptions.*

In general, it need not be the case that a system of natural deduction admits conversion to normal form, but it is the aim of structural proof theory to find systems that do. There is a series of properties of growing strength relating to normal form, the weakest being the **existence of normal form**. This property states that if a formula A is derivable in a system, there exists also a derivation of A in normal form. Secondly, we have the concept of **normalization**: A procedure is given for actually converting any given derivation into normal form. The next notion is **strong normalization:** The application of conversions to a non-normal derivation in any order whatsoever **terminates** after some number of steps in a derivation in normal form. Last, we have the notion of **uniqueness** of normal form: The process of normalization of a given derivation always terminates with the same normal proof. Note that it does not follow that there would be only one normal derivation of a formula, for different non-normal derivations would in general terminate in different normal derivations.

The conjunction and the disjunction introduction rules, as well as the special elimination rules for conjunction and implication, are simple one-step inferences. The rest of the rules are schematic, with "vertical dots" that indicate derivations with assumptions. The behavior of these assumptions is controlled by **discharge functions**: Each assumption gets a number and the discharge of assumptions is indicated by writing the number next to the inference line. Further, the discharge is optional, i.e., we can, and indeed sometimes must, leave an assumption open even if it could be discharged.

Some examples will illustrate the management of assumptions and point at some peculiarities of natural deduction derivations.

Example 1:

$$\frac{\overset{1.}{[A]}}{A \supset A} \supset I, 1.$$

The rule schemes of natural deduction display only the open assumptions that are **active** in the rule, but there may be any number of other assumptions. Thus the conclusion may depend on a whole set Γ of assumptions, which can be indicated by the notation $\Gamma \vdash A$. Now the rule of implication introduction can be written as

$$\frac{\Gamma \vdash B}{\Gamma - \{A\} \vdash A \supset B} \supset I$$

In words, if there is a derivation of B from the set of open assumptions Γ, there is a derivation of $A \supset B$ from assumptions Γ minus $\{A\}$. In this formulation there is a "compulsory" discharge of the assumption A. All the other rules of natural deduction can be written similarly. We give two examples:

$$\frac{\Gamma \vdash A \quad \Delta \vdash B}{\Gamma \cup \Delta \vdash A \& B} \&I \qquad \frac{\Gamma \vdash A \vee B \quad \Delta \cup \{A\} \vdash C \quad \Theta \cup \{B\} \vdash C}{\Gamma \cup \Delta \cup \Theta \vdash C} \vee E$$

The resulting system of inference, introduced by Gentzen in 1936, is usually known as "natural deduction in sequent calculus style." It can be used to clarify the strange-looking derivation of Example 1: The assumption of A is written as

$A \vdash A$, and we have the derivation

$$\frac{A \vdash A}{\vdash A \supset A} \supset I$$

The first occurrence of A has the set of assumptions $\Gamma = \{A\}$, and so (dropping the braces around singleton sets) the conclusion has the set of assumptions $A - A = \emptyset$.

Example 2 shows how superfluous assumptions can be added to **weaken** the consequent A of the first example into $B \supset A$.

Example 2:

$$\frac{\dfrac{\overset{1.}{[A]}}{\dfrac{B \supset A}{A \supset (B \supset A)}} \supset I}{} \supset I,1.$$

The first inference step is justified by the rule about sets of assumptions: $A - B = A$. There is a **vacuous discharge** of B in the first instance of $\supset I$, and discharge of A takes place only at the second instance of $\supset I$. Note that there is a problem here in the case of $B = A$, for compulsory discharge dictates that A is discharged at the first inference, the second becoming a vacuous discharge. The instance of the derivation with $B = A$ is not a syntactically correct one; therefore the original derivation cannot be correct either. Chapter 8 will give a method for handling the discharge of assumptions, the **unique discharge principle**, that does not lead to such problems.

In sequent calculus style, the derivation is

$$\frac{\dfrac{A \vdash A}{A \vdash B \supset A} \supset I}{\vdash A \supset (B \supset A)} \supset I$$

Example 3 gives a derivation that cannot be done with just a single use of assumption A.

Example 3:

$$\frac{\dfrac{\dfrac{\overset{2.}{[A \supset (A \supset B)]} \quad \overset{1.}{[A]}}{A \supset B} \supset E \quad \overset{1.}{[A]}}{\dfrac{B}{A \supset B} \supset I,1.} \supset E}{(A \supset (A \supset B)) \supset (A \supset B)} \supset I,2.$$

Assumption A had to be made twice, and there is correspondingly a **multiple discharge** at the first instance of $\supset I$ with both occurrences of assumption A discharged. Note the "nonlocality" of derivations in natural deduction: To control the correctness of inference steps in which assumptions can be discharged, we have to look higher up along derivation branches. (This will be crucial later with the variable restrictions in quantifier rules.) In sequent calculus style, instead, each step of inference is local:

$$\cfrac{\cfrac{\cfrac{A \supset (A \supset B) \vdash A \supset (A \supset B) \quad A \vdash A}{A \supset (A \supset B), A \vdash A \supset B} \supset E \quad A \vdash A}{\cfrac{A \supset (A \supset B), A \vdash B}{A \supset (A \supset B) \vdash A \supset B} \supset I} \supset E}{\vdash (A \supset (A \supset B)) \supset (A \supset B)} \supset I$$

In implication elimination, a rule with two premises, the assumptions from the left of the turnstile are collected together. At the second implication elimination of the derivation, a second occurrence of A in the assumption part is produced. The trace of this repetition disappears, however, when assumptions are collected into sets.

The above system of introduction and elimination rules for &, $\vee$, and $\supset$, together with the rule of assumption by which an assumption can be introduced at any stage in a derivation, is the system of natural deduction for **minimal** propositional logic. If we add rule $\perp E$ to it we have a system of natural deduction rules for **intuitionistic** propositional logic.

We obtain **classical** propositional logic by adding to the rules of intuitionistic logic a rule we call, in analogy to the law of excluded middle characteristic of classical logic in an axiomatic approach, the **rule of excluded middle**:

$$\cfrac{\begin{matrix} [A] & [\sim A] \\ \vdots & \vdots \\ C & C \end{matrix}}{C} \; Em$$

Both A and $\sim A$ are discharged at the inference. The law of excluded middle, $A \vee \sim A$, is derivable with the rule:

$$\cfrac{\cfrac{[A]}{A \vee \sim A} \vee I_1 \quad \cfrac{[\sim A]}{A \vee \sim A} \vee I_2}{A \vee \sim A} \; Em$$

The rule of excluded middle is a generalization of the **rule of indirect proof** ("reductio ad absurdum"),

$$\cfrac{\begin{matrix} [\sim A] \\ \vdots \\ \perp \end{matrix}}{A} \; Raa$$

The properties of the classical rules Em and Raa are presented in Chapter 8.

Rules of natural deduction can be categorized in a way similar to rules of proposition formation. This is based on the **propositions-as-sets** principle, and leads to **type systems**. We think of a proposition A as being the same as its set of formal proofs. Each such proof can be called a **proof-object** or **proof term** to emphasize that this special notion of proof is intended. Instead of an assertion of the form $\vdash A$ we have $a : A$, a is a proof-object for A. Rules of inference are categorized as functions operating on proof-objects.

Type-theoretical rules for proof-objects validate the BHK-explanations, by showing how proof-objects of compound propositions are constructed from proof-objects of their constituents. For example, the proof of an implication $A \supset B$ is a function that converts an arbitrary proof of A into some proof of B. In earlier times, the explanation of a proof of an implication $A \supset B$ was described as "a method that converts proofs of A into proofs of B," and this was thought to be circular or at least ill-founded through its reference to an arbitrary proof of A. However, in constructive type theory, the problem is solved.[1] The meaning explanations first concern only "canonical proofs," that is, the direct proofs of the forms given by the introduction rules. All other proofs, the "noncanonical" ones, are reduced to the canonical proofs through **computation rules** that correspond to the conversions in natural deduction. For this process to be well-founded, it is required that the conversion from noncanonical to canonical form terminate. These notions have deep connections to the structural properties of natural deduction derivations.

An exposition of type theory and its relation to natural deduction is given in Appendix B.

1.3. FROM NATURAL DEDUCTION TO SEQUENT CALCULUS

If our task is to derive $A \supset B$, rule $\supset I$ reduces the task to one of deriving B under the assumption A. So we assume A, but if B in turn is of the form $C \& D$, rule $\& I$ shows how the derivation of $C \& D$ is reduced to that of C and D. Thus we have to mentally decompose the goal $A \supset B$ into subgoals, but there is no formal way to keep track of the process. It is as if we had to construct a derivation backwards.

Sequent calculus corrects the lack of guidance of natural deduction. It has a notation for keeping track of open assumptions; moreover, this is local: Each formula C has the open assumptions Γ that it depends on listed on the same line, as follows:

$$\Gamma \Rightarrow C$$

---

[1] The explanation was rejected on these grounds by Gödel (1941), for example. The solution was given, in philosophical terms, by Dummett (1975) and more formally by Martin-Löf (1975).

Sequent calculus is a formal theory of the **derivability relation**. To make a difference to writing $\Gamma \vdash C$, where the turnstile is a metalevel expression, not part of the syntax as are the formulas, we use the now common symbol $\Rightarrow$. In $\Gamma \Rightarrow C$, the left side Γ is called the **antecedent** and C the **succedent**.

As mentioned, the rules of natural deduction are schematic and show only the active formulas, leaving implicit the set of remaining open assumptions. For example, the rule of conjunction introduction can be written more completely as follows, with a derivation of A with open assumptions Γ and a derivation of B with open assumptions Δ:

$$\begin{array}{cc} \Gamma & \Delta \\ \vdots & \vdots \\ A & B \\ \hline \multicolumn{2}{c}{A \& B} \end{array} \,\&I$$

Rule $\&I$ gives a derivation of $A \& B$ with open assumptions $\Gamma \cup \Delta$. With implication, we have a derivation of B from A and Γ, and the introduction rule gives a derivation of $A \supset B$ from Γ. The E-rules are similar; for example, disjunction elimination gives a derivation of C from $A \vee B, \Gamma, \Delta, \Theta$ if C is derived from A, Δ and from B, Θ. The management of sets of assumptions was already made explicit in the rules of natural deduction written in sequent calculus style. Sequent calculus maintains the introduction rules thus written, but the treatment of elimination rules is profoundly different.

The rules of sequent calculus are ordered in the same way as those of natural deduction, with the conclusion at the root. The introduction rules of natural deduction become **right rules** of sequent calculus, where a comma replaces set-theoretical union:

$$\frac{\Gamma \Rightarrow A \quad \Delta \Rightarrow B}{\Gamma, \Delta \Rightarrow A \& B} \, R\& \qquad \frac{A, \Gamma \Rightarrow B}{\Gamma \Rightarrow A \supset B} \, R\supset \qquad \frac{\Gamma \Rightarrow A}{\Gamma \Rightarrow A \vee B} \, R\vee_1 \qquad \frac{\Gamma \Rightarrow B}{\Gamma \Rightarrow A \vee B} \, R\vee_2$$

Rule $R\supset$ can also be read "root-first," and in this direction it shows how the derivation of an implication reduces to its components. Reduction here means that the premiss is derivable whenever the conclusion is.

In Gentzen's original formulation of 1934–35, the assumptions Γ, Δ, Θ were finite sequences, or **lists** as we would now say.[2] Gentzen had rules permitting the exchange of order of formulas in a sequence. However, matters are simplified

---

[2]Use of the word "sequent" as a noun was begun by Kleene. His *Introduction to Metamathematics* of 1952 (p. 441) explains the origin of the term as follows: "Gentzen says 'Sequenz', which we translate as 'sequent', because we have already used 'sequence' for any succession of objects, where the German is 'Folge'." This is the standard terminology now; Kleene's usage has even been adopted to some other languages. But Mostowski (1965) for example uses the literal translation "sequence."

if we treat assumptions as **finite multisets**, that is, lists with multiplicity but no order, and we shall do so from now on. Example 3 of Section 1.2 showed that if assumptions are treated simply as sets, control is lost over the number of times an assumption is made.

The elimination rules of natural deduction correspond to **left rules** of sequent calculus. In $\&E$, we have a derivation of C from A, B and some assumptions Γ, and we conclude that C follows from $A\&B$ and the assumptions Γ. In sequent calculus, this is written as

$$\frac{A, B, \Gamma \Rightarrow C}{A\&B, \Gamma \Rightarrow C} \; {}_{L\&}$$

The remaining two left rules are found similarly:

$$\frac{A, \Gamma \Rightarrow C \quad B, \Delta \Rightarrow C}{A \vee B, \Gamma, \Delta \Rightarrow C} \; {}_{L\vee} \qquad \frac{\Gamma \Rightarrow A \quad B, \Delta \Rightarrow C}{A \supset B, \Gamma, \Delta \Rightarrow C} \; {}_{L\supset}$$

The formula with the connective in a rule is the **principal** formula of that rule, and its components in the premises are the **active** formulas. The Greek letters denote possible additional assumptions that are not active in a rule; they are called the **contexts** of the rules.

In natural deduction elimination rules written in sequent calculus style, a formula disappears from the right; in sequent calculus, the same formula appears on the left. Inspection of sequent calculus rules shows what the effect of this change is.

> **Subformula property:** *All formulas in a sequent calculus derivation are subformulas of the endsequent of the derivation.*

The usual way to find derivations in sequent calculus is a "root-first proof search." However, in rules with two premises, we do not know how the context in the conclusion should be divided between the antecedents of the premises. Therefore we do not divide it at all but repeat it fully in both premises. The procedure can be motivated as follows: If assumptions Γ are permitted in the conclusion, it cannot do any harm to make the same assumptions elsewhere in the derivation. Rules $R\&$, $L\vee$, and $L\supset$ can be modified into

$$\frac{\Gamma \Rightarrow A \quad \Gamma \Rightarrow B}{\Gamma \Rightarrow A\&B} \; {}_{R\&} \qquad \frac{A, \Gamma \Rightarrow C \quad B, \Gamma \Rightarrow C}{A \vee B, \Gamma \Rightarrow C} \; {}_{L\vee} \qquad \frac{\Gamma \Rightarrow A \quad B, \Gamma \Rightarrow C}{A \supset B, \Gamma \Rightarrow C} \; {}_{L\supset}$$

The preceding two-premiss rules had **independent contexts**; the above rules instead have **shared contexts**.[3] It now follows that, given the endsequent to be

---

[3] Lately some authors have called these "additive" and "multiplicative" contexts, but these are not as easy to remember.

derived, once it is decided which formula of the endsequent is principal, the premisses are **uniquely determined**.

To show how derivations are found in sequent calculus, we derive the sequent

$$\Rightarrow (A \supset (A \supset B)) \supset (A \supset B)$$

corresponding to Example 3 of Section 1.2:

$$\cfrac{\cfrac{A \Rightarrow A \quad \cfrac{A \Rightarrow A \quad B, A \Rightarrow B}{A \supset B, A \Rightarrow B}\,L\supset}{\cfrac{A \supset (A \supset B), A \Rightarrow B}{\cfrac{A \supset (A \supset B) \Rightarrow A \supset B}{\Rightarrow (A \supset (A \supset B)) \supset (A \supset B)}\,R\supset}\,R\supset}\,L\supset}$$

Both instances of the two-premiss rule $L\supset$ have the shared context A. This root-first proof search is not completely deterministic: The last step can be only $R\supset$, but above that, there are choices in the order of application of rules. Further, proof search need not stop, but we stopped when we reached sequents with the same formula in the antecedent and succedent. This situation corresponds to the rule of assumption of natural deduction, by which we can start a derivation with any formula A as assumption. The rule is given in sequent calculus in the form of a **logical axiom**:

$$A \Rightarrow A$$

In the above derivation, proof search ended in one case with a sequent of the form $A, \Gamma \Rightarrow A$, with a superfluous extra assumption. Its presence was caused by the repetition of formulas in premisses when shared contexts are used.

The $\perp E$ rule of natural deduction gives a zero-premiss sequent calculus rule:

$$\cfrac{}{\perp \Rightarrow C}\,L\perp$$

Often this rule is also referred to as an axiom, but we want to emphasize its character as a left rule and do not call it so.

Formally, a sequent calculus derivation is defined inductively: Instances of axioms are derivations, and if instances of premisses of a rule are conclusions of derivations, an application of the rule will give a derivation. Thus sequent calculus derivations always begin with axioms or $L\perp$. But we depart in two ways from this "official" order of things:

First, note that the logical rules themselves are not derivations, for they have sequents as premisses that need not be axioms. The combination of logical rules likewise gives sequent calculus derivations with premisses. Each logical rule and each combination is correct in the sense that, given derivations of the premisses, the conclusion of the rule or of the combination becomes derivable.

Secondly, the usual root-first proof search procedure runs counter to the inductive generation of sequent calculus derivations. Proof search succeeds only when these two meet, i.e., when the root-first process reaches axioms or instances of $L\bot$.

We now come to the **structural rules** of sequent calculus. To derive the sequent $\Rightarrow A \supset (B \supset A)$ corresponding to Example 2 in Section 1.2, we use a rule of **weakening** that introduces an extra assumption in the antecedent:

$$\frac{\Gamma \Rightarrow C}{A, \Gamma \Rightarrow C} \; Wk$$

The rule is sometimes called "thinning." The derivation of Example 2 is

$$\frac{\dfrac{\dfrac{A \Rightarrow A}{A, B \Rightarrow A} \; Wk}{A \Rightarrow B \supset A} \; R\supset}{\Rightarrow A \supset (B \supset A)} \; R\supset$$

The derivation illustrates the role of weakening: Whenever there is a vacuous discharge in a natural deduction derivation, there is in a corresponding sequent calculus derivation an instance of a logical rule with an active formula that has been introduced in the derivation by weakening.

As noted, our example of proof search in sequent calculus led to a premiss that was not an axiom of the form $A \Rightarrow A$, but of the form $A, \Gamma \Rightarrow A$. These more general axioms are obtained from $A \Rightarrow A$ by repeated application of weakening. If instead we permit instances of axioms as well as the $L\bot$ rule to have an arbitrary context Γ in the antecedent, there is no need for a rule of weakening in sequent calculus.

Above we gave a derivation of the sequent corresponding to Example 3 of Section 1.2 using rules with shared contexts. We give another derivation, this time with the earlier rules that have independent contexts. A rule of **contraction** is now needed:

$$\frac{A, A, \Gamma \Rightarrow C}{A, \Gamma \Rightarrow C} \; Ctr$$

With this rule and axioms of the form $A \Rightarrow A$, the derivation is

$$\frac{\dfrac{\dfrac{A \Rightarrow A \quad \dfrac{A \Rightarrow A \quad B \Rightarrow B}{A \supset B, A \Rightarrow B} \; L\supset}{A \supset (A \supset B), A, A \Rightarrow B} \; L\supset}{\dfrac{A \supset (A \supset B), A \Rightarrow B}{A \supset (A \supset B) \Rightarrow A \supset B} \; R\supset} \; Ctr}{\Rightarrow (A \supset (A \supset B)) \supset (A \supset B)} \; R\supset$$

Contrary to the derivation with shared contexts, a **duplication** of A is produced on the fourth line from below. The meaning of contraction can be explained in terms of natural deduction: Whenever there is a multiple discharge in natural deduction, there is a contraction in a corresponding sequent calculus derivation.

If assumptions are treated as sets instead of multisets, contraction is in a way built into the system and cannot be expressed as a distinct rule.

As with weakening, the rule of contraction can be dispensed with, by use of rules with shared contexts and some additional modifications.

In Chapter 8 we show in a general way that weakening and contraction amount to vacuous and multiple discharge, respectively, in natural deduction, whenever the weakening or contraction formula is active in a derivation. Without this condition, weakening and contraction are purely formal matters produced by the separation of discharge of assumptions into independent structural and logical steps in sequent calculus.

We now come to the last and most important general rule of sequent calculus. In natural deduction, if two derivations $\Gamma \vdash A$ and $A, \Delta \vdash C$ are given, we can join them together into a derivation $\Gamma, \Delta \vdash C$, through a **substitution**. The sequent calculus rule corresponding to this is **cut**:

$$\frac{\Gamma \Rightarrow A \quad A, \Delta \Rightarrow C}{\Gamma, \Delta \Rightarrow C} \, Cut$$

Often cut is explained as follows: We break down the derivation of C from some assumptions into "lemmas," intermediate steps that are easier to prove and that are chained together in the way shown by the cut rule. In Chapter 8 we find a somewhat different explanation of cut: It arises, in terms of natural deduction, from non-normal instances of elimination rules. This points to an important analogy between normal derivations in natural deduction and **cut-free** derivations in sequent calculus, an analogy that will be made precise in Chapter 8.

As explained in Section 1.2, there is in natural deduction a series of concepts from the existence of normal form to strong normalization and uniqueness of normal form. In systems of sequent calculus, it is possible that two derivations $\Gamma \Rightarrow A$ and $A, \Delta \Rightarrow C$ are cut-free, but the derivation $\Gamma, \Delta \Rightarrow C$ obtained by cut need not, in general, have any such form. (This would correspond to the inexistence of normal form in natural deduction.) In the contrary case, we say that a system of sequent calculus is **closed with respect to cut**: If there is a derivation of the sequent $\Gamma \Rightarrow C$ that uses the rule of cut, there exists also a derivation without cut. A typical way of proving closure under cut is to show the **completeness** of a system that does not have the cut rule: All correct sequents are derivable in the system so that the addition of cut does not add any new derivable sequents. Next there is the notion of **elimination of cut** in which a procedure for the actual elimination of cuts in a derivation is given. It corresponds to normalization in natural deduction, but it is typical of sequent calculi that they do not admit of properties corresponding to strong normalization or uniqueness of normal form: in terms of sequent calculi, termination of cut elimination in any order whatsoever and uniqueness of the cut-free form.

An alternative formulation of the rule of cut is obtained if rule $R\supset$ is applied to its right premiss $A, \Delta \Rightarrow C$. The derivation

$$\cfrac{\Gamma \Rightarrow A \quad \cfrac{A, \Delta \Rightarrow C}{\Delta \Rightarrow A \supset C}R\supset}{\Gamma, \Delta \Rightarrow C}Mp$$

shows how cut is replaced by rule $R\supset$ and a sequent calculus version of modus ponens.

Weakening, contraction, and cut are the usual structural rules of sequent calculus. Cut has the effect of making a formula disappear during a derivation so that it need not be a subformula of the conclusion, whereas none of the other rules do this. If we wanted to determine whether a sequent $\Gamma \Rightarrow C$ is derivable, by using cut we could always try to reduce the task into $\Gamma \Rightarrow A$ and $A, \Gamma \Rightarrow C$ with a new formula A, with no end.

A main task of structural proof theory is to find systems that do not need the cut rule or use it in only some limited way. But note that contraction can be as "bad" as cut as concerns a root-first search for a derivation of a given sequent: Formulas in antecedents can be multiplied with no end if contraction cannot be dispensed with.

Two main types of sequent calculi arise: those with independent contexts, similar in many respects to calculi of natural deduction, and those with shared contexts, useful for proof search. Gentzen's original (1934–35) calculi for intuitionistic and classical logic had shared contexts for $R\&$ and $L\vee$ and independent ones for $L\supset$. Further, the left rule for $\&$ (as well as the $R\vee$ rule in the classical case) was given in the form of two rules

$$\cfrac{A, \Gamma \Rightarrow C}{A\&B, \Gamma \Rightarrow C} \qquad \cfrac{B, \Gamma \Rightarrow C}{A\&B, \Gamma \Rightarrow C}$$

that do not support proof search: It need not be the case that $A, \Gamma \Rightarrow C$ is derivable even if $A\&B, \Gamma \Rightarrow C$ is. The single $L\&$ rule we use is due to Ketonen (1944). He also improved the classical $R\vee$ rule in an analogous way and found a classical $L\supset$ rule with shared contexts. With these changes, the sequent calculus for classical propositional logic is **invertible**: From the derivability of a sequent of any of the forms given in the conclusions of the logical rules, the derivability of its premisses follows. Starting with the endsequent, decomposition by invertible rules gives a terminating method of proof search for classical propositional logic.

For intuitionistic logic, a sequent calculus with shared contexts was found by Kleene (1952). The rule of cut can be eliminated in calculi with independent as well as shared contexts. In calculi of the latter kind, also weakening and contraction can be eliminated, so that derivations contain logical rules only. Chapters 2–4 are mainly devoted to the development and study of such calculi. Calculi with independent contexts are studied in Chapter 5.

1.4. THE STRUCTURE OF PROOFS

Given a system of rules **G**, we say that a rule with premisses $S_1, \ldots, S_n$ and conclusion S is **admissible** in **G** if, whenever an instance of $S_1, \ldots, S_n$ is derivable in **G**, the corresponding instance of S is derivable in **G**. Structural proof theory has as its first task the study of admissibility of rules such as weakening, contraction, and cut. Our methods for establishing such results will be thoroughly elementary: In part we show that the addition of a structural rule has no effect on derivability (as for weakening), or we give explicit transformations of derivations that use structural rules into ones that do not use them (as for cut). A major difficulty is to find the correct rules in the first place. Even if the proof methods are all elementary, the proofs often depend on the right combination of many details and are much easier to read than write.

If the cut rule has been shown admissible for a system of rules, we see by inspection of all the rules of inference that no formula disappears in a derivation. Thus cut-free derivations have the subformula property: Each formula in the derivation of a sequent $\Gamma \Rightarrow C$ is a subformula of this endsequent. Later we shall relax on this a bit, by letting atomic formulas disappear, and then the subformula property becomes the statement that each formula in a derivation is a subformula of the endsequent or an atomic formula. Such a **weak** subformula property is still adequate for structural proof-analysis.

Standard applications of cut elimination include elementary syntactic proofs of consistency, the disjunction property for intuitionistic systems, interpolation theorems, and so on. For the first, assume a system to be inconsistent, i.e., assume the sequent $\Rightarrow \perp$ is derivable in it. Each logical rule adds a logical constant, and the axioms and weakening and contraction are rules that have formulas in the antecedent. Therefore there cannot be any derivation of $\Rightarrow \perp$; a cut-free system is consistent. Similarly, assuming that $\Rightarrow A \lor B$ is derivable in a system of rules, it can be the case that the only way by which it can be concluded is by the rules for right disjunction. Thus either $\Rightarrow A$ or $\Rightarrow B$ can be derived, and we say that the system of rules has the **disjunction property**. If a system is both cut-free and contraction-free, it can have the property that the premisses are proper parts of the conclusion, i.e., at least some formula is reduced to a subformula. In this case, we have a root-first proof search resulting in a tree that **terminates**. If the leaves of the tree thus reached are axioms or instances of $L\perp$, we can read it top-down as a derivation of the endsequent. But to show that a sequent is **underivable**, we have to be able to survey all possible derivations. For example, assume that

$\Rightarrow P \lor \sim P$ is derivable in a cut-free intuitionistic system. Then the last rule is one of the two right disjunction rules, and either $\Rightarrow P$ or $\Rightarrow \sim P$ is derivable. No logical rule concludes $\Rightarrow P$, and if $\Rightarrow \sim P$ were derivable, the last rule would have had to be $R\supset$. Again, no logical rule concludes the premiss $P \Rightarrow \perp$.

Above we found a way that led to the rules of sequent calculus from those of natural deduction. Often the structure of cut-free sequent calculus derivations is seen more clearly if it is translated back into natural deduction. This can be made algorithmic, as shown in Chapter 8. Not all sequent calculus derivations can be translated, but only those that do not have "useless" weakening or contraction steps. The translation is such that the order of application of logical rules is reflected in the natural deduction derivation. The meaning of a cut-free derivation is that all major premisses of elimination rules turn into assumptions.

The connection between sequent calculus and natural deduction is straightforward for single succedent sequent calculi, i.e., those with just one formula in the succedent to the right of the sequent arrow. However, there are also systems with a whole multiset as succedent. It can be shown that systems of intuitionistic logic are obtained from classical multisuccedent systems by some innocent-looking restrictions on the succedents. In Chapter 5 we show that the converse is also true, at least for propositional logic: We obtain classical logic from intuitionistic single succedent sequent calculus by the addition of a suitable rule corresponding to the classical law of excluded middle.

Most of the research on sequent calculus has been on systems of pure logic. Considering that the original aim of proof theory was to show the consistency of mathematics, this is rather unfortunate. It is commonly believed that there is nothing to be done: that the main tool of structural proof theory, cut elimination, does not apply if mathematical axioms are added to the purely logical systems of derivation of sequent calculus. In Chapter 6 we show that these limitations can be overcome. A simple example of the failure of cut elimination in the presence of axioms is given by Girard (1987, p. 125): Let the axioms have the forms $\Rightarrow A \supset B$ and $\Rightarrow A$. The sequent $\Rightarrow B$ is derived from these axioms by

$$
\cfrac{\Rightarrow A \qquad \cfrac{\Rightarrow A \supset B \quad \cfrac{A \Rightarrow A \quad B \Rightarrow B}{A \supset B, A \Rightarrow B}\, L\supset}{A \Rightarrow B}\, Cut}{\Rightarrow B}\, Cut
$$

Inspection of sequent calculus rules shows that there is no cut-free derivation of $\Rightarrow B$, which leads Girard to conclude that "the *Hauptsatz* fails for systems with proper axioms" (ibid.). More generally stated, the cut elimination theorem does not apply to sequent calculus derivations with premisses.

We shall give a way of adding axioms to sequent calculus in the form of **nonlogical rules of inference** and show that cut elimination need not be lost by such addition. This depends critically on formulating the rules of inference in a particular way. It then follows that the resulting systems of sequent calculus are both contraction- and cut-free. A limitation, not of the method, but one that is due to the nature of the matter, is that in constructive systems there will be some special forms of axioms, notably $(P \supset Q) \supset R$, that cannot be treated through cut-free rules. For classical systems, our method works uniformly. Gentzen's original subformula property is lost, but typical consequences of that property, such as consistency or the disjunction property for constructive systems, usually follow from the weaker subformula property.

To give an idea of the method, consider again the above example. With P and Q atomic formulas and C an arbitrary formula, $P \supset Q$ is turned into a rule by requiring that if C follows from Q, then it follows from P and P is turned into a rule by requiring that if C follows from P, then C follows:

$$\frac{Q \Rightarrow C}{P \Rightarrow C} \qquad \frac{P \Rightarrow C}{\Rightarrow C}$$

The sequent $\Rightarrow Q$ now has the cut-free derivation

$$\frac{\dfrac{Q \Rightarrow Q}{P \Rightarrow Q}}{\Rightarrow Q}$$

The method of converting axioms into cut-free systems of rules has many applications in mathematics; for example, it can be used in syntactic proofs of consistency and mutual independence for axiom systems. If we use classical logic, we can convert a theorem to be proved into a finite number of sequents that have no logical structure but only atomic formulas and falsity. By cut elimination, their derivation uses only the nonlogical rules, and a very strong control on structure of derivations is achieved. In typical cases such as affine geometry, an axiom can be proved underivable from the rest of an axiom system by showing its underivability by the rules corresponding to the latter.

The aim of proof theory, as envisaged by Hilbert in 1904, was to give a consistency proof of arithmetic and analysis and thereby to resolve the foundational problems of mathematics for good. There had been earlier consistency proofs, such as those for non-Euclidean geometries, in which a **model** was given for an axiom system. However, such proofs are relative: They assume the consistency of the theory in which the model is given. Hilbert's aim instead was an absolute consistency proof, carried through by elementary means. The results of Gödel in 1931 are usually taken to show such proofs to be an impossibility as soon as a system contains the principles of arithmetic. However, we shall see in Chapter 6 that, when this is not the case, purely syntactic and elementary consistency proofs can be obtained as corollaries to cut elimination.

A whole branch of logical research is devoted to the study of **intermediate** logical systems. These are by definition systems that stand between intuitionistic and classical logic in deductive power. In Chapter 7, we shall study the structure of proofs in intermediate logical systems by presenting them as extensions of the basic intuitionistic calculus. One method of extension follows the model of extending this calculus by the rule of excluded middle. Such extension works perfectly for the logical system obeying the **weak** law of excluded middle, $\sim A \vee \sim\sim A$. A limit is reached here, too, for in order to have a subformula property, the characteristic law of an intermediate logic is restricted to instances of the law with atomic formulas, as for the law of excluded middle. If the law for arbitrary formulas cannot be proved from the law for atoms, there is no good structural proof theory under this approach. Such is the case for **Dummett logic**, characterized by the law $(A \supset B) \vee (B \supset A)$. Another method that has been used is to start with the multisuccedent intuitionistic calculus and to relax the intuitionistic restriction on the $R\supset$ rule. This approach will lead to a satisfactory system for Dummett logic.

Our approach to structural proof theory is mainly based on contraction- and cut-free sequent calculi. However, we also present, in Chapter 5, calculi in which weakening and contraction are explicit rules and only cut is eliminated. The sequent calculus rules of the previous section are precisely the propositional and structural rules of the first such calculus, in Section 5.1(a). Further, we also present a calculus in which there is no explicit weakening or contraction, but these are built into the logical rules. This calculus, studied in Section 5.2, can be described as a "sequent calculus in natural deduction style." Sequent calculi with independent contexts are useful for relating derivations in sequent calculus to derivations in natural deduction. The use of special elimination rules in natural deduction brings problems that vanish only if the general elimination rules are taken into use. In Chapter 8 we show that this change will give an isomorphism between sequent calculus derivations and natural deduction derivations. The analysis of proofs by means of natural deduction can often provide insights it would be hard to obtain by the use of sequent calculus only.

Notes to Chapter 1

The definition of languages through categorial grammars, and predicate logic especially, is treated at length in Ranta's *Type-Theoretical Grammar*, 1994. A discussion of logical systems from the point of view of constructive type theory is given in Martin-Löf's *Intuitionistic Type Theory*, 1984, but see also Ranta's book for later developments.

An illuminating discussion of the nature of logical rules and the justification of introduction rules in terms of constructive meaning explanations is given in Martin-Löf (1985). Dummett's views on these matters are collected in his *Truth & Other Enigmas* of 1978.

Our treatment of the elimination rules of natural deduction for propositional logic comes from von Plato (1998) and differs from the usual one that recognizes only the special elimination rules, as in Gentzen's original paper *Untersuchungen über das logische Schliessen* (in two parts, 1934–35) or Prawitz' influential book *Natural Deduction: A Proof-Theoretical Study* of 1965. The change is due to our formulation of the inversion principle in terms of arbitrary consequences of the direct grounds of the corresponding introduction rule, instead of just these direct grounds. The general elimination rule for conjunction is presented in Schroeder-Heister (1984). Chapter 8 will show what the effect of general elimination rules is for the structure of derivations in natural deduction.

Natural deduction in sequent calculus style is used systematically in Dummett's book *Elements of Intuitionism* of 1977.

Our way of obtaining classical propositional logic from the intuitionistic one uses the rule of excluded middle. It appears in this form, as a rule for arbitrary propositions, in Tennant (1978) and Ungar (1992), but the first one to propose the rule was Gentzen (1936). The rule has not been popular, for the obvious reason that it does not have the subformula property. Prawitz (1965) uses the rule of indirect proof and shows that its restriction to atomic formulas will give a satisfactory normal form and subformula property for derivations in the ∨-free fragment of classical propositional logic. We restrict in Chapter 8 the rule of excluded middle to atomic formulas and show that this gives a complete system of natural deduction rules and a full normal form for classical propositional logic. We also show that the rule can be restricted to atoms of the conclusion, thereby obtaining the full subformula property.

The long survey article by Prawitz, *Ideas and results in proof theory*, 1971, offers valuable insights into the development of structural proof theory. The notes to the chapters of Troelstra and Schwichtenberg's *Basic Proof Theory*, 1996, also contain many historical comments. Feferman's *Highlights in proof theory*, 2000, discusses Hilbert's program, sequent calculi, and the proof theory of arithmetic. Finally, the life story of the founder of structural proof theory is given in Menzler-Trott's *Gentzen's Problem: Mathematische Logik im nationalsozialistischen Deutschland*, 2001.

2

Sequent Calculus for Intuitionistic Logic

We present a system of sequent calculus for intuitionistic propositional logic. In later chapters we obtain stronger systems by adding rules to this basic system, and we therefore go through its proof-theoretical properties in detail, in particular the admissibility of structural rules and the basic consequences of cut elimination. Many of these properties can then be verified in a routine fashion for extensions of the system. We begin with a discussion of the significance of constructive reasoning.

2.1. CONSTRUCTIVE REASONING

Intuitionistic logic, and intuitionism more generally, used to be philosophically motivated, but today the grounds for using intuitionistic logic can be completely neutral philosophically. Intuitionistic or constructive reasoning, which are the same thing, systematically supports computability: If the initial data in a problem or theorem are computable and if one reasons constructively, logic will never make one committed to an infinite computation. Classical logic, instead, does not make the distinction between the computable and the noncomputable. We illustrate these phenomena by an example:

A mathematical colleague comes with an algorithm for generating a decimal expansion $0.a_1a_2a_3\ldots$, and also gives a proof that if none of the decimals a_i is greater than zero, a contradiction follows. Then you are asked to find the first decimal a_k such that $a_k > 0$. But you are out of luck in this task, for several hours and days of computation bring forth only 0's. . . .

Given two real numbers a and b, if it happens to be true that they are equal, a and b would have to be computed to infinite precision to verify $a = b$. Obviously the truth of the proposition $a = b$ is not continuous in its two arguments; to see this, think of a and b as points on the real line, assume that $a = b$ is true, and then "move" one of the points just a bit. In a constructive approach, we start with the relation of **apartness**, or distinctness, of two real numbers, written as $a \neq b$. Apartness can be proved by showing that the difference $|a - b|$ has a positive lower bound. This time the proposition is continuous in its arguments: a finite computation can verify $a \neq b$.

What are the axioms of an apartness relation? First, irreflexivity; call it AP1:

AP1. $\sim a \neq a$.

Second, assume $a \neq b$, and take any third number c. If you are unable to decide whether $a \neq c$, it must be the case that $b \neq c$, and similarly with deciding $b \neq c$. This property, the **splitting** of the apartness $a \neq b$ into two cases $a \neq c$ and $b \neq c$, is the second axiom:

AP2. $a \neq b \supset a \neq c \vee b \neq c$.

The principle is intuitively very clear if the points a, b, c are depicted geometrically, as points on the real line.

We now obtain equality as a defined notion:

Definition 2.1.1: $a = b \equiv \sim a \neq b$.

Thus equality is a negative notion, and an infinitistic one also: To prove $a = b$, we have to show how to convert any of the infinitely many a priori possible proofs of $a \neq b$ into an impossibility.

From AP1 we get at once

EQ1. $a = a$,

and from the contraposition of AP2

EQ2. $a = c \,\&\, b = c \supset a = b$.

Substitution of a for c in AP2 gives $a \neq b \supset a \neq a \vee b \neq a$, so by AP1, $b \neq a$ follows from $a \neq b$. Symmetry of equality is obtained by contraposition. Thus the negation of an apartness relation is an equivalence relation.

Let us denote by a the number $0.a_1 a_2 a_3 \ldots$ of our mathematical colleague. From the proof that $a = 0$ leads to a contradiction, $\sim a = 0$ can be concluded. However, this proof does not give any lower bound for $|a - 0|$; thus we have **not** concluded $a \neq 0$. Logically, the difference is one between $\sim\sim a \neq 0$ and $a \neq 0$. The former says that it is impossible that a is equal to zero, the latter says that a positively is distinct from zero.

Classical logic contains the principle of **indirect proof**: If $\sim A$ leads to a contradiction, A can be inferred. Axiomatically expressed, this principle is contained in the law of **double negation**, $\sim\sim A \supset A$. The law of **excluded middle**, $A \vee \sim A$, is a somewhat stronger way of expressing the same principle.

In constructive logic, the connectives and quantifiers obtain a meaning different from the one of classical logic in terms of absolute truth. The constructive "BHK meaning explanations" for propositional logic were given in Section 1.2, and those for quantifiers will be presented in Section 4.1. One particular feature in these explanations is that a direct proof of a disjunction consists of a proof of one of the disjuncts. However, the classical law of excluded middle $A \vee \sim A$ cannot be proved in this way, as there is no method of proving any proposition or its negation. Under the constructive interpretation, the law of excluded middle is not an empty "tautology," but expresses the **decidability** of proposition A. Similarly, a direct proof of an existential proposition $\exists x\, A$ consists of a proof of A for some a. Classically, we can prove existence indirectly by assuming that there is no x such that A, then deriving a contradiction, and concluding that such an x exists. Here the classical law of double negation is used for deriving $\exists x\, A$ from $\sim\sim \exists x\, A$.

More generally, the inference pattern, if something leads to a contradiction the contrary follows, is known as the principle of **reductio ad absurdum**. Dictionary definitions of this principle rarely make the distinction into a genuine indirect proof and a proof of a negative proposition: If A leads to a contradiction, then $\sim A$ can be inferred. Mathematical and even logical literature are full of examples in which the latter inference, a special case of a constructive proof of an implication, is confused with a genuine reductio. A typical example is the proof of irrationality of a real number x: Assume that x is rational, derive a contradiction, and conclude that x is irrational. The fallacy in claiming that this is an indirect proof stems from not realizing that to be an irrational number is a negative property: There do not exist integers n, m such that $x = n/m$.

The effect of constructive reasoning on logic is captured by **intuitionistic logic**. From the point of view of classical logic, it is no limitation not to use the law of excluded middle, or the principle of indirect proof, for the following reason: Given a formula C, there is a translation giving a formula C^* such that C and C^* are classically equivalent and C^* is intuitionistically derivable if C is classically derivable. For example, a disjunction can be translated by $(A \vee B)^* = \sim (\sim A^* \,\&\, \sim B^*)$. The translation gives an interpretation of classical logic in intuitionistic logic. Another intuitionistic interpretation will be given in Chapter 5: For propositional logic, if a formula C is classically derivable, the formula

$$(P_1 \vee \sim P_1)\& \cdots \&(P_n \vee \sim P_n) \supset C$$

where $P_1, \ldots, P_n$ are the atoms of C, is intuitionistically derivable. By this translation, classical propositional logic can be interpreted intuitionistically as a logic in which the proofs of theorems are relativized to decisions on their atoms.

The method of interpreting classical logic in intuitionistic logic through a suitable translation applies to predicate logic and axiomatic theories formalizable in it, such as arithmetic. For such theories, constructive reasoning will only apparently decrease the deductive strength of the theory.

The essential difference between classical and constructive reasoning concerns **predicativity**: The idea, advocated by Poincaré and by Russell, is that "anything involving a totality must not be defined in terms of that totality." Poincaré wanted mathematical objects and structures to be generated from the rock bottom of natural numbers. In Russell, predicativity was a response to the set-theoretical paradoxes, particularly Russell's paradox that arises from defining a "set of all sets that are not members of themselves." In this **impredicative** definition, the totality of all sets is presupposed. Another traditional example of an impredicative definition is the definition of the set of real numbers through complete ordered fields. Impredicativity is met in **second-order logic** in which quantifiers range over propositions. With $X, Y, Z \ldots$ standing as variables for propositions, we can form second-order propositions such as $(\forall X)X$ and $(\forall X)((A \supset (B \supset X)) \supset X)$. Assertion of the first proposition means that for any proposition X, it is the case that X. Then $(\forall X)X$ must be false in a consistent system, and falsity has a second-order definition as $\bot \equiv (\forall X)X$. The rule of falsity elimination becomes a special case of universal instantiation,

$$\frac{(\forall X)X}{C}$$

The second example of a second-order proposition defines the proposition $A \& B$, as can be seen by deriving the rules of conjunction introduction and elimination from the definition, using only the rules for implication and second-order universal quantification. Through the propositions-as-sets principle, we see that second-order quantification amounts to quantification over sets.

2.2. INTUITIONISTIC SEQUENT CALCULUS

In this section, we present an intuitionistic sequent calculus with the remarkable property that all structural rules, weakening, contraction, and cut, are admissible in it. Classical sequent calculi are obtained by removing certain intuitionistic restrictions, and admissibility of structural rules carries over to the classical calculi, as shown in the next chapter. Other extensions of the basic calculus will be studied in later chapters. Sequents are of the form $\Gamma \Rightarrow C$, where Γ is a finite, possibly empty, multiset. The rules of the calculus **G3ip** for intuitionistic propositional logic are the following:

G3ip

Logical axiom:

$P, \Gamma \Rightarrow P$

Logical rules:

$$\frac{A, B, \Gamma \Rightarrow C}{A \& B, \Gamma \Rightarrow C} L\& \qquad \frac{\Gamma \Rightarrow A \quad \Gamma \Rightarrow B}{\Gamma \Rightarrow A \& B} R\&$$

$$\frac{A, \Gamma \Rightarrow C \quad B, \Gamma \Rightarrow C}{A \vee B, \Gamma \Rightarrow C} L\vee \qquad \frac{\Gamma \Rightarrow A}{\Gamma \Rightarrow A \vee B} R\vee_1 \qquad \frac{\Gamma \Rightarrow B}{\Gamma \Rightarrow A \vee B} R\vee_2$$

$$\frac{A \supset B, \Gamma \Rightarrow A \quad B, \Gamma \Rightarrow C}{A \supset B, \Gamma \Rightarrow C} L\supset \qquad \frac{A, \Gamma \Rightarrow B}{\Gamma \Rightarrow A \supset B} R\supset$$

$$\frac{}{\bot, \Gamma \Rightarrow C} L\bot$$

The axiom is restricted to atomic formulas. It is essential that $\bot$ is not considered an atomic formula, but a zero-place logical operation.

Each rule has a **context** designated by Γ in the above rules, **active** formulas designated by A and B, and a **principal** formula that is introduced on the left or the right by the rule in question.

The above calculus differs in three respects from the sequent calculus rules presented in Section 1.3: Only atoms appear in axioms, and the formula $A \supset B$ is repeated in the left premiss of the $L\supset$ rule. The reason for the latter will become apparent later, when admissibility of contraction is proved. Third, the rules have shared contexts.

The calculus has been developed by Troelstra, as a single succedent variant of the calculus of Dragalin (1988). None of the usual structural rules of sequent calculus, weakening, contraction, and cut, need be assumed in it. Exchange rules are absent because of properties of multisets and the other structural rules; those of weakening, contraction, and cut will be proved admissible. The structural rules we consider are

$$\frac{\Gamma \Rightarrow C}{A, \Gamma \Rightarrow C}\, Wk \qquad \frac{A, A, \Gamma \Rightarrow C}{A, \Gamma \Rightarrow C}\, Ctr$$

$$\frac{\Gamma \Rightarrow A \quad A, \Delta \Rightarrow C}{\Gamma, \Delta \Rightarrow C}\, Cut$$

In Gentzen's original calculus of 1934–35, the structural rules were first assumed, and then it was shown how to eliminate applications of the cut rule. A calculus for intuitionistic logic of the above type, with no structural rules, was first developed by Kleene in 1952 for the purpose of proof search. In Gentzen, negation is primitive, but this does not make a great difference. It has the simplifying effect that derivations begin with axioms only, not $L\bot$. Gentzen's calculus maintained the rule of weakening; therefore axioms were of the form $A \Rightarrow A$, with no context since it could be added by weakening. In the calculus **G3ip**, weakening is admissible because it is built into the axiom and the $L\bot$ rule.

The logical rules of the calculus are intuitionistic versions of the rules of Ketonen (1944); In Gentzen's calculus, there were two left rules for conjunction, one for each premiss of the form $A, \Gamma \Rightarrow C$ and $B, \Gamma \Rightarrow C$, and the conclusion as in the above rule. Further, the left implication rule was as follows:

$$\frac{\Gamma \Rightarrow A \quad B, \Delta \Rightarrow C}{A \supset B, \Gamma, \Delta \Rightarrow C}$$

There are two contexts that are joined in the conclusion, so that the rule has independent contexts. In the calculus **G3ip**, instead, all two-premiss logical rules are context-sharing, or have the same context. A shared context is needed for having a contraction-free calculus. Further, the principal formula in $L\supset$ is repeated in the left premiss for the same purpose, a device invented by Kleene in 1952.

(He repeated the principal formula in all the left rules, but such repetition is needed only for noninvertible rules.)

2.3. PROOF METHODS FOR ADMISSIBILITY

Our task in the next two sections is to establish the admissibility of structural rules for the calculus **G3ip**. Proofs of admissibility will use induction on **weight** of formulas and **height** of derivations. Formula weight can be defined in different ways, depending on what is needed in a proof. For the next few chapters a simple definition, amounting to the length of a formula, will be sufficient. In Section 5.5, we shall encounter more complicated formula weights.

Definition 2.3.1: *The weight $w(A)$ of a formula A is defined inductively by*

$w(\perp) = 0$,
$w(P) = 1$ *for atoms P*,
$w(A \circ B) = w(A) + w(B) + 1$ *for conjunction, disjunction, and implication.*

It follows that $w(\sim A) = w(A) + w(\perp) + 1 = w(A) + 1$.

Definition 2.3.2: *A* derivation *in* **G3ip** *is either an axiom, an instance of $L\perp$, or an application of a logical rule to derivations concluding its premisses. The* height *of a derivation is the greatest number of successive applications of rules in it, where an axiom and $L\perp$ have height* 0.

Lemma 2.3.3: *The sequent $C, \Gamma \Rightarrow C$ is derivable for an arbitrary formula C and arbitrary context Γ.*

Proof: The proof is by induction on weight of C. If $w(C) \leqslant 1$, either $C = \perp$ or $C = P$ for some atom P or $C = \perp \supset \perp$. In the first case, $C, \Gamma \Rightarrow C$ is an instance of $L\perp$; in the second it is an axiom. If $C = \perp \supset \perp$, then $C, \Gamma \Rightarrow C$ is derived by

$$\frac{\overline{\perp, \perp \supset \perp, \Gamma \Rightarrow \perp}^{\ L\perp}}{\perp \supset \perp, \Gamma \Rightarrow \perp \supset \perp}^{\ R\supset}$$

The inductive hypothesis is that $C, \Gamma \Rightarrow C$ is derivable for all formulas C with $w(C) \leqslant n$, and we have to show that $D, \Gamma \Rightarrow D$ is derivable for formulas D of weight $\leqslant n + 1$. There are three cases:

$D = A \& B$. By the definition of weight, $w(A) \leqslant n$ and $w(B) \leqslant n$. Noting that the context is arbitrary, we have that $A, \Gamma' \Rightarrow A$ and $B, \Gamma'' \Rightarrow B$ are derivable, where $\Gamma' = B, \Gamma$ and $\Gamma'' = A, \Gamma$. We now derive $A\&B, \Gamma \Rightarrow A\&B$ by

$$\frac{\dfrac{A, B, \Gamma \Rightarrow A}{A\&B, \Gamma \Rightarrow A}\,L\&\quad\dfrac{A, B, \Gamma \Rightarrow B}{A\&B, \Gamma \Rightarrow B}\,L\&}{A\&B, \Gamma \Rightarrow A\&B}\,R\&$$

$D = A \vee B$. As before $w(A) \leqslant n$, $w(B) \leqslant n$, and we have the derivation

$$\frac{\dfrac{A, \Gamma \Rightarrow A}{A, \Gamma \Rightarrow A \vee B}\,R\vee_1\quad\dfrac{B, \Gamma \Rightarrow B}{B, \Gamma \Rightarrow A \vee B}\,R\vee_2}{A \vee B, \Gamma \Rightarrow A \vee B}\,L\vee$$

$D = A \supset B$. As before $w(A) \leqslant n$, $w(B) \leqslant n$, and we have the derivation

$$\frac{\dfrac{A, A \supset B, \Gamma \Rightarrow A\quad B, A, \Gamma \Rightarrow B}{A, A \supset B, \Gamma \Rightarrow B}\,L\supset}{A \supset B, \Gamma \Rightarrow A \supset B}\,R\supset$$

Here $A, A \supset B, \Gamma \Rightarrow A$ and $B, \Gamma \Rightarrow B$ are derivable by the inductive hypothesis.   QED.

Proof by induction on height of derivation is a usual method, often as a subinduction in an inductive proof on formula weight. In the following, the notation

$$\vdash_n \Gamma \Rightarrow C$$

will stand for: the sequent $\Gamma \Rightarrow C$ in **G3ip** is derivable with a height of derivation **at most** n.

When proving the admissibility of a rule by induction on height of derivation, we prove it for subderivations ending in a topmost occurrence of the rule in question, then generalize by induction on the number of applications of the rule to arbitrary derivations. Therefore it can be assumed that in a derivation there is only one instance of the rule in question, the last one.

Theorem 2.3.4: Height-preserving weakening. *If $\vdash_n \Gamma \Rightarrow C$, then $\vdash_n D, \Gamma \Rightarrow C$ for arbitrary D.*

Proof: The proof is by induction on height of derivation. If $n = 0$, then $\Gamma \Rightarrow C$ is an axiom or conclusion of $L\bot$ and either C is an atom and a formula in Γ or $\bot$ is a formula in Γ. In either case, also $D, \Gamma \Rightarrow C$ is an axiom or concluded by $L\bot$. Assume now that height-preserving weakening is admissible up to derivations of height $\leqslant n$, and let $\vdash_{n+1} \Gamma \Rightarrow C$. If the last rule applied is $L\&$, $\Gamma = A\&B, \Gamma'$ and the last step is

$$\frac{A, B, \Gamma' \Rightarrow C}{A\&B, \Gamma' \Rightarrow C}\,L\&$$

so the premiss $A, B, \Gamma' \Rightarrow C$ is derivable in $\leqslant n$ steps. By inductive hypothesis, also $D, A, B, \Gamma' \Rightarrow C$ is derivable in $\leqslant n$ steps. Then an application of $L\&$ gives a derivation of $D, A\&B, \Gamma \Rightarrow C$ in $\leqslant n + 1$ steps.

A similar argument applies to all the other logical rules.   QED.

A more direct way of obtaining height-preserving weakening is to transform the given derivation by adding the weakening formula to the antecedents of all its sequents. Two-premiss rules of **G3ip** have the same context in both premisses, and the conclusion inherits only one copy of these. In the proof transformation showing admissibility of height-preserving weakening, the weakening formula is added always into the contexts of axioms or $L\perp$, and therefore no multiplication of the weakening formula is produced. By repeating weakening, we find weakening admissible for an arbitrary context Γ': If $\vdash_n \Gamma \Rightarrow C$, then $\vdash_n \Gamma, \Gamma' \Rightarrow C$.

For proving the admissibility of contraction, we will need the following **inversion lemma:**

Lemma 2.3.5:
  (i)   *If $\vdash_n A\&B, \Gamma \Rightarrow C$, then $\vdash_n A, B, \Gamma \Rightarrow C$,*
  (ii) *If $\vdash_n A \vee B, \Gamma \Rightarrow C$, then $\vdash_n A, \Gamma \Rightarrow C$ and $\vdash_n B, \Gamma \Rightarrow C$,*
  (iii) *If $\vdash_n A \supset B, \Gamma \Rightarrow C$, then $\vdash_n B, \Gamma \Rightarrow C$.*

Proof: By induction on n.

  (i) If $A\&B, \Gamma \Rightarrow C$ is an axiom or conclusion of $L\perp$; then, $A\&B$ not being atomic or $\perp$, also $A, B, \Gamma \Rightarrow C$ is an axiom or conclusion of $L\perp$.

Assume height-preserving inversion up to height n, and let $\vdash_{n+1} A\&B, \Gamma \Rightarrow C$.

If $A\&B$ is the principal formula, the premiss $A, B, \Gamma \Rightarrow C$ has a derivation of height n.

If $A\&B$ is not principal in the last rule, it has one or two premisses $A\&B, \Gamma' \Rightarrow C', A\&B, \Gamma'' \Rightarrow C''$, of derivation height $\leqslant n$; so by inductive hypothesis, $\vdash_n A, B, \Gamma' \Rightarrow C''$ and $\vdash_n A, B, \Gamma'' \Rightarrow C''$. Now apply the last rule to these premisses to conclude $A, B, \Gamma \Rightarrow C$ in at most $n + 1$ steps.

  (ii) As in (i), if $A \vee B, \Gamma \Rightarrow C$ is an axiom, also $A, \Gamma \Rightarrow C$ and $B, \Gamma \Rightarrow C$ are axioms.

If $A \vee B$ is the principal formula, the two premisses $A, \Gamma \Rightarrow C$ and $B, \Gamma \Rightarrow C$ are derivable in n steps.

If $A \vee B$ is not principal in the last rule, it has one or two premisses $A \vee B, \Gamma' \Rightarrow C', A \vee B, \Gamma'' \Rightarrow C''$, of derivation height $\leqslant n$, so by inductive hypothesis, $\vdash_n A, \Gamma' \Rightarrow C'$ and $\vdash_n B, \Gamma' \Rightarrow C'$ and $\vdash_n A, \Gamma'' \Rightarrow C''$ and $\vdash_n B, \Gamma'' \Rightarrow C''$: Now apply the last rule to the first and the third to conclude $A, \Gamma \Rightarrow C$ and to the second and the fourth to conclude $B, \Gamma \Rightarrow C$ in at most $n + 1$ steps.

(iii) As above for the case that $A \supset B, \Gamma \Rightarrow C$ is an axiom.

If $A \supset B$ is the principal formula, the premiss $B, \Gamma \Rightarrow C$ has a derivation of height n.

If $A \supset B$ is not principal in the last rule, it has one or two premisses $A \supset B, \Gamma' \Rightarrow C'$, $A \supset B, \Gamma'' \Rightarrow C''$, of derivation height $\leqslant n$, so by inductive hypothesis, $\vdash_n B, \Gamma' \Rightarrow C'$ and $\vdash_n B, \Gamma'' \Rightarrow C''$: Now apply the last rule to these premisses to conclude $B, \Gamma \Rightarrow C$ in at most $n + 1$ steps.   QED.

If a rule is invertible, we often indicate use of the inverse rule by writing *Inv* at the inference line. Similarly, if a step is permitted by an inductive hypothesis, we write *Ind* next to the inference line.

The following example shows that $L\supset$ is not invertible with respect to its first premiss: The sequent $\bot \supset \bot \Rightarrow \bot \supset \bot$ is derivable in **G3ip** by

$$\frac{\dfrac{}{\bot, \bot \supset \bot \Rightarrow \bot} \, L\bot}{\bot \supset \bot \Rightarrow \bot \supset \bot} \, R\supset$$

If $L\supset$ were invertible with respect to its first premiss, from the derivability of a sequent with an implication in the antecedent would follow the derivability of its first premiss as determined by the $L\supset$ rule. For the sequent $\bot \supset \bot \Rightarrow \bot \supset \bot$, this first premiss would be $\bot \supset \bot \Rightarrow \bot$. The sequent $\bot \Rightarrow \bot$ is an instance of $L\bot$, and $R\supset$ gives $\Rightarrow \bot \supset \bot$. An application of the cut rule now gives $\Rightarrow \bot$, which would make the system **G3ip** inconsistent. (The formula $\bot \supset \bot$ of this example is the "standard" true formula, abbreviated as $\top = \bot \supset \bot$.)

2.4. ADMISSIBILITY OF CONTRACTION AND CUT

Next we prove the admissibility of the rule of contraction in **G3ip**:

Theorem 2.4.1: Height-preserving contraction. *If* $\vdash_n D, D, \Gamma \Rightarrow C$, *then* $\vdash_n D, \Gamma \Rightarrow C$.

Proof: The proof is by induction on the height of derivation n. If $n = 0$, $D, D, \Gamma \Rightarrow C$ is an axiom or conclusion of $L\bot$ and either C is an atom in the antecedent or the antecedent contains $\bot$. In either case, also $D, \Gamma \Rightarrow C$ is an axiom or conclusion of $L\bot$.

Let contraction be admissible up to derivation height n. We have two cases according to whether the contraction formula is not principal or is principal in the last inference step.

If the contraction formula D is not principal in the last (one-premiss) rule concluding the premiss of contraction we have

$$\frac{D, D, \Gamma' \Rightarrow C'}{D, D, \Gamma \Rightarrow C}$$

which has a derivation height $\leqslant n$, so by inductive hypothesis we obtain $\vdash_n D, \Gamma' \Rightarrow C'$ and by applying the last rule $\vdash_{n+1} D, \Gamma \Rightarrow C$. Two-premiss rules have two occurrences of D in both premisses and the same argument applies.

If the contraction formula D is principal in the last rule, we have three cases according to the form of D:

$D = A\&B$. Then the last step is $L\&$ and we have $\vdash_n A, B, A\&B, \Gamma \Rightarrow C$. By Lemma 2.3.5, we obtain $\vdash_n A, B, A, B, \Gamma \Rightarrow C$ and by inductive hypothesis applied twice, $\vdash_n A, B, \Gamma \Rightarrow C$. Application of $L\&$ now gives $\vdash_{n+1} A\&B, \Gamma \Rightarrow C$.

$D = A \vee B$. Then the last step is $L\vee$ and we have $\vdash_n A, A \vee B, \Gamma \Rightarrow C$ and $\vdash_n B, A \vee B, \Gamma \Rightarrow C$. Lemma 2.3.5 gives $\vdash_n A, A, \Gamma \Rightarrow C$ and $\vdash_n B, B, \Gamma \Rightarrow C$ so by inductive hypothesis, $\vdash_n A, \Gamma \Rightarrow C$ and $\vdash_n B, \Gamma \Rightarrow C$, so by $L\vee, \vdash_{n+1} A \vee B, \Gamma \Rightarrow C$.

$D = A \supset B$. Then the last step is $L\supset$ and we have $\vdash_n A \supset B, A \supset B, \Gamma \Rightarrow A$ and $\vdash_n B, A \supset B, \Gamma \Rightarrow C$. By inductive hypothesis, the first gives $\vdash_n A \supset B, \Gamma \Rightarrow A$. By Lemma 2.3.5, the second gives $\vdash_n B, B, \Gamma \Rightarrow C$ so by inductive hypothesis, $\vdash_n B, \Gamma \Rightarrow C$. Application of $L\supset$ now gives $\vdash_{n+1} A \supset B, \Gamma \Rightarrow C$. QED.

Remarkably, the weaker result of admissibility of contraction without preservation of height is more difficult to prove than admissibility of height-preserving contraction, for its proof requires a double induction on formula weight with a subinduction on height of derivation.

The repetition of the principal formula in the first premiss of rule $L\supset$ is needed in order to apply the inductive hypothesis that permits contraction in a derivation of less height. In classical sequent calculus with shared contexts, all rules are invertible and there is no need for such repetition. The same is true in **G3ip** in the sense that the rule without repetition,

$$\frac{\Gamma \Rightarrow A \quad B, \Gamma \Rightarrow C}{A \supset B, \Gamma \Rightarrow C}$$

is admissible in **G3ip**. This follows by the application of weakening with $A \supset B$ to the left premiss $\Gamma \Rightarrow A$.

We now come to the main result of this chapter, the admissibility of cut for the calculus **G3ip**. Gentzen called his cut elimination theorem the "Hauptsatz," the main theorem, and this is how cut elimination is often called today also. The proof uses, explicitly or implicitly, all the preceding lemmas and theorems to show that cuts can be **permuted** upward in a derivation until they reach the axioms and conclusions of $L\bot$ the derivation started with. When both premisses of a cut are axioms or conclusions of $L\bot$, the conclusion also is an axiom or conclusion of $L\bot$: If the first premiss is $\bot, \Gamma \Rightarrow C$, the conclusion has $\bot$ in the antecedent, and if the first premiss is $P, \Gamma \Rightarrow P$, the second premiss is $P, \Delta \Rightarrow C$. This is an axiom only if $C = P$ or C is an atom in Δ, and it is a conclusion of $L\perp$ only if $\perp$ is in Δ. In each case, the conclusion of cut $P, \Gamma, \Delta \Rightarrow C$ is an axiom or conclusion of $L\perp$. As a consequence, when cut has reached axioms and instances of $L\perp$, the derivation can be transformed into one beginning with the conclusion of the cut, by just deleting the premisses.

The proof of admissibility of cut for **G3ip** is by induction on the weight of the cut formula and a subinduction on the sum of heights of derivations of the two premisses. This sum is called **cut-height**:

Definition 2.4.2: Cut-height. *The* cut-height *of an instance of the rule of cut in a derivation is the sum of heights of derivation of the two premisses of cut.*

We give transformations that always reduce the weight of cut formula or cut-height. Actually, what happens is that cut-height is reduced in all cases in which the cut formula is not principal in both premisses of cut. In the contrary case, cut is reduced to formulas of lesser weight. This process terminates since atoms can never be principal in logical rules.

Cut-height is not monotone as we go down in a derivation; that is, a cut below another one can have a lesser cut-height: In the derivation of one of its premisses there is the first cut, and this derivation has a greater height than either of the premisses of the first cut. But the other premiss may have a height much shorter than either premiss of the first cut, making the sum less than the sum in the first cut. It follows that the permutation of a cut upward does not always reduce cut-height but can increase it. For this reason, we shall explicitly calculate the height of each cut in what follows. As with weakening and contraction, we may assume that there is only one occurrence of the rule of cut, as the last step.

Theorem 2.4.3: *The rule of cut,*

$$\frac{\Gamma \Rightarrow D \quad D, \Delta \Rightarrow C}{\Gamma, \Delta \Rightarrow C} \; Cut$$

is admissible in **G3ip**.

Proof: The proof is organized as follows: We consider first the case that at least one premiss in a cut is an axiom or conclusion of $L\perp$ and show how cut is eliminated. For the rest there are three cases: 1. The cut formula is not principal in either premiss of cut. 2. The cut formula is principal in just one premiss of cut. 3. The cut formula is principal in both premisses of cut.

Cut with an axiom or conclusion of $L\perp$ as premiss: If at least one of the premisses of cut is an axiom or conclusion of $L\perp$, we distinguish two cases:

1. The **left premiss** $\Gamma \Rightarrow D$ of cut is an axiom or conclusion of $L\perp$. There are two subcases:

1.1. The cut formula D is in Γ. In this case we derive $\Gamma, \Delta \Rightarrow C$ from $D, \Delta \Rightarrow C$ by weakening.

1.2. $\perp$ is a formula in Γ. Then $\Gamma, \Delta \Rightarrow C$ is a conclusion of $L\perp$.

2. The **right premiss** $D, \Delta \Rightarrow C$ is an axiom or conclusion of $L\perp$. There are four subcases:

2.1. C is in Δ. Then $\Gamma, \Delta \Rightarrow C$ is an axiom.

2.2. $C = D$. Then the first premiss is $\Gamma \Rightarrow C$ and $\Gamma, \Delta \Rightarrow C$ follows by weakening.

2.3. $\perp$ is in Δ. Then $\Gamma, \Delta \Rightarrow C$ is a conclusion of $L\perp$.

2.4. $D = \perp$. Then either the first premiss $\Gamma \Rightarrow \perp$ is an axiom and $\Gamma, \Delta \Rightarrow C$ follows as in case *1*, or $\Gamma \Rightarrow \perp$ has been derived by a left rule. There are three cases according to the rule used. These are transformed into derivations with less cut-height. Since the transformations are special cases of the transformations *3.1–3.3* below, with $D = \perp$, we do not write them out here.

Cut with neither premiss an axiom: We have three cases:

3. Cut formula D is **not principal** in the left premiss, that is, not derived by an R-rule. We have three subcases according to the rule used to derive the left premiss. In the derivations, it is assumed that the topsequents, from left to right, have derivation heights $n, m, k, \ldots$.

3.1. $L\&$, with $\Gamma = A\&B, \Gamma'$. The derivation with a cut of cut-height $n + 1 + m$ is

$$\frac{\dfrac{A, B, \Gamma' \Rightarrow D}{A\&B, \Gamma' \Rightarrow D} \, L\& \quad D, \Delta \Rightarrow C}{A\&B, \Gamma', \Delta \Rightarrow C} \, Cut$$

and it is transformed by permuting the order $L\&,Cut$ into the order $Cut,L\&$. The result is the derivation with a cut of cut-height $n + m$:

$$\frac{\dfrac{A, B, \Gamma' \Rightarrow D \quad D, \Delta \Rightarrow C}{A, B, \Gamma', \Delta \Rightarrow C} \, Cut}{A\&B, \Gamma', \Delta \Rightarrow C} \, L\&$$

3.2. $L\vee$, with $\Gamma = A \vee B, \Gamma'$. The derivation with a cut of cut-height $max(n, m) + 1 + k$

$$\frac{\dfrac{A, \Gamma' \Rightarrow D \quad B, \Gamma' \Rightarrow D}{A \vee B, \Gamma' \Rightarrow D} \, L\vee \quad D, \Delta \Rightarrow C}{A \vee B, \Gamma', \Delta \Rightarrow C} \, Cut$$

is transformed into the derivation with two cuts of heights $n + k$ and $m + k$:

$$\frac{\dfrac{A, \Gamma' \Rightarrow D \quad D, \Delta \Rightarrow C}{A, \Gamma', \Delta \Rightarrow C}\ Cut \qquad \dfrac{B, \Gamma' \Rightarrow D \quad D, \Delta \Rightarrow C}{B, \Gamma', \Delta \Rightarrow C}\ Cut}{A \vee B, \Gamma', \Delta \Rightarrow C}\ L\vee$$

3.3. $L\supset$, with $\Gamma = A \supset B, \Gamma'$. The derivation with a cut of cut-height $max(n, m) + 1 + k$

$$\frac{\dfrac{A \supset B, \Gamma' \Rightarrow A \quad B, \Gamma' \Rightarrow D}{A \supset B, \Gamma' \Rightarrow D}\ L\supset \qquad D, \Delta \Rightarrow C}{A \supset B, \Gamma', \Delta \Rightarrow C}\ Cut$$

is transformed into the derivation with a cut of cut-height $m + k$:

$$\frac{\dfrac{A \supset B, \Gamma' \Rightarrow A}{A \supset B, \Gamma', \Delta \Rightarrow A}\ Wk \qquad \dfrac{B, \Gamma' \Rightarrow D \quad D, \Delta \Rightarrow C}{B, \Gamma', \Delta \Rightarrow C}\ Cut}{A \supset B, \Gamma', \Delta \Rightarrow C}\ L\supset$$

We observe that cut-height is reduced in each transformation.

4. Cut formula D is **principal in the left premiss only**, and the derivation is transformed into one with a cut of lesser cut-height according to the derivation of the right premiss. We have six subcases according to the rule used:

4.1. $L\&$, with $\Delta = A\&B, \Delta'$, and the derivation with a cut of cut-height $n + m + 1$

$$\frac{\Gamma \Rightarrow D \qquad \dfrac{D, A, B, \Delta' \Rightarrow C}{D, A\&B, \Delta' \Rightarrow C}\ L\&}{\Gamma, A\&B, \Delta' \Rightarrow C}\ Cut$$

is transformed into the derivation with a cut of cut-height $n + m$:

$$\frac{\dfrac{\Gamma \Rightarrow D \quad D, A, B, \Delta' \Rightarrow C}{\Gamma, A, B, \Delta' \Rightarrow C}\ Cut}{\Gamma, A\&B, \Delta' \Rightarrow C}\ L\&$$

4.2. $L\vee$, with $\Delta = A \vee B, \Delta'$, and the derivation with a cut of cut-height $n + max(m, k) + 1$

$$\frac{\Gamma \Rightarrow D \qquad \dfrac{D, A, \Delta' \Rightarrow C \quad D, B, \Delta' \Rightarrow C}{D, A \vee B, \Delta' \Rightarrow C}\ L\vee}{\Gamma, A \vee B, \Delta' \Rightarrow C}\ Cut$$

is transformed into the derivation with two cuts of heights $n + m$ and $n + k$:

$$\dfrac{\dfrac{\Gamma \Rightarrow D \quad D, A, \Delta' \Rightarrow C}{\Gamma, A, \Delta' \Rightarrow C} \, Cut \quad \dfrac{\Gamma \Rightarrow D \quad D, B, \Delta' \Rightarrow C}{\Gamma, B, \Delta' \Rightarrow C} \, Cut}{\Gamma, A \vee B, \Delta' \Rightarrow C} \, L\vee$$

4.3. $L\supset$, with $\Delta = A \supset B, \Delta'$, and the derivation with a cut of cut-height $n + max(m, k) + 1$

$$\dfrac{\Gamma \Rightarrow D \quad \dfrac{D, A \supset B, \Delta' \Rightarrow A \quad D, B, \Delta' \Rightarrow C}{D, A \supset B, \Delta' \Rightarrow C} \, L\supset}{\Gamma, A \supset B, \Delta' \Rightarrow C} \, Cut$$

is transformed into the derivation with two cuts of heights $n + m$ and $n + k$:

$$\dfrac{\dfrac{\Gamma \Rightarrow D \quad D, A \supset B, \Delta' \Rightarrow A}{\Gamma, A \supset B, \Delta' \Rightarrow A} \, Cut \quad \dfrac{\Gamma \Rightarrow D \quad D, B, \Delta' \Rightarrow C}{\Gamma, B, \Delta' \Rightarrow C} \, Cut}{\Gamma, A \supset B, \Delta' \Rightarrow C} \, L\supset$$

4.4. $R\&$, with $C = A\&B$, and the derivation with a cut of cut-height $n + max(m, k) + 1$

$$\dfrac{\Gamma \Rightarrow D \quad \dfrac{D, \Delta \Rightarrow A \quad D, \Delta \Rightarrow B}{D, \Delta \Rightarrow A\&B} \, R\&}{\Gamma, \Delta \Rightarrow A\&B} \, Cut$$

is transformed into the derivation with two cuts of heights $n + m$ and $n + k$:

$$\dfrac{\dfrac{\Gamma \Rightarrow D \quad D, \Delta \Rightarrow A}{\Gamma, \Delta \Rightarrow A} \, Cut \quad \dfrac{\Gamma \Rightarrow D \quad D, \Delta \Rightarrow B}{\Gamma, \Delta \Rightarrow B} \, Cut}{\Gamma, \Delta \Rightarrow A\&B} \, R\&$$

4.5. $R\vee$, with $C = A \vee B$, and the derivations with cuts of cut-heights $n + m + 1$ and $n + k + 1$, respectively,

$$\dfrac{\Gamma \Rightarrow D \quad \dfrac{D, \Delta \Rightarrow A}{D, \Delta \Rightarrow A \vee B} \, R\vee_1}{\Gamma, \Delta \Rightarrow A \vee B} \, Cut \qquad \dfrac{\Gamma \Rightarrow D \quad \dfrac{D, \Delta \Rightarrow B}{D, \Delta \Rightarrow A \vee B} \, R\vee_2}{\Gamma, \Delta \Rightarrow A \vee B} \, Cut$$

are transformed into the derivations with cuts of cut-heights $n + m$ and $n + k$:

$$\dfrac{\dfrac{\Gamma \Rightarrow D \quad D, \Delta \Rightarrow A}{\Gamma, \Delta \Rightarrow A} \, Cut}{\Gamma, \Delta \Rightarrow A \vee B} \, R\vee_1 \qquad \dfrac{\dfrac{\Gamma \Rightarrow D \quad D, \Delta \Rightarrow B}{\Gamma, \Delta \Rightarrow B} \, Cut}{\Gamma, \Delta \Rightarrow A \vee B} \, R\vee_2$$

4.6. $R\supset$, with $C = A \supset B$, and the derivation with a cut of cut-height $n + m + 1$

$$\cfrac{\Gamma \Rightarrow D \quad \cfrac{D, A, \Delta \Rightarrow B}{D, \Delta \Rightarrow A \supset B} R\supset}{\Gamma, \Delta \Rightarrow A \supset B} Cut$$

is transformed into the derivation with a cut of cut-height $n + m$:

$$\cfrac{\cfrac{\Gamma \Rightarrow D \quad D, A, \Delta \Rightarrow B}{\Gamma, A, \Delta \Rightarrow B} Cut}{\Gamma, \Delta \Rightarrow A \supset B} R\supset$$

In each case, cut-height is reduced.

5. Cut formula D is **principal in both premisses**, and we have three subcases:

5.1. $D = A\&B$, and the derivation with a cut of cut-height $max(n, m) + 1 + k + 1$ is

$$\cfrac{\cfrac{\Gamma \Rightarrow A \quad \Gamma \Rightarrow B}{\Gamma \Rightarrow A\&B} R\& \quad \cfrac{A, B, \Delta \Rightarrow C}{A\&B, \Delta \Rightarrow C} L\&}{\Gamma, \Delta \Rightarrow C} Cut$$

This is transformed into the derivation with two cuts of heights $n + k$ and $m + max(n, k) + 1$:

$$\cfrac{\cfrac{\Gamma \Rightarrow B \quad \cfrac{\Gamma \Rightarrow A \quad A, B, \Delta \Rightarrow C}{\Gamma, B, \Delta \Rightarrow C} Cut}{\Gamma, \Gamma, \Delta \Rightarrow C} Cut}{\Gamma, \Delta \Rightarrow C} Ctr$$

Note that cut-height can increase in the transformation, but the cut formula is reduced.

5.2. $D = A \vee B$, and the derivation is either

$$\cfrac{\cfrac{\Gamma \Rightarrow A}{\Gamma \Rightarrow A \vee B} R\vee_1 \quad \cfrac{A, \Delta \Rightarrow C \quad B, \Delta \Rightarrow C}{A \vee B, \Delta \Rightarrow C} L\vee}{\Gamma, \Delta \Rightarrow C} Cut$$

with cut-height $n + 1 + max(m, k) + 1$ or

$$\cfrac{\cfrac{\Gamma \Rightarrow B}{\Gamma \Rightarrow A \vee B} R\vee_2 \quad \cfrac{A, \Delta \Rightarrow C \quad B, \Delta \Rightarrow C}{A \vee B, \Delta \Rightarrow C} L\vee}{\Gamma, \Delta \Rightarrow C} Cut$$

with the same cut-height. These are transformed into derivations with cuts of cut-heights $n + m$ and $n + k$,

$$\frac{\Gamma \Rightarrow A \quad A, \Delta \Rightarrow C}{\Gamma, \Delta \Rightarrow C} \, Cut \qquad \frac{\Gamma \Rightarrow B \quad B, \Delta \Rightarrow C}{\Gamma, \Delta \Rightarrow C} \, Cut$$

where both cut-height and weight of cut formula are reduced.

5.3. $D = A \supset B$, and the derivation with a cut of cut-height $n + 1 + max(m, k) + 1$ is

$$\frac{\dfrac{A, \Gamma \Rightarrow B}{\Gamma \Rightarrow A \supset B} \, R\supset \quad \dfrac{A \supset B, \Delta \Rightarrow A \quad B, \Delta \Rightarrow C}{A \supset B, \Delta \Rightarrow C} \, L\supset}{\Gamma, \Delta \Rightarrow C} \, Cut$$

This is transformed into the derivation with three cuts of heights $n + 1 + m, n + k$ and $max(n + 1, m) + 1 + max(n, k) + 1$

$$\frac{\dfrac{\dfrac{A, \Gamma \Rightarrow B}{\Gamma \Rightarrow A \supset B} \, R\supset \quad A \supset B, \Delta \Rightarrow A}{\Gamma, \Delta \Rightarrow A} \, Cut \quad \dfrac{A, \Gamma \Rightarrow B \quad B, \Delta \Rightarrow C}{A, \Gamma, \Delta \Rightarrow C} \, Cut}{\Gamma, \Delta \Rightarrow C} \, Cut$$

In the first and second cut, cut-height is reduced; in the second and third, weight of cut formula.   QED.

In many of the permutations of cut upward in a derivation, the number of cuts increases exponentially.

In contrast to the logical rules, the contexts in the two premises of the cut rule are independent. However, by the admissibility of structural rules, we can show that also the cut rule with a shared context,

$$\frac{\Gamma \Rightarrow D \quad D, \Gamma \Rightarrow C}{\Gamma \Rightarrow C}$$

is admissible. To see this, first apply the usual cut rule to the two premises to derive $\Gamma, \Gamma, \Rightarrow C$, then contract the duplication of Γ in its conclusion.

2.5. SOME CONSEQUENCES OF CUT ELIMINATION

(a) **The subformula property:** Since structural rules can be dispensed with in **G3ip**, we find by inspection of its rules of inference that no formulas disappear from derivations:

Theorem 2.5.1: *If $\Gamma \Rightarrow C$ has a derivation in* **G3ip**, *all formulas in the derivation are subformulas of Γ, C.*

Similarly, a connective that has once appeared in a derivation cannot disappear. From this it follows in particular that $\Rightarrow \perp$ is not derivable, i.e., the calculus is syntactically **consistent**.

Theorem 2.5.2: *If* $\Rightarrow A \vee B$ *is derivable in* **G3ip**, *then* $\Rightarrow A$ *or* $\Rightarrow B$ *is derivable.*

Proof: Only right rules can conclude sequents with an empty antecedent so the last rule can only be $R\vee$.   QED.

This theorem establishes the **disjunction property** of the calculus for intuitionistic propositional logic. That such a property should hold follows from the constructive meaning of disjunction as given in Section 2.1. The result can be strengthened into a disjunction property under suitable hypotheses:

Definition 2.5.3: *The class of* Harrop formulas *is defined by*

   (i) $P, Q, R, \ldots$, *and* $\perp$ *are Harrop formulas,*
   (ii) $A\&B$ *is a Harrop formula whenever A and B are Harrop formulas,*
   (iii) $A \supset B$ *is a Harrop formula whenever B is a Harrop formula.*

Theorem 2.5.4: *If* $\Gamma \Rightarrow A \vee B$ *is derivable in* **G3ip** *and* Γ *consists of Harrop formulas, then* $\Gamma \Rightarrow A$ *or* $\Gamma \Rightarrow B$ *is derivable.*

Proof: The proof is by induction on the height of derivation. For the base case, $\Gamma \Rightarrow A \vee B$ is not an axiom, and if it is the conclusion of $L\perp$ also $\Gamma \Rightarrow A$ is. If the last rule in the derivation of $\Gamma \Rightarrow A \vee B$ is $R\vee$, the premiss is either $\Gamma \Rightarrow A$ or $\Gamma \Rightarrow B$. If the last rule is $L\&$, then $\Gamma = C\&D, \Gamma'$ and the premiss is $C, D, \Gamma' \Rightarrow A \vee B$. Since $C\&D$ is a Harrop formula, also C and D are and the inductive hypothesis applies to the premiss. If the last rule is $L\supset$, then $\Gamma = C \supset D, \Gamma'$ and the inductive hypothesis applies to the right premiss $D, \Gamma' \Rightarrow A \vee B$. The last rule cannot be $L\vee$.   QED.

Proof by induction on the height of derivation in a system with no structural rules is remarkably simple compared with the original proof of the result in Harrop (1960).

(b) Hilbert-style systems: We show that, from **G3ip**, the more traditional Hilbert-style axiomatic formulation of intuitionistic propositional logic follows. In a Hilbert-style system, formulas rather than sequents are derived, starting with instances of axioms and using in the propositional case only one rule of inference, modus ponens. The axioms are given schematically as

1. $\perp \supset A$,
2. $A \supset (B \supset A\&B)$,   3. $A\&B \supset A$,   4. $A\&B \supset B$,
5. $A \supset A \vee B$,   6. $B \supset A \vee B$,   7. $(A \supset C) \supset ((B \supset C) \supset (A \vee B \supset C))$,
8. $A \supset (B \supset A)$,   9. $(A \supset (B \supset C)) \supset ((A \supset B) \supset (A \supset C))$.

In Hilbert-style systems, substitution of formulas is done in the schematic axioms to obtain the top formulas of derivations. These systems are next to impossible to use for the actual derivation of formulas because of the difficulty of locating the substitution instances that are needed. A notorious example is the derivation of $A \supset A$ by substitutions in axioms 8 and 9:

$$\dfrac{\dfrac{(A \supset ((A \supset A) \supset A)) \supset ((A \supset (A \supset A)) \supset (A \supset A)) \quad A \supset ((A \supset A) \supset A)}{(A \supset (A \supset A)) \supset (A \supset A)} \quad A \supset (A \supset A)}{A \supset A}$$

There seems to be very little relation between the simplicity of the conclusion and the complexity of its derivation. In order to translate derivations in the Hilbert-style system into **G3ip** we shall write axiom schemes as sequents with empty antecedents and the rule of modus ponens as the sequent calculus rule

$$\dfrac{\Rightarrow A \supset B \quad \Rightarrow A}{\Rightarrow B} Mp$$

We show that this translation of derivations in the Hilbert-style system gives derivations in **G3ip**:

Theorem 2.5.5: *If formula C is derivable in the Hilbert-style system, then $\Rightarrow C$ is derivable in* **G3ip**.

Proof: In a derivation of C, each instance A of an axiom is replaced by a derivation of the sequent $\Rightarrow A$. All the axiom schemes as sequents with empty antecedents are easily derived in **G3ip**, and we show only the first two:

$$\dfrac{\bot \Rightarrow A}{\Rightarrow \bot \supset A} R\supset \qquad \dfrac{\dfrac{\dfrac{A, B \Rightarrow A \quad A, B \Rightarrow B}{A, B \Rightarrow A\&B} R\&}{A \Rightarrow B \supset A\&B} R\supset}{\Rightarrow A \supset (B \supset A\&B)} R\supset$$

Each application of modus ponens in the derivation of C is replaced by its sequent calculus version. We note that a rule is admissibile in **G3ip** if it is derivable using also structural rules. Modus ponens as a sequent calculus rule concluding $\Rightarrow B$ from $\Rightarrow A \supset B$ and $\Rightarrow A$ is derived by

$$\dfrac{\Rightarrow A \quad \dfrac{\Rightarrow A \supset B \quad \dfrac{A \supset B, A \Rightarrow A \quad B \Rightarrow B}{A \supset B, A \Rightarrow B} L\supset}{A \Rightarrow B} Cut}{\Rightarrow B} Cut$$

By cut elimination, a derivation of $\Rightarrow C$ in **G3ip** is obtained.   QED.

Hilbert-style systems are widely used in model theory and related fields, but in proof theory they are more of historical interest. It is possible, although very laborious, to prove a converse to Theorem 2.5.5, by the following translations: Sequents $A_1, \ldots, A_m \Rightarrow B$ are translated into formulas $A_1 \& \ldots \& A_m \supset B$, and instances of sequent calculus rules, say

$$\frac{A_1, \ldots, A_m \Rightarrow A \quad B_1, \ldots, B_n \Rightarrow B}{C_1, \ldots, C_k \Rightarrow C}$$

into implications

$$(A_1 \& \ldots \& A_m \supset A) \& (B_1 \& \ldots \& B_n \supset B) \supset (C_1 \& \ldots \& C_k \supset C).$$

(c) Underivability results: Certain sequents require for their derivation logical systems stronger in deductive strength than intuitionistic logic. Examples of such sequents are

$\Rightarrow A \vee \sim A$, the law of excluded middle,
$\Rightarrow \sim A \vee \sim\sim A$, the weak law of excluded middle,
$\Rightarrow \sim\sim A \supset A$, the law of double-negation,
$\Rightarrow (A \supset B) \vee (B \supset A)$, Dummett's law,
$\Rightarrow ((A \supset B) \supset A) \supset A$, Peirce's law,
$\Rightarrow (A \supset B \vee C) \supset (A \supset B) \vee (A \supset C)$, disjunction property under hypothesis,
$\Rightarrow (\sim A \supset B \vee C) \supset (\sim A \supset B) \vee (\sim A \supset C)$, disjunction property under negative hypothesis.

The underivability of these sequents in intuitionistic logic is usually established by model-theoretical means. We show their underivability proof-theoretically by the elementary method of contraction- and cut-free derivability. We note that if a sequent is underivable for atomic formulas, such as $\Rightarrow P \vee \sim P$, then the corresponding sequent $\Rightarrow A \vee \sim A$ with arbitrary formulas is also underivable. Whenever in a root-first proof search a premiss is found that is equal to some previous sequent, proof search on that branch is stopped. One says that a **loop** obtains in the search tree. Stopping the proof search is justified by the fact that a continuation from the repeated sequent succeeds if and only if a search from its first occurrence succeeds.

Theorem 2.5.6: *The following sequents are not derivable in* **G3ip**:

(i) $\Rightarrow P \vee \sim P$,
(ii) $\Rightarrow \sim P \vee \sim\sim P$,
(iii) $\Rightarrow ((P \supset Q) \supset P) \supset P$,
(iv) $\Rightarrow (P \supset Q \vee R) \supset (P \supset Q) \vee (P \supset R)$.

Proof: (i) Assume there is a derivation of $\Rightarrow P \vee \sim P$. By the disjunction property, either $\Rightarrow P$ or $\Rightarrow \sim P$ is derivable. No rule concludes $\Rightarrow P$ for an atom P, and only $R\supset$ concludes $\Rightarrow \sim P$, so by invertibility of $R\supset$, $P \Rightarrow \bot$ is derivable. But no rule concludes such a sequent.

(ii) For $\Rightarrow \sim P \vee \sim\sim P$ to be derivable, by proof of (i), $\Rightarrow \sim\sim P$ must be derivable. Proceeding root-first, the last three steps must be

$$
\cfrac{\cfrac{P \supset \bot \Rightarrow P \quad \bot \Rightarrow P}{P \supset \bot \Rightarrow P}\,{\scriptstyle L\supset} \quad \cfrac{}{\bot \Rightarrow \bot}\,{\scriptstyle L\bot}}{\cfrac{P \supset \bot \Rightarrow \bot}{\Rightarrow (P \supset \bot) \supset \bot}\,{\scriptstyle R\supset}}\,{\scriptstyle L\supset}
$$

Since the left premiss of the uppermost instance of $L\supset$ is equal to its conclusion, this proof search does not terminate. Therefore there is no derivation of $\Rightarrow \sim\sim P$.

(iii) With $\Rightarrow ((P \supset Q) \supset P) \supset P$, the last two steps must be

$$
\cfrac{\cfrac{(P \supset Q) \supset P \Rightarrow P \supset Q \quad P \Rightarrow P}{(P \supset Q) \supset P \Rightarrow P}\,{\scriptstyle L\supset}}{\Rightarrow ((P \supset Q) \supset P) \supset P}\,{\scriptstyle R\supset}
$$

If we continue by $R\supset$ we get

$$
\cfrac{\cfrac{P, (P \supset Q) \supset P \Rightarrow P \supset Q \quad P, P \Rightarrow Q}{P, (P \supset Q) \supset P \Rightarrow Q}\,{\scriptstyle L\supset}}{(P \supset Q) \supset P \Rightarrow P \supset Q}\,{\scriptstyle R\supset}
$$

but the right premiss is not derivable by any rule. If we apply $L\supset$ instead we get

$$
\cfrac{(P \supset Q) \supset P \Rightarrow P \supset Q \quad \cfrac{P, P \Rightarrow Q}{P \Rightarrow P \supset Q}\,{\scriptstyle R\supset}}{(P \supset Q) \supset P \Rightarrow P \supset Q}\,{\scriptstyle L\supset}
$$

This proof search fails because the sequent $P, P \Rightarrow Q$ is not derivable. Therefore $\Rightarrow ((P \supset Q) \supset P) \supset P$ is not derivable.

(iv) Derivations of $\Rightarrow (P \supset Q \vee R) \supset (P \supset Q) \vee (P \supset R)$ must end with $R\supset$. The premiss is $P \supset Q \vee R \Rightarrow (P \supset Q) \vee (P \supset R)$, and if the last rule was $R\vee_1$, the derivation has either $R\supset$ or $L\supset$. If it is the former, we have the steps

$$
\cfrac{\cfrac{\cfrac{P, P \supset Q \vee R \Rightarrow P \quad \cfrac{P, Q \Rightarrow Q \quad P, R \Rightarrow Q}{P, Q \vee R \Rightarrow Q}\,{\scriptstyle L\vee}}{P, P \supset Q \vee R \Rightarrow Q}\,{\scriptstyle L\supset}}{P \supset Q \vee R \Rightarrow P \supset Q}\,{\scriptstyle R\supset}}{P \supset Q \vee R \Rightarrow (P \supset Q) \vee (P \supset R)}\,{\scriptstyle R\vee_1}
$$

But the premiss $P, R \Rightarrow Q$ is not derivable. Else the last steps are

$$\dfrac{\dfrac{P \supset Q \vee R \Rightarrow P \quad Q \vee R \Rightarrow P}{P \supset Q \vee R \Rightarrow P} \, L\supset \quad Q \vee R \Rightarrow P \supset Q}{\dfrac{P \supset Q \vee R \Rightarrow P \supset Q}{P \supset Q \vee R \Rightarrow (P \supset Q) \vee (P \supset R)} \, R\vee_1} \, L\supset$$

Conclusion of $L\supset$ is repeated in the premiss. If the last rule was $R\vee_2$, underivability follows in an entirely similar way. The only remaining possibility is that the rule was $L\supset$, and we have as the last steps

$$\dfrac{\dfrac{P \supset Q \vee R \Rightarrow P \quad Q \vee R \Rightarrow P}{P \supset Q \vee R \Rightarrow P} \, L\supset \quad Q \vee R \Rightarrow (P \supset Q) \vee (P \supset R)}{P \supset Q \vee R \Rightarrow (P \supset Q) \vee (P \supset R)} \, L\supset$$

Again the conclusion of the upper $L\supset$ was repeated in the premiss. We do not need to analyze the right premiss, since the proof search fails in any case.   QED.

(d) Independence of the intuitionistic connectives: None of the standard interdefinabilities of classical propositional logic obtain in intuitionistic logic. By arguments similar to those above, it is shown that the following sequents are underivable:

   (i) $\sim(\sim A \& \sim B) \Rightarrow A \vee B$,
   (ii) $\sim A \supset B \Rightarrow A \vee B$,
   (iii) $\sim(A \& \sim B) \Rightarrow A \supset B$.

(e) Decidability of intuitionistic propositional logic: In the above examples, we were able to survey all possible derivations and found by various arguments that none turned out to be good. This depended essentially on having all derivations contraction- and cut-free.

Theorem 2.5.7: *Derivability of a sequent* $\Gamma \Rightarrow C$ *in the calculus* **G3ip** *is decidable.*

Proof: We generate all possible finite derivation trees with endsequent $\Gamma \Rightarrow C$ and show them to be bounded in number. Starting with $\Gamma \Rightarrow C$, we write all instances of rules that conclude it, then do the same for all the premisses of the last step. All rules except $L\supset$ reduce the sequent to be derived into ones with less weight, where the weight of a sequent is the sum of the weights of its formulas. If in a proof search we arrive at a sequent that does not reduce by any rule, then if it is not an axiom or conclusion of $L\bot$, we terminate the proof search. If in a proof search we have two applications of $L\supset$ that conclude the same sequent, we also terminate the proof search. Application of $L\supset$ root-first can produce only a bounded number of different sequents as premisses. Therefore each proof search tree terminates. If there is one tree all leaves of which are axioms or conclusions of $L\perp$, the endsequent is derivable; if not, it is underivable.   QED.

This algorithm of proof search is not very efficient, as one can see by trying, say, the disjunction property under negative hypothesis. There are sequent calculi for intuitionistic propositional logic that are much better in this respect. One such calculus will be studied in Section 5.5.

NOTES TO CHAPTER 2

Constructive real numbers and constructive analysis is treated in Bishop and Bridges (1985). The two-volume book of Troelstra and van Dalen (1988) is an encyclopedia of metamathematical studies on constructive logic and formal systems of constructive mathematics. A discussion of predicativity, with references to original papers by Poincaré and Russell, is found in Kleene (1952, p. 42). The same reference also discusses the background and development of intuitionism (ibid., p. 46).

The calculus **G3ip** is the propositional part of a single succedent version of Dragalin's (1988) calculus and is presented as such in Troelstra and Schwichtenberg (1996). The proofs of admissibility of contraction and cut follow the method of Dragalin, with inversion lemmas and induction on height of derivation. The proof in Dragalin (1988) is an outline; a detailed presentation is given in Dyckhoff (1997).

3

Sequent Calculus for Classical Logic

There are many formulations of sequent calculi. Historically, Gentzen first found systems of natural deduction for intuitionistic and classical logic, denoted by **NJ** and **NK**, respectively, but was not able to find a normal form for derivations in **NK**. To this purpose, he developed the classical sequent calculus **LK** that had sequences of formulas also in the succedent part. In our notation, such **multisuccedent** sequents are written as $\Gamma \Rightarrow \Delta$, where both Γ and Δ are multisets of formulas. Gentzen (1934–35) gives what is now called the **denotational** interpretation of multisuccedent sequents: The conjunction of formulas in Γ implies the disjunction of formulas in Δ. But the **operational** interpretation of single succedent sequents $\Gamma \Rightarrow C$, as expressing that from assumptions Γ, conclusion C **can be derived**, does not extend to multiple succedents.

Gentzen's somewhat later explanation of the multisuccedent calculus is that it is a natural representation of the **division into cases** often found in mathematical proofs (1938, p. 21). Proofs by cases are met in natural deduction in disjunction elimination, where a common consequence C of the two disjuncts A and B is sought, permitting to conclude C from $A \vee B$. There is a generalization of natural deduction into a **multiple conclusion** calculus that includes this mode of inference. Gentzen suggests such a multiple conclusion rule for disjunction (ibid., p. 21):

$$\frac{A \vee B}{A \quad B}$$

Disjunction elimination corresponds to arriving at the same formula C along both downward branches.

Along these lines, we may read a sequent $\Gamma \Rightarrow \Delta$ as consisting of the **open assumptions** Γ and the **open cases** Δ. Logical rules change and combine open assumptions and cases: $L\&$ replaces the open assumptions A, B by the open assumption $A\&B$, and there will be a dual multisuccedent rule $R\vee$ that changes the open cases A, B into the open case $A \vee B$, and so on. If there is just one case, we have the situation of an ordinary conclusion from open assumptions. Finally, we can have, as a dual to an empty assumption, an **empty case** representing impossibility, with nothing on the right of the sequent arrow.

In an axiomatic formulation, classical logic is obtained from intuitionistic logic by the addition of the principle of excluded third to the logical axioms (Gentzen 1934–35, p. 117). In natural deduction, one adds that derivations may start from instances of the law $A \vee \sim A$ (Gentzen, ibid., p. 81). Alternatively, one may add either the rule $\frac{\sim\sim A}{A}$ (Gentzen, ibid.) or the rule of indirect proof (Prawitz 1965, p. 20):

In sequent calculus, in the words of Gentzen (ibid., p. 80), "the difference is characterized by the restriction on the succedent," that is, a calculus for intuitionistic logic is obtained from the classical calculus **LK** by restricting the succedent to be one (alternatively, at most one) formula. The essential point here is that the classical $R\supset$ rule

$$\frac{A, \Gamma \Rightarrow \Delta, B}{\Gamma \Rightarrow \Delta, A \supset B}$$

becomes

$$\frac{A, \Gamma \Rightarrow B}{\Gamma \Rightarrow A \supset B}$$

An instance of the former is

$$\frac{A \Rightarrow A, \bot}{\Rightarrow A, A \supset \bot}$$

By the multisuccedent $R\vee$ rule, the cases $A, A \supset \bot$ can be replaced by the disjunction $A \vee (A \supset \bot)$, a derivation of the law of excluded middle that gets barred in the intuitionistic calculus.

It is, however, possible to give an operational interpretation to a restricted multisuccedent calculus corresponding precisely to intuitionistic derivability, as will be shown in Chapter 5. Therefore, it is not the feature of having a multiset as a succedent that leads to classical logic, but the unrestricted $R\supset$ rule. If only one formula is permitted in the succedent of its premiss, comma on the right can be interpreted as an intuitionistic disjunction.

3.1. AN INVERTIBLE CLASSICAL CALCULUS

We give the rules for a calculus **G3cp** of classical propositional logic and show that they are all invertible. Then we describe a variant of the calculus with negation as a primitive connective.

(a) The calculus G3cp: Sequents are of the form $\Gamma \Rightarrow \Delta$, where Γ and Δ are finite multisets and Γ and Δ can be empty. In contrast to the single succedent calculus, it is possible to have sequents of the form $\Gamma \Rightarrow$ and even $\Rightarrow$. One of the admissible structural rules of the multisuccedent calculus will be right weakening, from which it follows that if $\Gamma \Rightarrow$ is derivable, then also $\Gamma \Rightarrow \bot$ is derivable.

G3cp

Logical axiom:

$P, \Gamma \Rightarrow \Delta, P$

Logical rules:

$$\frac{A, B, \Gamma \Rightarrow \Delta}{A \& B, \Gamma \Rightarrow \Delta} \, L\& \qquad\qquad \frac{\Gamma \Rightarrow \Delta, A \quad \Gamma \Rightarrow \Delta, B}{\Gamma \Rightarrow \Delta, A \& B} \, R\&$$

$$\frac{A, \Gamma \Rightarrow \Delta \quad B, \Gamma \Rightarrow \Delta}{A \vee B, \Gamma \Rightarrow \Delta} \, L\vee \qquad\qquad \frac{\Gamma \Rightarrow \Delta, A, B}{\Gamma \Rightarrow \Delta, A \vee B} \, R\vee$$

$$\frac{\Gamma \Rightarrow \Delta, A \quad B, \Gamma \Rightarrow \Delta}{A \supset B, \Gamma \Rightarrow \Delta} \, L\supset \qquad\qquad \frac{A, \Gamma \Rightarrow \Delta, B}{\Gamma \Rightarrow \Delta, A \supset B} \, R\supset$$

$$\frac{}{\bot, \Gamma \Rightarrow \Delta} \, L\bot$$

The logical rules display the perfect duality of left and right rules for conjunction and disjunction, of which only the duality $L\vee - R\&$ could be observed in the intuitionistic calculus. Here there is only one right disjunction rule, and it is invertible, and also the left implication rule is invertible, with no need to repeat the principal formula in the left premiss, which has profound consequences for the structure of derivations and for proof search.

Theorem 3.1.1: Height-preserving inversion. *All rules of* **G3cp** *are invertible, with height-preserving inversion.*

Proof: For $L\&$, $L\vee$, and the second premiss of $L\supset$, the proof goes through as in Lemma 2.3.5, with Δ in place of C. We proceed from there with a proof by induction on height of derivation:

If the endsequent is $A \supset B, \Gamma \Rightarrow \Delta$ with $A \supset B$ not principal, the last rule has one or two premisses $A \supset B, \Gamma' \Rightarrow \Delta'$ and $A \supset B, \Gamma'' \Rightarrow \Delta''$, of derivation height $\leqslant n$, so by inductive hypothesis, $\Gamma' \Rightarrow \Delta', A$ and $\Gamma'' \Rightarrow \Delta'', A$ have derivations of height $\leqslant n$: Now apply the last rule to these premisses to conclude $\Gamma \Rightarrow \Delta, A$ with height of derivation $\leqslant n + 1$.

If $A \supset B$ is principal in the last rule, the premiss $\Gamma \Rightarrow \Delta, A$ has a derivation of height $\leqslant n$.

We now prove invertibility of the right rules:

If $\Gamma \Rightarrow \Delta, A\&B$ is an axiom or conclusion of $L\perp$, then, $A\&B$ not being atomic, also $\Gamma \Rightarrow \Delta, A$ and $\Gamma \Rightarrow \Delta, B$ are axioms or conclusions of $L\perp$. Assume height-preserving inversion up to height n and let $\vdash_{n+1} \Gamma \Rightarrow \Delta, A\&B$. There are two cases:

If $A\&B$ is not principal in the last rule, it has one or two premises, $\Gamma' \Rightarrow \Delta', A\&B$ and $\Gamma'' \Rightarrow \Delta'', A\&B$, of derivation height $\leqslant n$, so by inductive hypothesis, $\vdash_n \Gamma' \Rightarrow \Delta', A$ and $\vdash_n \Gamma' \Rightarrow \Delta', B$ and $\vdash_n \Gamma'' \Rightarrow \Delta'', A$ and $\vdash_n \Gamma'' \Rightarrow \Delta'', B$. Now apply the last rule to these premises to conclude $\Gamma \Rightarrow \Delta, A$ and $\Gamma \Rightarrow \Delta, B$ with a height of derivation $\leqslant n + 1$.

If $A\&B$ is principal in the last rule, the premises $\Gamma \Rightarrow \Delta, A$ and $\Gamma \Rightarrow \Delta, B$ have derivations of height $\leqslant n$.

If $\Gamma \Rightarrow \Delta, A \vee B$ is an axiom or conclusion of $L\perp$, then, $A \vee B$ not being atomic, also $\Gamma \Rightarrow \Delta, A, B$ is an axiom or conclusion of $L\perp$. Assume height-preserving inversion up to height n and let $\vdash_{n+1} \Gamma \Rightarrow \Delta, A \vee B$. There are again two cases:

If $A \vee B$ is not principal in the last rule, it has one or two premises $\Gamma' \Rightarrow \Delta', A \vee B$ and $\Gamma'' \Rightarrow \Delta'', A \vee B$, of derivation height $\leqslant n$, so by inductive hypothesis, $\vdash_n \Gamma' \Rightarrow \Delta', A, B$ and $\vdash_n \Gamma'' \Rightarrow \Delta'', A, B$. Now apply the last rule to these premises to conclude $\Gamma \Rightarrow \Delta, A, B$ with a height of derivation $\leqslant n + 1$.

If $A \vee B$ is principal in the last rule, the premiss $\Gamma \Rightarrow \Delta, A, B$ has a derivation of height $\leqslant n$.

If $\Gamma \Rightarrow \Delta, A \supset B$ is an axiom or conclusion of $L\perp$, then, $A \supset B$ not being atomic, also $A, \Gamma \Rightarrow \Delta, B$ is an axiom or conclusion of $L\perp$. Assume height-preserving inversion up to height n and let $\vdash_{n+1} \Gamma \Rightarrow \Delta, A \supset B$. As above, there are two cases:

If $A \supset B$ is not principal in the last rule, it has one or two premises $\Gamma' \Rightarrow \Delta', A \supset B$ and $\Gamma'' \Rightarrow \Delta'', A \supset B$, of derivation height $\leqslant n$, so by inductive hypothesis, $\vdash_n A, \Gamma' \Rightarrow \Delta', B$ and $\vdash_n A, \Gamma'' \Rightarrow \Delta'', B$. Now apply the last rule to these premises to conclude $A, \Gamma \Rightarrow \Delta, B$ with a derivation of height $\leqslant n + 1$.

If $A \supset B$ is principal in the last rule, the premiss $A, \Gamma \Rightarrow \Delta, B$ has a derivation of height $\leqslant n$.   QED.

Given a sequent $\Gamma \Rightarrow \Delta$, each step of a root-first proof search is a reduction that removes a connective and it follows that proof search terminates. The leaves are topsequents of form

$$\perp, \ldots, \perp, P_1, \ldots, P_m \Rightarrow Q_1, \ldots, Q_n, \perp, \ldots, \perp$$

where the number of $\perp$'s in the antecedent or succedent as well as m or n can be 0.

Lemma 3.1.2: *The decomposition of a sequent* $\Gamma \Rightarrow \Delta$ *into topsequents in* **G3cp** *is unique.*

Proof: By noting that successive application of any two logical rules in **G3cp** commutes.   QED.

Root-first proof search gives a method for finding a representation of formulas of propositional logic in a certain **normal form**: Given a formula C, apply the decomposition to $\Rightarrow C$, and after having reduced all connectives, remove those topsequents that are axioms or conclusions of $L\bot$, i.e., those that have the same atom in the antecedent and succedent or $\bot$ in the antecedent.

Definition 3.1.3: *A regular sequent is a sequent of the form* $P_1, \ldots, P_m \Rightarrow Q_1, \ldots, Q_n, \bot, \ldots, \bot$ *where* $P_i \neq Q_j$, *the antecedent is empty if* $m = 0$, *and the succedent is* $\bot$ *if* $n = 0$. *The* trace formula *of a regular sequent is*

1. $P_1 \& \ldots \& P_m \supset Q_1 \vee \ldots \vee Q_n$ *for* $m, n > 0$,
2. $Q_1 \vee \ldots \vee Q_n$ *for* $m = 0, n > 0$,
3. $\sim(P_1 \& \ldots \& P_m)$ *for* $m > 0, n = 0$,
4. $\bot$ *for* $m, n = 0$,

where possible repetitions of the P_i *or* Q_j *in the regular sequent are deleted.*

Regular sequents correspond to Gentzen's (1934–35) "basic mathematical sequents," except that Gentzen did not have $\bot$ as a primitive. The term "regular" is explained in Chapter 6. Trace formulas of regular sequents are unique up to the order in the disjunctions and conjunctions. By the invertibility of the rules of **G3cp**, a regular sequent with trace formula C is derivable if and only if the sequent $\Rightarrow C$ is derivable. It follows that a formula is equivalent to the conjunction of its trace formulas:

Theorem 3.1.4: *A formula C is equivalent to the conjunction of the trace formulas of its decomposition into regular sequents.*

Proof: Let the topsequents of the decomposition of $\Rightarrow C$ be $\Gamma_1 \Rightarrow \Delta_1, \ldots, \Gamma_m \Rightarrow \Delta_m$, with the n first giving the trace formulas $C_1, \ldots, C_n$ and the rest, if $m > n$, having $\bot$ in the antecedent or the same atom in the antecedent and succedent. We have to show that $\Rightarrow C \supset\subset C_1 \& \ldots \& C_n$ is derivable. We have a derivation of $\Rightarrow C$ from $\Gamma_1 \Rightarrow \Delta_1, \ldots, \Gamma_m \Rightarrow \Delta_m$ that uses invertible rules. By adding the formula C to the antecedent of each sequent in the derivation, we obtain a derivation of $C \Rightarrow C$ from $C, \Gamma_1 \Rightarrow \Delta_1, \ldots, C, \Gamma_m \Rightarrow \Delta_m$ by the same invertible rules. Therefore each step in each root-first path, from $C \Rightarrow C$ to $C, \Gamma_i \Rightarrow \Delta_i$, is admissible. Since $C \Rightarrow C$ is derivable, each $C, \Gamma_i \Rightarrow \Delta_i$ is derivable. It follows that, for each trace formula, up to n, the sequent $C \Rightarrow C_i$ is derivable.

Therefore, by repeated application of $R\&$, $C \Rightarrow C_1 \& \ldots \& C_n$ is derivable, and by $R \supset$, $\Rightarrow C \supset C_1 \& \ldots \& C_n$ is derivable.

Conversely, starting from the given derivation of $\Rightarrow C$ from topsequents $\Gamma_1 \Rightarrow \Delta_1, \ldots, \Gamma_m \Rightarrow \Delta_m$, add the formulas $C_1, \ldots, C_n$ to the antecedent of each sequent in the derivation to obtain a derivation of $C_1, \ldots, C_n \Rightarrow C$ from new topsequents of the form $C_1, \ldots, C_n, \Gamma_i \Rightarrow \Delta_i$. For $i > n$, such sequents are axioms since they have $\bot$ in the antecedent or the same atom in the antecedent and succedent. For $i \leqslant n$ they are derivable since each $C_1, \ldots, C_n \Rightarrow C_i$ is derivable. Application of $L\&$ and $R\supset$ to $C_1, \ldots, C_n \Rightarrow C$ now gives a derivation of $\Rightarrow C_1 \& \ldots \& C_n \supset C$.   QED.

As a consequence of Lemma 3.1.2, the representation given by the theorem is unique up to order in the conjunction and the conjunctions and disjunctions in the trace formulas. Each trace formula $P_1 \& \ldots \& P_m \supset Q_1 \vee \ldots \vee Q_n$ is classically equivalent to $\sim P_1 \vee \ldots \vee \sim P_m \vee Q_1 \vee \ldots \vee Q_n$; the representation is in effect a variant of the **conjunctive normal form** of formulas of classical propositional logic.

(b) Negation as a primitive connective: In Gentzen's original classical sequent calculus **LK** of 1934–35, negation was a primitive, with two rules that make a negation appear on the left and the right part of the conclusion, respectively:

$$\frac{\Gamma \Rightarrow \Delta, A}{\sim A, \Gamma \Rightarrow \Delta} L\sim \qquad \frac{A, \Gamma \Rightarrow \Delta}{\Gamma \Rightarrow \Delta, \sim A} R\sim$$

Now negation displays the same elegant symmetry of left and right rules as the other connectives. Some years later, Gentzen commented on this property of the multisuccedent calculus as follows (1938, p. 25): "The special role of negation, an annoying exception in the natural deduction calculus, has been completely removed, in a way approaching magic. I should be permitted to express myself thus since I was, when putting up the calculus **LK** for the first time, greatly surprised that it had such a property."

Gentzen's rules for negation, with the definition $\sim A \equiv A \supset \bot$, are admissible in **G3cp**, the first one by

$$\frac{\Gamma \Rightarrow \Delta, A \quad \overline{\bot, \Gamma \Rightarrow \Delta}^{L\bot}}{A \supset \bot, \Gamma \Rightarrow \Delta} L\supset$$

and the second one by

$$\frac{\dfrac{A, \Gamma \Rightarrow \Delta}{A, \Gamma \Rightarrow \Delta, \bot} RW}{\Gamma \Rightarrow \Delta, A \supset \bot} R\supset$$

where RW is **right weakening**, to be proved admissible in the next section.

3.2. ADMISSIBILITY OF STRUCTURAL RULES

We shall prove admissibility of weakening, contraction, and cut for the calculus **G3cp**. There will be two weakening rules, a **left** one for weakening in the antecedent and a **right** one for weakening in the succedent, and similarly for contraction. The rules are as follows:

$$\frac{\Gamma \Rightarrow \Delta}{A, \Gamma \Rightarrow \Delta} LW \qquad \frac{\Gamma \Rightarrow \Delta}{\Gamma \Rightarrow \Delta, A} RW \qquad \frac{A, A, \Gamma \Rightarrow \Delta}{A, \Gamma \Rightarrow \Delta} LC \qquad \frac{\Gamma \Rightarrow \Delta, A, A}{\Gamma \Rightarrow \Delta, A} RC$$

The proofs of admissibility of left and right weakening are similar to the proof of height-preserving weakening for **G3ip** in Theorem 2.3.4:

Theorem 3.2.1: Height-preserving weakening. *If* $\vdash_n \Gamma \Rightarrow \Delta$, *then* $\vdash_n A, \Gamma \Rightarrow \Delta$. *If* $\vdash_n \Gamma \Rightarrow \Delta$, *then* $\vdash_n \Gamma \Rightarrow \Delta, A$.

Proof: The addition of formula A to the antecedent and consequent, respectively, of each sequent in the derivation of $\Gamma \Rightarrow \Delta$ will produce derivations of $A, \Gamma \Rightarrow \Delta$ and $\Gamma \Rightarrow \Delta, A$. QED.

It follows that if a sequent $\Gamma \Rightarrow$ with an empty succedent is derivable, the sequent $\Gamma \Rightarrow \bot$ also is derivable.

Theorem 3.2.2: Height-preserving contraction. *If* $\vdash_n C, C, \Gamma \Rightarrow \Delta$, *then* $\vdash_n C, \Gamma \Rightarrow \Delta$. *If* $\vdash_n \Gamma \Rightarrow \Delta, C, C$, *then* $\vdash_n \Gamma \Rightarrow \Delta, C$.

Proof: The proof of admissibility of left and right contraction is done simultaneously by induction on height of derivation of the premiss. For $n = 0$, if the premiss is an axiom or conclusion of $L\bot$, the conclusion also is an axiom or conclusion of $L\bot$, whether contraction was applied on the left or right. For the inductive case, assume height-preserving left and right contraction up to derivations of height n. As in the proof of contraction for the single succedent calculus, Theorem 2.4.1, we distinguish two cases: If the contraction formula is not principal in the last rule applied, we apply the inductive hypothesis to the premisses and then the rule. If the contraction formula is principal, we have six subcases according to the last rule applied.

If the last rule is $L\&$ or $L\vee$, the proof proceeds as in Theorem 2.4.1. If the last rule is $R\&$, the premisses are $\vdash_n \Gamma \Rightarrow \Delta, A\&B, A$ and $\vdash_n \Gamma \Rightarrow \Delta, A\&B, B$. By height-preserving invertibility, we obtain $\vdash_n \Gamma \Rightarrow \Delta, A, A$ and $\vdash_n \Gamma \Rightarrow \Delta, B, B$, and the inductive hypothesis gives $\vdash_n \Gamma \Rightarrow \Delta, A$ and $\vdash_n \Gamma \Rightarrow \Delta, B$. The conclusion $\vdash_{n+1} \Gamma \Rightarrow \Delta, A\&B$ follows by $R\&$. If the last rule is $R\vee$, the premiss is $\vdash_n \Gamma \Rightarrow \Delta, A \vee B, A, B$ and we apply height-preserving invertibility to conclude $\vdash_n \Gamma \Rightarrow \Delta, A, B, A, B$, then the inductive hypothesis twice to obtain $\vdash_n \Gamma \Rightarrow \Delta, A, B$, and last $R\vee$.

If the last rule is $R\supset$, the premiss is $\vdash_n A, \Gamma \Rightarrow \Delta, A \supset B, B$ and we apply height-preserving invertibility to conclude $\vdash_n A, A, \Gamma \Rightarrow \Delta, B, B$, then the inductive hypothesis to conclude $\vdash_n A, \Gamma \Rightarrow \Delta, B$ and then $R\supset$. If $L\supset$ was applied, we have the derivation of the premiss of contraction,

$$\frac{A \supset B, \Gamma \Rightarrow \Delta, A \quad B, A \supset B, \Gamma \Rightarrow \Delta}{A \supset B, A \supset B, \Gamma \Rightarrow \Delta} L\supset$$

By height-preserving inversion, we have $\vdash_n \Gamma \Rightarrow \Delta, A, A$ and $\vdash_n B, B, \Gamma \Rightarrow \Delta$. By the inductive hypothesis, we have $\vdash_n \Gamma \Rightarrow \Delta, A$ and $\vdash_n B, \Gamma \Rightarrow \Delta$, and obtain a derivation of $A \supset B, \Gamma \Rightarrow \Delta$ in at most $n + 1$ steps.   QED.

A proof by separate induction on left and right contraction will not go through if the last rule is $L\supset$ or $R\supset$.

Theorem 3.2.3: *The rule of cut,*

$$\frac{\Gamma \Rightarrow \Delta, D \quad D, \Gamma' \Rightarrow \Delta'}{\Gamma, \Gamma' \Rightarrow \Delta, \Delta'} Cut$$

is admissible in **G3cp**.

Proof: The proof is organized as that of Theorem 2.4.3, with the same numbering of cases.

Cut with an axiom or conclusion of $L\perp$ as premiss: If at least one of the premisses of cut is an axiom, we distinguish two cases:

1. The **left premiss** $\Gamma \Rightarrow \Delta, D$ of cut is an axiom or conclusion of $L\perp$. There are three subcases:

1.1. The cut formula D is in Γ. In this case we derive $\Gamma, \Gamma' \Rightarrow \Delta, \Delta'$ from the right premiss $D, \Gamma' \Rightarrow \Delta'$ by weakening.

1.2. Γ and Δ have a common atom. Then $\Gamma, \Gamma' \Rightarrow \Delta, \Delta'$ is an axiom.

1.3. $\perp$ is a formula in Γ. Then $\Gamma, \Gamma' \Rightarrow \Delta, \Delta'$ is a conclusion of $L\perp$.

2. The **right premiss** $D, \Gamma' \Rightarrow \Delta'$ is an axiom or conclusion of $L\perp$. There are four subcases:

2.1. D is in Δ'. Then $\Gamma, \Gamma' \Rightarrow \Delta, \Delta'$ follows from the first premiss by weakening.

2.2. Γ' and Δ' have a common atom. Then $\Gamma, \Gamma' \Rightarrow \Delta, \Delta'$ is an axiom.

2.3. $\perp$ is in Γ'. Then $\Gamma, \Gamma' \Rightarrow \Delta, \Delta'$ is a conclusion of $L\perp$.

2.4. $D = \perp$. Then either the first premiss is an axiom or conclusion of $L\perp$ and $\Gamma, \Gamma' \Rightarrow \Delta, \Delta'$ follows as in case 1, or $\Gamma \Rightarrow \Delta, \perp$ has been derived. There are six cases according to the rule used. These are transformed into derivations with cuts of lesser cut-height. Since $\perp$ is never principal in a rule and the transformations are special cases of transformations 3.1–3.6 below, with $D = \perp$, they need not be written out here.

Cut with neither premiss an axiom: We have three cases:

3. Cut formula D is **not principal** in the left premiss. We have six subcases according to the rule used to derive the left premiss. For $L\&$ and $L\vee$, the transformations are analogous to those of cases 3.1 and 3.2 of Theorem 2.4.3. For implication, we have

3.3. $L\supset$, with $\Gamma = A \supset B, \Gamma''$. The derivation

$$\cfrac{\cfrac{\Gamma'' \Rightarrow \Delta, D, A \quad B, \Gamma'' \Rightarrow \Delta, D}{A \supset B, \Gamma'' \Rightarrow \Delta, D}\ L\supset \quad D, \Gamma' \Rightarrow \Delta'}{A \supset B, \Gamma'', \Gamma' \Rightarrow \Delta, \Delta'}\ Cut$$

is transformed into the derivation

$$\cfrac{\cfrac{\Gamma'' \Rightarrow \Delta, D, A \quad D, \Gamma' \Rightarrow \Delta'}{\Gamma'', \Gamma' \Rightarrow \Delta, \Delta', A}\ Cut \quad \cfrac{B, \Gamma'' \Rightarrow \Delta, D \quad D, \Gamma' \Rightarrow \Delta'}{B, \Gamma'', \Gamma' \Rightarrow \Delta, \Delta'}\ Cut}{A \supset B, \Gamma'', \Gamma' \Rightarrow \Delta, \Delta'}\ L\supset$$

with two cuts of lower cut-height.

3.4. $R\&$, with $\Delta = A\&B, \Delta''$. The derivation

$$\cfrac{\cfrac{\Gamma \Rightarrow \Delta'', A, D \quad \Gamma \Rightarrow \Delta'', B, D}{\Gamma \Rightarrow \Delta'', A\&B, D}\ R\& \quad D, \Gamma' \Rightarrow \Delta'}{\Gamma, \Gamma' \Rightarrow \Delta'', A\&B, \Delta'}\ Cut$$

is transformed into the derivation with two cuts of lower height

$$\cfrac{\cfrac{\Gamma \Rightarrow \Delta'', A, D \quad D, \Gamma' \Rightarrow \Delta'}{\Gamma, \Gamma' \Rightarrow \Delta'', A, \Delta'}\ Cut \quad \cfrac{\Gamma \Rightarrow \Delta'', B, D \quad D, \Gamma' \Rightarrow \Delta'}{\Gamma, \Gamma' \Rightarrow \Delta'', B, \Delta'}\ Cut}{\Gamma, \Gamma' \Rightarrow \Delta'', A\&B, \Delta'}\ R\&$$

3.5. $R\vee$, with $\Delta = A \vee B, \Delta''$. The derivation

$$\cfrac{\cfrac{\Gamma \Rightarrow \Delta'', A, B, D}{\Gamma \Rightarrow \Delta'', A \vee B, D}\ R\vee \quad D, \Gamma' \Rightarrow \Delta'}{\Gamma, \Gamma' \Rightarrow \Delta'', A \vee B, \Delta'}\ Cut$$

is transformed into the derivation with a cut of lower cut-height:

$$\cfrac{\cfrac{\Gamma \Rightarrow \Delta'', A, B, D \quad D, \Gamma' \Rightarrow \Delta'}{\Gamma, \Gamma' \Rightarrow \Delta'', A, B, \Delta'}\ Cut}{\Gamma, \Gamma' \Rightarrow \Delta'', A \vee B, \Delta'}\ R\vee$$

3.6. $R \supset$, with $\Delta = A \supset B, \Delta''$. The derivation

$$\dfrac{\dfrac{\Gamma, A \Rightarrow \Delta'', B, D}{\Gamma \Rightarrow \Delta'', A \supset B, D} {\scriptstyle R \supset} \qquad D, \Gamma' \Rightarrow \Delta'}{\Gamma, \Gamma' \Rightarrow \Delta'', A \supset B, \Delta'} {\scriptstyle Cut}$$

is transformed into the derivation with a cut of lower cut-height:

$$\dfrac{\dfrac{\Gamma, A \Rightarrow \Delta'', B, D \qquad D, \Gamma' \Rightarrow \Delta'}{\Gamma, \Gamma', A \Rightarrow \Delta'', B} {\scriptstyle Cut}}{\Gamma, \Gamma' \Rightarrow \Delta'', A \supset B, \Delta'} {\scriptstyle R \supset}$$

4. Cut formula D is **principal in the left premiss only**, and the derivation is transformed in one with a cut of lower cut-height according to derivation of the right premiss. We have six subcases according to the rule used. Only the cases of $L \supset$ and $R \vee$ are significantly different from the cases of Theorem 2.4.3:

4.3. $L \supset$, with $\Delta = A \supset B, \Delta'$. The derivation and its transformation are similar to those of previous case 3.3.

4.5. $R \vee$, with $\Delta = A \vee B, \Delta''$. The derivation

$$\dfrac{\Gamma \Rightarrow \Delta, D \qquad \dfrac{D, \Gamma' \Rightarrow A, B, \Delta}{D, \Gamma' \Rightarrow A \vee B, \Delta''} {\scriptstyle R \vee}}{\Gamma, \Gamma' \Rightarrow \Delta, A \vee B, \Delta''} {\scriptstyle Cut}$$

is transformed into the derivation with a cut of lower cut-height

$$\dfrac{\dfrac{\Gamma \Rightarrow \Delta, D \qquad D, \Gamma' \Rightarrow A, B, \Delta''}{\Gamma, \Gamma' \Rightarrow \Delta, A, B, \Delta''} {\scriptstyle Cut}}{\Gamma, \Gamma' \Rightarrow \Delta, A \vee B, \Delta''} {\scriptstyle R \vee}$$

5. Cut formula D is **principal in both premisses**, and we have three subcases, of which conjunction is very similar to that of case 5.1 of Theorem 2.4.3.

5.2. $D = A \vee B$, and the derivation

$$\dfrac{\dfrac{\Gamma \Rightarrow \Delta, A, B}{\Gamma \Rightarrow \Delta, A \vee B} {\scriptstyle R \vee} \qquad \dfrac{A, \Gamma' \Rightarrow \Delta' \qquad B, \Gamma' \Rightarrow \Delta'}{A \vee B, \Gamma' \Rightarrow \Delta'} {\scriptstyle L \vee}}{\Gamma, \Gamma' \Rightarrow \Delta, \Delta'} {\scriptstyle Cut}$$

is transformed into

$$\dfrac{\dfrac{\dfrac{\Gamma \Rightarrow \Delta, A, B \qquad A, \Gamma' \Rightarrow \Delta'}{\Gamma, \Gamma' \Rightarrow \Delta, \Delta', B} {\scriptstyle Cut} \qquad B, \Gamma' \Rightarrow \Delta'}{\Gamma, \Gamma', \Gamma' \Rightarrow \Delta, \Delta', \Delta'} {\scriptstyle Cut}}{\Gamma, \Gamma' \Rightarrow \Delta, \Delta'} {\scriptstyle Ctr}$$

with two cuts of lower cut-height.

5.3. $D = A \supset B$, and the derivation

$$\dfrac{\dfrac{A, \Gamma \Rightarrow \Delta, B}{\Gamma \Rightarrow \Delta, A \supset B} R\supset \quad \dfrac{\Gamma' \Rightarrow \Delta', A \quad B, \Gamma' \Rightarrow \Delta'}{A \supset B, \Gamma' \Rightarrow \Delta'} L\supset}{\Gamma, \Gamma' \Rightarrow \Delta, \Delta'} Cut$$

is transformed into the derivation with two cuts of lower cut-heights:

$$\dfrac{\dfrac{\Gamma' \Rightarrow \Delta', A \quad A, \Gamma \Rightarrow \Delta, B}{\Gamma, \Gamma' \Rightarrow \Delta, \Delta', B} Cut \quad B, \Gamma' \Rightarrow \Delta'}{\dfrac{\Gamma, \Gamma', \Gamma' \Rightarrow \Delta, \Delta', \Delta'}{\Gamma, \Gamma' \Rightarrow \Delta, \Delta'} Ctr} Cut$$

QED.

We obtain, just as for the calculus **G3ip**, the following subformula property.

Corollary 3.2.4: *Each formula in the derivation of* $\Gamma \Rightarrow \Delta$ *in* **G3cp** *is a subformula of* Γ, Δ.

It follows in particular that the sequent $\Rightarrow$ is not derivable. We concluded from the admissibility of weakening that if $\Gamma \Rightarrow$ is derivable, then also $\Gamma \Rightarrow \perp$ is derivable. We now obtain the converse by applying cut to $\Gamma \Rightarrow \perp$ and $\perp \Rightarrow$; thus an empty succedent behaves like $\perp$.

In intuitionistic logic, all connectives are needed, but in classical logic, negation and one of &, $\vee$, $\supset$ can express the remaining two. How does the interdefinability of connectives affect proof analysis? Gentzen says that one could replace some rules by others in classical sequent calculus, but that if this were done, the cut elimination theorem would not be provable anymore (1934–35, III. 2.1).

If we consider, say, the $\supset$, $\perp$ fragment of **G3cp**, the cut elimination theorem remains valid. Conjunction and disjunction can be defined in terms of implication and falsity; thus for any formula A there is a translated formula A^* in the fragment classically equivalent to it. Similarly, sequents $\Gamma \Rightarrow \Delta$ of **G3cp** have translations $\Gamma^* \Rightarrow \Delta^*$ derivable in the fragment if and only if the original sequent is derivable in **G3cp**. By the admissibility of cut, the derivation uses only the logical rules for implication and falsity.

Gentzen's statement about losing the cut elimination theorem is probably based on considerations of the following kind: According to Hilbert's program, logic and mathematics had to be represented as formal manipulations of **concrete signs**. In propositional logic, the signs are the connectives, atomic formulas, and parentheses. Once these are given, there is no question of **defining** one sign by another. However, it is permitted to reduce or change the set of formal axioms and rules by which the signs are manipulated. Thus one gets along in propositional logic with just one rule, modus ponens. The axioms for conjunction and disjunction in Hilbert-style, in Section 2.5(b) above, could in classical logic be replaced by the axioms $(\sim A \supset B) \supset A \vee B$, $A \vee B \supset (\sim A \supset B)$ and $\sim (A \supset \sim B) \supset A \& B$, $A \& B \supset \sim (A \supset \sim B)$. If these axioms are added to the fragment of **G3cp** in the same way as in Section 2.5(b), as sequents with empty antecedents, they can be put to use only by the rule of cut, and it is this phenomenon that Gentzen seems to have had in mind.

Later on Gentzen admitted, however, the possibility of "dispensing with the sign $\supset$ in the classical calculus **LK** by considering $A \supset B$ as an abbreviation for $\sim A \vee B$; it is easy to prove that rules $R\supset$ and $L\supset$ can be replaced by the rules for $\vee$ and $\sim$" (1934–35, III. 2.41).[1]

3.3. COMPLETENESS

The decomposability of formulas in **G3cp** can be turned into a proof of completeness of the calculus. For this purpose, we have to define the basic semantical concepts of classical propositional logic:

Definition 3.3.1: *A valuation is a function v from formulas of propositional logic to the values* 0, 1 *assumed to be given for all atoms,*

$$v(P) = 0 \quad or \quad v(P) = 1,$$

and extended inductively to all formulas,

$$v(\bot) = 0,$$
$$v(A \& B) = min(v(A), v(B)),$$
$$v(A \vee B) = max(v(A), v(B)),$$
$$v(A \supset B) = max(1 - v(A), v(B)).$$

Observe that, by definition of v, $v(A \supset B) = 1$ if and only if $v(A) \leqslant v(B)$. Valuations are extended to multisets Γ by taking conjunctions $\bigwedge(\Gamma)$ and disjunctions $\bigvee(\Gamma)$ of formulas in Γ, with $\bigwedge(\ ) = \bot$ and $\bigvee(\ ) = \top$ for the empty multiset and by setting

$$v \bigwedge(\Gamma) \equiv min(v(C)) \text{ for formulas } C \text{ in } \Gamma,$$
$$v \bigvee(\Gamma) \equiv max(v(C)) \text{ for formulas } C \text{ in } \Gamma.$$

Definition 3.3.2: *A sequent $\Gamma \Rightarrow \Delta$ is* refutable *if there is a valuation v such that $v \bigwedge(\Gamma) > v \bigvee(\Delta)$. Sequent $\Gamma \Rightarrow \Delta$ is* valid *if it is not refutable.*

It follows that $\Gamma \Rightarrow \Delta$ is valid if for all valuations v, $v \bigwedge(\Gamma) \leqslant v \bigvee(\Delta)$. For proving the soundness of **G3cp**, we need the following lemma about valuations:

---

[1] The text has "**NK**" (also in the English translation) that is Gentzen's name for classical natural deduction, but this must be a misprint since he expressly refers to rules of sequent calculus.

Lemma 3.3.3: *For a valuation* v, $min(v(A), v(B)) \leqslant v(C)$ *if and only if* $v(A) \leqslant v(B \supset C)$.

Proof: If $v(A) = 0$ the claim trivially holds. Else $min(v(A), v(B)) = v(B)$; thus $min(v(A), v(B)) \leqslant v(C)$ if and only if $v(B) \leqslant v(C)$, if and only if $v(B \supset C) = 1$, i.e., $v(A) \leqslant v(B \supset C)$. QED.

Corollary 3.3.4: $min(v(A \supset B), v(A)) \leqslant v(B)$.

Proof: Immediate by Lemma 3.3.3. QED.

Theorem 3.3.5: Soundness. *If a sequent* $\Gamma \Rightarrow \Delta$ *is derivable in* **G3cp**, *it is valid.*

Proof: Assume that $\Gamma \Rightarrow \Delta$ is derivable. We prove by induction on height of derivation that it is valid. If it is an axiom or conclusion of $L\bot$, it is valid since we always have $v \bigwedge(P, \Gamma) \leqslant v \bigvee(\Delta, P)$ and $v \bigwedge(\bot, \Gamma) \leqslant v \bigvee(\Delta)$.

If the last rule is $L\&$, we have by inductive hypothesis for all valuations v that $v \bigwedge(A, B, \Gamma) \leqslant v \bigvee(\Delta)$, and $v \bigwedge(A\&B, \Gamma) \leqslant v \bigvee(\Delta)$ follows by $v \bigwedge(A\&B, \Gamma) = v \bigwedge(A, B, \Gamma)$. The case for $R\vee$ is dual to this. For $L\vee$, we have $v \bigwedge(A, \Gamma) \leqslant v \bigvee(\Delta)$ and $v \bigwedge(B, \Gamma) \leqslant v \bigvee(\Delta)$. Then

$$v \bigwedge(A \vee B, \Gamma) = max(v \bigwedge(A, \Gamma), v \bigwedge(B, \Gamma)) \leqslant v \bigvee(\Delta).$$

The case of $R\&$ is dual to this. If the last rule is $L\supset$, suppose

$$v \bigwedge(\Gamma) \leqslant max(v \bigvee(\Delta), v(A)) \text{ and } min(v(B), v \bigwedge(\Gamma)) \leqslant v \bigvee(\Delta).$$

There are two cases: If $v \bigvee(\Delta) = 1$, then the conclusion is trivial. If $v \bigvee(\Delta) = 0$, then $v \bigwedge(\Gamma) \leqslant v(A)$ and $min(v(B), v \bigwedge(\Gamma)) \leqslant 0$. From the former follows

$$min(v(A \supset B), v \bigwedge(\Gamma)) \leqslant min(min(v(A \supset B), v(A)), v \bigwedge(\Gamma))$$

and therefore, by using Corollary 3.3.4, we obtain

$$min(v(A \supset B), v \bigwedge(\Gamma)) \leqslant min(v(B), v \bigwedge(\Gamma)) \leqslant 0.$$

If the last rule is $R\supset$, we have

$$min(v(A), v \bigwedge(\Gamma)) \leqslant max(v \bigvee(\Delta), v(B))$$

and there are two cases: If $v \bigvee(\Delta) = 1$, then the conclusion is trivial. Otherwise we have $min(v(A), v \bigwedge(\Gamma)) \leqslant v(B)$: hence by Lemma 3.3.3, $v \bigwedge(\Gamma) \leqslant v(A \supset B)$ and *a fortiori* $v \bigwedge(\Gamma) \leqslant max(v \bigvee(\Delta), v(A \supset B))$. QED.

Theorem 3.3.6: Completeness. *If a sequent* $\Gamma \Rightarrow \Delta$ *is valid, it is derivable in* **G3cp**.

Proof: Apply root-first the rules of **G3cp** to the sequent $\Gamma \Rightarrow \Delta$, obtaining leaves that are either axioms, conclusions of $L\perp$, or regular sequents. We prove that if $\Gamma \Rightarrow \Delta$ is valid, then the set of regular sequents is empty, and therefore $\Gamma \Rightarrow \Delta$ is derivable. Suppose that the set of regular sequents consists of $\Gamma_1 \Rightarrow \Delta_1, \ldots, \Gamma_n \Rightarrow \Delta_n$, with $n > 0$, and let C_i be their corresponding trace formulas. We have, by Theorem 3.1.4, $\Rightarrow C \supset\subset C_1 \& \ldots \& C_n$, where C is $\bigwedge(\Gamma) \supset \bigvee(\Delta)$. Since $\Gamma \Rightarrow \Delta$ is valid, by definition $v(C) = 1$ for every valuation v, and since $C \Rightarrow C_1 \& \ldots \& C_n$, by soundness $v(C) \leqslant v(C_1 \& \ldots \& C_n)$, which gives $v(C_i) = 1$ for each C_i and every valuation v. No C_i is $\perp$, since no valuation validates it. No C_i is $\sim(P_1 \& \ldots \& P_m)$ since the valuation with $v(P_j) = 1$ for all $j \leqslant m$ does not validate it. Finally no C_i is $P_1 \& \ldots \& P_m \supset Q_i \vee \ldots \vee Q_r$ or $Q_i \vee \ldots \vee Q_r$ since it is refuted by the valuation with $v(P_j) = 1$ for all $j \leqslant m$ and $v(Q_k) = 0$ for all $k \leqslant r$. QED.

Decomposition into regular sequents gives a syntactic **decision method** for formulas of classical propositional logic: A formula C is valid if and only if no topsequent is a regular sequent.

Notes to Chapter 3

The logical rules of the calculus **G3cp** first appear in Ketonen (1944, p. 14), the main results of whom were made known through the long review by Bernays (1945). Negation is a primitive connective, derivations start with axioms of the form $A \Rightarrow A$, and only cut is eliminated, the proof being similar to that of Gentzen. Invertibility is proved by structural rules.

Direct proofs of invertibility were given by Schütte (1950) and Curry (1963). The proofs of admissibility of structural rules we give follow the method of Dragalin, similarly to the intuitionistic calculus. Normal form by means of decomposition through invertible rules and the related completeness theorem are due to Ketonen (1944). He seems to have found his calculus by making systematic the necessity that anyone trying root-first proof search experiences, namely, that one has to repeat the contexts of the conclusion in both premises of two-premise rules. In an earlier expository paper, he gives an example of proof search and states that, because of invertibility of the propositional rules, the making of derivations consists of purely mechanical decomposition (1943, pp. 138–139).

The idea of validity as a negative notion, as in Definition 3.3.2, was introduced in Negri and von Plato (1998a).

4

The Quantifiers

In this chapter, we give the language and rules for intuitionistic and classical predicate logic. Proofs of admissibility of structural rules are extensions of the previous proofs for the propositional calculi **G3ip** and **G3cp**. We then present some basic consequences of cut elimination, such as the existence property and the lack of prenex normal form in intuitionistic logic. The invertible rules for the classical sequent calculus **G3c** are exploited to give a, possibly nonterminating, procedure of proof search. This procedure, called construction of the reduction tree for a given sequent, is the basis of Schütte's method for proving completeness of classical predicate logic. We give a completeness proof, using the reduction tree, but define validity through valuations, as an extension of the definition of validity of classical propositional logic in Section 3.3.

4.1. Quantifiers in Natural Deduction and in Sequent Calculus

(a) The language of predicate logic: The **language** of first-order logic contains constants $a, b, \ldots$, variables $x, y, \ldots$, n-place functions $f^n, g^n, \ldots$, and predicates $P^n, Q^n, \ldots$, for any $n \geqslant 0$, the zero-place connective $\bot$, the two-place connectives $\&, \vee, \supset$, and the quantifiers $\forall, \exists$. The arity of functions and predicates is often left unwritten. Constants and variables are sometimes written as $a_1, a_2, \ldots, x_1, x_2, \ldots$, or $a, a', \ldots, x, x', \ldots$. Constants can be thought of as zero-place functions, and there can be, analogously, zero-place constant propositions.

Terms are denoted by $t, u, \ldots$ or $t_1, t_2, \ldots$ or $t, t', \ldots$ and are defined inductively by the clauses:

1. Constants are terms,

2. Variables are terms,

3. Application of an n-ary function f^n to terms $t_1, \ldots, t_n$ gives a term $f^n(t_1, \ldots, t_n)$.

Formulas are defined inductively by the clauses:

1. $\perp$ is a formula,
2. Application of an n-ary predicate P^n to terms $t_1,\ldots,t_n$ gives a formula $P(t_1,\ldots,t_n)$,
3. If A and B are formulas, $A\&B$, $A \vee B$, and $A \supset B$ are formulas,
4. If A is a formula, $\forall x A$ and $\exists x A$ are formulas.

The set of **free variables** $FV(t)$ in a term t is defined inductively by:

1. For $t = a$, $FV(a) = \emptyset$,
2. For $t = x$, $FV(x) = \{x\}$,
3. For $t = f^n(t_1,\ldots,t_n)$, $FV(f(t_1,\ldots,t_n)) = FV(t_1)\cup\ldots\cup FV(t_n)$.

The set of **free variables** $FV(A)$ in a formula A is defined inductively by:

1. $FV(\perp) = \emptyset$,
2. $FV(P(t_1,\ldots,t_n)) = FV(t_1)\cup\ldots\cup FV(t_n)$,
3. $FV(A\&B) = FV(A \vee B) = FV(A \supset B) = FV(A)\cup FV(B)$,
4. $FV(\forall x A) = FV(\exists x A) = FV(A) - \{x\}$.

A term or formula that has free variables is **open**; otherwise it is **closed**. In 4, x is a **bound variable**. In first-order logic, a **principle of renaming of bound variables**, or α-**conversion**, is often assumed: It consists in identifying formulas differing only in the names of bound variables, usually expressed as $\forall x A(x) \equiv \forall y A(y)$ and $\exists x A(x) \equiv \exists y A(y)$. Such a principle is intuitively justified by the role of bound variables as placeholders, as in $\int_a^b f(x)\,dx$, which is the same as $\int_a^b f(y)\,dy$. We shall not need to assume this principle here, as it will be formally derivable once the quantifier rules are given an appropriate formulation. Also, we shall not use the notation $A(x)$ to indicate that A contains, or may contain, the free variable x. The parenthesis notation is used for **application** of a function or predicate, as in the definition of terms and formulas above. If a variable does not occur in a formula, sequent, or derivation, we say it is **fresh** for that formula, sequent, or derivation.

In terms t, as well as in formulas A, a variable x can be substituted by a term t'. To identify what is substituted for what, the notation $[t'/x]$ is used. The result of substitution $[t'/x]$ is written as $t(t'/x)$ for a term t and as $A(t'/x)$ for a formula A. Substitution is defined by induction on the terms and formulas in which the substitution is performed:

Substitution $[t/x]$ in a term:

1. $a(t/x) = a$,
2. $y(t/x) = y$ if $y \neq x$ and $y(t/x) = t$ if $y = x$,
3. $f^n(t_1, \ldots, t_n)(t/x) = f^n(t_1(t/x), \ldots, t_n(t/x))$.

Substitution $[t/x]$ in a formula:

1. $\bot(t/x) = \bot$,
2. $(P^n(t_1, \ldots, t_n))(t/x) = P^n(t_1(t/x), \ldots, t_n(t/x))$,
3. $(A \circ B)(t/x) = A(t/x) \circ B(t/x)$, for $\circ = \&, \vee, \supset$,
4. $(\forall y A)(t/x) = \forall y A(t/x)$ if $y \neq x$, $(\forall y A)(t/x) = \forall y A$ if $y = x$,
5. $(\exists y A)(t/x) = \exists y A(t/x)$ if $y \neq x$, $(\exists y A)(t/x) = \exists y A$ if $y = x$.

The last two clauses in the above definition guarantee that substitution does not act on bound variables. We shall call $A(t/x)$ a **substitution instance** of A. **Simultaneous** substitution of n terms $t_1, \ldots, t_n$ for n variables $x_1, \ldots, x_n$ is written as $[t_1/x_1, \ldots, t_n/x_n]$, and its result in a term t is written as $t(t_1/x_1, \ldots, t_n/x_n)$ and in a formula A as $A(t_1/x_1, \ldots, t_n/x_n)$.

When a term t is substituted for a variable x in a formula A, it may happen that some variables of the term t "get caught" in the substitution, by some quantifiers in A. If this happens, the validity of substitution instances of a universal formula is no longer guaranteed. As an example, consider the formula $\forall y \exists x (y < x)$ that holds in a linearly ordered set without greatest element. Dropping the first quantifier and substituting x for y produces $\exists x (x < x)$, which is not satisfiable in the same domain. However, if we rename the bound variable x by z before performing the substitution, we obtain $\exists z (x < z)$ that is satisfiable.

We say that a term t is **free for x in** A if no variable of t becomes bound as an effect of the substitution of t for x in A. The binding may happen if some variables of t are in the scope of quantifiers in A. However, instead of controlling the condition for each substitution, we observe that the condition can always be met by appropriate renaming of bound variables in the formula A: If A is, say, $\forall x B$ and y is a variable not occurring in A, α-conversion guarantees that we can identify A with $\forall y B(y/x)$. In this way we can ensure that a variable does not occur both free and bound in a formula. When considering a substitution, we shall assume that this condition is satisfied, if necessary by renaming of bound variables, and shall sometimes omit recalling the condition.

(b) Quantifiers in natural deduction: The intuitionistic meaning explanations for quantified formulas $\forall x A$ and $\exists x A$ are as follows:

1. A direct proof of $\forall x A$ consists of a proof of $A(y/x)$ for an arbitrary y.

2. A direct proof of $\exists x A$ consists of a proof of $A(a/x)$ for some individual a.

In natural deduction, the premiss for the introduction rule of a universal proposition $\forall x A$ is that $A(y/x)$ has been derived for an **arbitrary** y. To be arbitrary means that nothing more than the range of possible values is assumed known about y. This condition is expressed as a **variable restriction** on the introduction rule for the universal quantifier. The premiss for an existential proposition is that A has been derived for some individual a.

The **introduction rules** for the quantifiers are:

$$\frac{A(y/x)}{\forall x A} \, \forall I \qquad \frac{A(a/x)}{\exists x A} \, \exists I$$

The rule of universal introduction has the variable restriction that y must not occur free in any assumption that $A(y/x)$ depends on nor in $\forall x A$. The latter condition can be equivalently expressed by requiring that y is equal to x or else y is not free in A. The variable restriction guarantees that y stands for an "arbitrary individual" for which A holds, which is the direct ground for asserting the universal proposition.

Usually rule $\forall I$ is written as $\frac{A}{\forall x A}$ and the restriction is that x is not free in any of the assumptions that A depends on, where one must keep in mind that if A is an assumption it depends on itself. With this rule, α-conversion has to be postulated as a principle to be added to the system. In the rule we use, instead, this conversion is built in. We also modify rule $\exists I$ for the same reason: Instead of having the premiss for some individual, the premiss will have an arbitrary term t, thus the rule we use for the existential quantifier is:

$$\frac{A(t/x)}{\exists x A} \, \exists I$$

To determine the general elimination rule corresponding to rule $\forall I$, assume a derivation of $A(y/x)$ for an arbitrary y. In deriving consequences from $A(y/x)$ for y arbitrary, any instances $A(t/x)$ may be used; thus the auxiliary derivation of the general elimination rule leads to some consequence C from assumptions $A(t_1/x), \ldots, A(t_n/x)$. We simplify this situation by admitting only one assumption $A(t/x)$ and obtain the rule

$$
\frac{\forall x A \qquad \begin{array}{c} [A(t/x)] \\ \vdots \\ C \end{array}}{C} \ \forall E
$$

The rule with the assumptions $A(t_1/x), \ldots, A(t_n/x)$ is admissible, by the repetition of the above rule n times.

Rules $\forall I$ and $\forall E$ satisfy the inversion principle: Given a derivation of $A(y/x)$ for an arbitrary y, a derivation of $A(t/x)$ is obtained for any given term t by substitution. Therefore, the derivation

$$
\frac{\dfrac{\begin{array}{c}\vdots\\A(y/x)\end{array}}{\forall x A}\forall I \qquad \begin{array}{c} [A(t/x)] \\ \vdots \\ C \end{array}}{C} \ \forall E
$$

converts through the use of the substitution into a derivation of C without rules $\forall I$ and $\forall E$:

$$
\begin{array}{c}
\vdots \\
A(t/x) \\
\vdots \\
C
\end{array}
$$

The standard elimination rule for universal quantification is obtained when $C = A(t/x)$:

$$
\frac{\forall x A}{A(t/x)} \ \forall E
$$

The standard elimination rule for the existential quantifier already is of the form of general elimination rules:

$$
\frac{\exists x A \qquad \begin{array}{c} [A(y/x)] \\ \vdots \\ C \end{array}}{C} \ \exists E
$$

It has the restriction that y must not occur free in $\exists x A$, C nor in any assumption C depends on except $A(y/x)$. The rule is in accordance with our inversion principle: The direct grounds for deriving $\exists x A$ can consist in deriving A for any one individual in the domain of the bound variable. For C to be derivable from $\exists x A$, in order to take all possible cases into account, it must be required that C follow from $A(y/x)$ for an arbitrary y, and this is what the variable restriction in rule

$\exists E$ expresses. If in a derivation $\exists x A$ was inferred by $\exists I$ from $A(t/x)$ and $\exists x A$ is the major premiss of $\exists E$, then C follows in particular from $A(t/x)$. The inversion principle is satisfied, for the derivation

$$
\begin{array}{cc}
\vdots & [A(y/x)] \\
\dfrac{A(t/x)}{\exists x A}\ {\exists I} & \vdots \\
 & C \\
\end{array}
$$
$$
\dfrac{}{C}\ {\exists E}
$$

converts into a derivation of C without rules $\exists I$ and $\exists E$:

$$
\vdots \\
A(t/x) \\
\vdots \\
C
$$

The addition of the above introduction and elimination rules to the system of natural deduction for intuitionistic propositional logic will result in the system of natural deduction for **intuitionistic predicate logic**.

The introduction rule for existence is in accordance with the intuitionistic or constructive notion of existence. Further, the definability of existence in terms of negation and universal quantification, $\exists x A \equiv {\sim} \forall x \sim A$, fails in intuitionistic predicate logic, and all four quantifier rules are needed.

In classical logic, indirect existence proofs are permitted and existence cannot have the same meaning as in intuitionistic logic, but no finitary system of natural deduction with normalization and subformula property has been found for full classical predicate logic. There is a natural deduction system with good structural properties only for the $\vee$- and $\exists$-free fragment, to be presented in Chapter 8. The idea is to translate formulas with $\vee$ or $\exists$ into formulas known to be classically equivalent but not containing these operations: For any formula C that should be derivable but is not, there is a translated formula C^* that is derivable. An example of such a translation is Prawitz' system of natural deduction for **stable logic**, i.e., a system of propositional logic in which the law of double negation is derivable. The translation gives a $\vee$-free fragment by translating disjunctions $A \vee B$ into implications $\sim A \supset B$. However, such a translation is not suited for representing the structure of derivations in full classical logic, disjunction included.

(c) Quantifiers in sequent calculus: As mentioned, there is at present no finitary normalizing system of natural deduction for the full language of classical predicate logic. In sequent calculus, instead, cut-free calculi for intuitionistic as well as full classical predicate logic were found already by Gentzen. To obtain a sequent calculus for intuitionistic predicate logic, quantifier rules are added to the intuitionistic propositional calculus **G3ip**. The rules are, with repetition of the principal formula in $L\forall$ similarly to $L\supset$,

G3i

$$\frac{A(t/x), \forall x A, \Gamma \Rightarrow C}{\forall x A, \Gamma \Rightarrow C} \, L\forall \qquad\qquad \frac{\Gamma \Rightarrow A(y/x)}{\Gamma \Rightarrow \forall x A} \, R\forall$$

$$\frac{A(y/x), \Gamma \Rightarrow C}{\exists x A, \Gamma \Rightarrow C} \, L\exists \qquad\qquad \frac{\Gamma \Rightarrow A(t/x)}{\Gamma \Rightarrow \exists x A} \, R\exists$$

The restriction in $R\forall$ is that y must not occur free in $\Gamma, \forall x A$. The restriction in $L\exists$ is that y must not occur free in $\exists x A, \Gamma, C$.

Unlike in natural deduction, in sequent calculus all propositional rules are local. In quantifier rules, fulfillment of the variable restrictions is controlled in the same local way.

We obtain the rules of sequent calculus for classical predicate logic by adding to the propositional calculus **G3cp** the rules

G3c

$$\frac{A(t/x), \forall x A, \Gamma \Rightarrow \Delta}{\forall x A, \Gamma \Rightarrow \Delta} \, L\forall \qquad\qquad \frac{\Gamma \Rightarrow \Delta, A(y/x)}{\Gamma \Rightarrow \Delta, \forall x A} \, R\forall$$

$$\frac{A(y/x), \Gamma \Rightarrow \Delta}{\exists x A, \Gamma \Rightarrow \Delta} \, L\exists \qquad\qquad \frac{\Gamma \Rightarrow \Delta, \exists x A, A(t/x)}{\Gamma \Rightarrow \Delta, \exists x A} \, R\exists$$

The restriction in $R\forall$ is that y must not occur free in $\Gamma, \Delta, \forall x A$. The restriction in $L\exists$ is that y must not occur free in $\exists x A, \Gamma, \Delta$. We may summarize these conditions by the requirement that y must not occur free in the conclusion of the two rules. In the propositional part of **G3c**, because of invertibility of all rules there was no need to repeat principal formulas in premises of rules, but to obtain admissibility of contraction for **G3c**, repetition is needed in $L\forall$ and $R\exists$.

The weight of quantified formulas is defined as

$$w(\forall x A) = w(A) + 1,$$
$$w(\exists x A) = w(A) + 1.$$

Height of derivation is defined as before.

The following lemma is a formal version of the principle of renaming of bound variables:

Lemma 4.1.1: Height-preserving α-conversion. *Given a derivation $\mathcal{D}$ of $\Gamma \Rightarrow C$ in **G3i** (of $\Gamma \Rightarrow \Delta$ in **G3c**), it can be transformed into a derivation $\mathcal{D}'$ of $\Gamma' \Rightarrow C'$ (of $\Gamma' \Rightarrow \Delta'$) where Γ', Δ', C', and $\mathcal{D}'$ differ from Γ, Δ, C, and $\mathcal{D}$ only by fresh renamings of bound variables.*

Proof: We shall give the proof for **G3i**, the proof for **G3c** being similar, by induction on the height n of the derivation. If $n = 0$, $\Gamma \Rightarrow C$ is an axiom or conclusion of $L\bot$, then also $\Gamma' \Rightarrow C'$, where the bound variables have been renamed by fresh variables, is an axiom or conclusion of $L\bot$. Else $\Gamma \Rightarrow C$ has derivation height > 0. If all the rules in the derivation are propositional ones, the renaming is inherited from the premisses (axioms) to the conclusion of the derivation. Otherwise we consider the last quantifier rule in the derivation. If it is $L\forall$, with conclusion $\forall x A, \Gamma'' \Rightarrow C$ and premiss $A(t/x), \forall x A, \Gamma'' \Rightarrow C$, and x has to be renamed by the fresh variable y, by inductive hypothesis from the premiss we obtain a derivation of the same height of $A(t/x), \forall y A(y/x), \Gamma' \Rightarrow C'$, that is, of $A(y/x)(t/y), \forall y A(y/x), \Gamma' \Rightarrow C'$, and therefore, by applying $L\forall$, we obtain a derivation of the same height of the conclusion $\forall y A(y/x), \Gamma' \Rightarrow C'$. If the last quantifier rule is $R\forall$, with conclusion $\Gamma \Rightarrow \forall x A$ from the premiss $\Gamma \Rightarrow A(z/x)$, by inductive hypothesis we have a derivation of the same height of $\Gamma' \Rightarrow A'(z/x)$, where bound occurrences of x have been renamed by y. This is the same as $\Gamma' \Rightarrow A'(y/x)(z/y)$; thus we obtain by $R\forall$, $\Gamma' \Rightarrow \forall y A'(y/x)$, with the same bound on the derivation height as the sequent $\Gamma \Rightarrow \forall x A$. The cases of $L\exists$ and $R\exists$ are treated symmetrically to $R\forall$ and $L\exists$.   QED.

If in a sequent $\Gamma \Rightarrow \Delta$ a term t is substituted for a variable x, derivability of the sequent is maintained, with the same derivation height. The proviso for the substitution is that the term t be free for x in every formula of the sequent $\Gamma \Rightarrow \Delta$ (or, for short, t is free for x in $\Gamma \Rightarrow \Delta$). Substitution of free occurrences of x by t in all formulas of Γ is denoted by $\Gamma(t/x)$.

Lemma 4.1.2: Substitution lemma.
(i) If $\Gamma \Rightarrow C$ is derivable in **G3i** and t is free for x in Γ, C, then $\Gamma(t/x) \Rightarrow C(t/x)$ is derivable in **G3i**, with the same derivation height.
(ii) If $\Gamma \Rightarrow \Delta$ is derivable in **G3c** and t is free for x in Γ, Δ, then $\Gamma(t/x) \Rightarrow \Delta(t/x)$ is derivable in **G3c**, with the same derivation height.

Proof: We only give the proof of (i), the proof of (ii) being similar. The proof is by induction on height of derivation.

By Lemma 4.1.1, it is not restrictive to suppose that in the derivation of $\Gamma \Rightarrow C$ the bound variables have been renamed so that the sets of free and bound variables are disjoint. With this assumption, some cases in the proof can be avoided. Furthermore, by the choice of fresh variables not occurring in the term to be substituted, the condition of being free for x in the sequent where the substitution occurs is maintained.

If $\Gamma \Rightarrow C$ is an axiom or conclusion of $L\bot$, then $\Gamma(t/x) \Rightarrow C(t/x)$ also is an axiom or conclusion of $L\bot$. Else $\Gamma \Rightarrow C$ has derivation height $n > 0$, and we consider the last rule in the derivation. If $\Gamma \Rightarrow C$ has been derived by a propositional rule, we observe that if t is free for x in the conclusion of any such rule, then it is free for x in the premisses, since there is no alteration in the sets of free and bound variables. Therefore the inductive hypothesis can be applied to the premisses, and the conclusion follows by application of the rule.

If $\Gamma \Rightarrow C$ has been derived by $L\forall$, we can exclude the case in which x is the quantified variable in the rule since in this case x would not be free and the substitution would be vacuous. Therefore we can assume that the premiss is

$$A(t'/y), \forall y A, \Gamma' \Rightarrow C$$

with $y \neq x$. Since t is free for x in $A(t'/y)$, by inductive hypothesis we have a derivation of height $\leqslant n - 1$ of

$$(A(t'/y))(t/x), (\forall y A)(t/x), \Gamma'(t/x) \Rightarrow C(t/x)$$

Observe that by definition of substitution and the fact that $x \neq y$, we have

$$(\forall y A)(t/x) = \forall y A(t/x).$$

The two substitutions in $(A(t'/y))(t/x)$ can be given as one simultaneous substitution $A(t'(t/x)/y, t/x)$; Since t is free for x in $\forall y A$, the term t does not contain the variable y, so the latter is equal to $(A(t/x))(t'(t/x)/y)$. Summing up, we have a derivation of height $\leqslant n - 1$ of

$$(A(t/x))(t''/y), \forall y A(t/x), \Gamma'(t/x) \Rightarrow C(t/x)$$

where $t'' = t'(t/x)$; so by $L\forall$ we obtain a derivation of height $\leqslant n$ of

$$\forall y A(t/x), \Gamma'(t/x) \Rightarrow C(t/x)$$

that is, of

$$(\forall y A)(t/x), \Gamma'(t/x) \Rightarrow C(t/x)$$

If the last rule is $R\forall$, we can exclude, as above, the case in which x is the quantified variable. Therefore the derivation ends with

$$\frac{\Gamma \Rightarrow A(z/y)}{\Gamma \Rightarrow \forall y A}$$

where $y \neq x$, and z is not free in Γ, and $z = y$ or z is not free in A. By inductive hypothesis we can replace z by a fresh variable v not in t. So we have a derivation of height $\leqslant n - 1$ of $\Gamma \Rightarrow A(v/y)$. Again by inductive hypothesis we obtain a derivation of height $\leqslant n - 1$ of

$$\Gamma(t/x) \Rightarrow A(v/y)(t/x)$$

By the choice of v and the fact that t does not contain the variable y (as it is free for x in $\forall y A$) we can switch the order of substitutions and obtain a derivation of height $\leqslant n - 1$ of

$$\Gamma(t/x) \Rightarrow A(t/x)(v/y)$$

where the variable conditions for applying $R\forall$ are met and we can infer $\Gamma(t/x) \Rightarrow \forall y A(t/x)$, and since $x \neq y$ this is the same as $\Gamma(t/x) \Rightarrow (\forall y A)(t/x)$.

The cases of $L\exists$ and $R\exists$ are treated symmetrically to $R\forall$ and $L\forall$. QED.

4.2. ADMISSIBILITY OF STRUCTURAL RULES

We first prove the admissibility of structural rules for the intuitionistic calculus. The corresponding proofs for the classical calculus are very close to this because of the similarity of the quantifier rules.

(a) Admissibility of structural rules for G3i: The proofs extend those for the propositional calculus in Chapter 2.

Lemma 4.2.1: *Sequents of the form $C, \Gamma \Rightarrow C$ are derivable in* **G3i**.

Proof: As for **G3ip** in Lemma 2.3.3, by induction on weight of C. The new cases are for the quantified formulas. If $C = \forall x A$, by inductive hypothesis $A(y/x), \forall x A, \Gamma \Rightarrow A(y/x)$ is derivable, where y is a fresh variable. By application of $L\forall$ and $R\forall$, $\forall x A, \Gamma \Rightarrow \forall x A$ is derivable. If $C = \exists x A$, by inductive hypothesis we have a derivation of $A(y/x), \Gamma \Rightarrow A(y/x)$ and the conclusion follows by application of $R\exists$ and $L\exists$. QED.

Theorem 4.2.2: Height-preserving weakening for G3i.

   If $\vdash_n \Gamma \Rightarrow C$, then $\vdash_n D, \Gamma \Rightarrow C$.

Proof: By induction on height of derivation, as in weakening for **G3ip**, Theorem 2.3.4. For applications of $L\forall$ and $R\exists$, the weakening formula can be added to the context of the premiss. For $R\forall$ and $L\exists$, we have to consider the effect of variable restrictions.

   If the last rule applied is $R\forall$, the premiss is $\Gamma \Rightarrow A(y/x)$. If y is not free in D, by inductive hypothesis we get $D, \Gamma \Rightarrow A(y/x)$ and hence $D, \Gamma \Rightarrow \forall x A(x)$ by $R\forall$. If y is free in D, we choose a fresh variable z and apply Lemma 4.1.2 to $\Gamma \Rightarrow A(y/x)$ to obtain $\Gamma \Rightarrow A(z/x)$. The inductive hypothesis gives $D, \Gamma \Rightarrow A(z/x)$, so that by $R\forall$ we derive $D, \Gamma \Rightarrow \forall x A$.

   If the last rule is $L\exists$, with premiss $A(y/x), \Gamma' \Rightarrow C$ and y is not free in D, we derive the conclusion by applying the inductive hypothesis to the premiss and then the rule. If y is free in D, we choose a fresh variable z for substitution in the premiss and obtain $A(z/x), \Gamma' \Rightarrow C$. By inductive hypothesis $D, A(z/x), \Gamma' \Rightarrow C$ is derivable, and by $L\exists$ the conclusion $D, \exists x A, \Gamma' \Rightarrow C$ follows. QED.

Lemma 4.2.3: Height-preserving inversion of $L\exists$ for G3i.

If $\vdash_n \exists x A, \Gamma \Rightarrow C$, then $\vdash_n A(y/x), \Gamma \Rightarrow C$.

Proof: By induction on height of derivation. If $n = 0$ and if $\exists x A, \Gamma \Rightarrow C$ is an axiom or conclusion of $L\bot$, then also $A(y/x), \Gamma \Rightarrow C$ is an axiom or conclusion of $L\bot$.

For the inductive case, if $\exists x A$ is not principal in the last rule and y is not free in its premisses, we have one or two premisses, $\exists x A, \Gamma' \Rightarrow C'$ and $\exists x A, \Gamma'' \Rightarrow C''$. Now apply inductive hypothesis and then the rule again to conclude $A(y/x), \Gamma \Rightarrow C$. If instead the last rule is a rule with a variable restriction on y, we need a substitution before applying the inductive hypothesis to the premiss of the rule, as the substitution $[y/x]$ could bring in free occurrences of y that would then prevent applying the rule again. Suppose for instance that the derivation ends with

$$\frac{\exists x A, \Gamma \Rightarrow B(y/z)}{\exists x A, \Gamma \Rightarrow \forall z B} \ R\forall$$

By the substitution lemma we can replace y by a fresh variable v in the premiss and obtain, using $B(y/z)(v/y) \equiv B(v/z)$, the derivation,

$$\frac{\dfrac{\exists x A, \Gamma \Rightarrow B(v/z)}{A(y/x), \Gamma \Rightarrow B(v/z)} \ Ind}{A(y/x), \Gamma \Rightarrow \forall z B} \ R\forall$$

If $\exists x A$ is principal in the last rule, the premiss gives a derivation of $A(z/x), \Gamma \Rightarrow C$, where z is not free in Γ, C. By Lemma 4.1.2 we obtain a derivation with the same height of $A(y/x), \Gamma \Rightarrow C$. QED.

We can now prove that contraction is admissible and height-preserving in **G3i**:

Theorem 4.2.4: Height-preserving contraction for G3i.

If $\vdash_n D, D, \Gamma \Rightarrow C$, then $\vdash_n D, \Gamma \Rightarrow C$.

Proof: Continuing the proof of admissibility for **G3ip**, Theorem 2.4.1, with $n > 0$ and $D = \forall x A$ principal, we have as the last step

$$\frac{A(t/x), \forall x A, \forall x A, \Gamma \Rightarrow C}{\forall x A, \forall x A, \Gamma \Rightarrow C} \ L\forall$$

and this is transformed into

$$\frac{\dfrac{A(t/x), \forall x A, \forall x A, \Gamma \Rightarrow C}{A(t/x), \forall x A, \Gamma \Rightarrow C} \ Ind}{\forall x A, \Gamma \Rightarrow C} \ L\forall$$

With $D = \exists x A$ as principal formula, we have as the last step

$$\frac{A(y/x), \exists x A, \Gamma \Rightarrow C}{\exists x A, \exists x A, \Gamma \Rightarrow C} \; L\exists$$

where y is not free in Γ, C. By Lemma 4.2.3, we have the derivation

$$\frac{\dfrac{\dfrac{A(y/x), \exists x A, \Gamma \Rightarrow C}{A(y/x), A(y/x), \Gamma \Rightarrow C} \; Inv}{A(y/x), \Gamma \Rightarrow C} \; Ind}{\exists x A, \Gamma \Rightarrow C} \; L\exists$$

QED.

Theorem 4.2.5: *The rule of cut is admissible in* **G3i**.

Proof: Continuing the proof of admissibility of cut for **G3ip** with its numbering of cases, Theorem 2.4.3, we have to consider only the case that neither premiss is an axiom. There are three such cases:

3. The cut formula is **not principal** in the left premiss. There are two additional subcases:

3.4. The left premiss has been concluded by $L\forall$. Then $\Gamma = \forall x A, \Gamma'$ and we have the derivation

$$\frac{\dfrac{A(t/x), \forall x A, \Gamma' \Rightarrow D}{\forall x A, \Gamma' \Rightarrow D} \; L\forall \qquad D, \Delta \Rightarrow C}{\forall x A, \Gamma', \Delta \Rightarrow C} \; Cut$$

This is transformed into

$$\frac{\dfrac{A(t/x), \forall x A, \Gamma' \Rightarrow D \qquad D, \Gamma \Rightarrow C}{A(t/x), \forall x A, \Gamma', \Delta \Rightarrow C} \; Cut}{\forall x A, \Gamma', \Delta \Rightarrow C} \; L\forall$$

3.5. If the left premiss has been concluded by the $L\exists$ rule, $\Gamma = \exists x A, \Gamma'$ and we have

$$\frac{\dfrac{A(y/x), \Gamma' \Rightarrow D}{\exists x A, \Gamma' \Rightarrow D} \; L\exists \qquad D, \Delta \Rightarrow C}{\exists x A, \Gamma', \Delta \Rightarrow C} \; Cut$$

The cut cannot simply be permuted with $L\exists$ as it can bring in formulas that do not meet the variable restriction necessary for the application of $L\exists$, and a variable substitution has to be performed first. Let z be a fresh variable. We have the derivation, where the left premiss is derivable by Lemma 4.1.2,

$$\frac{\dfrac{A(z/x), \Gamma' \Rightarrow D \quad D, \Delta \Rightarrow C}{A(z/x), \Gamma', \Delta \Rightarrow C} \; Cut}{\exists x A, \Gamma', \Delta \Rightarrow C} \; L\exists$$

4. If D is **principal** in the left premiss only, the derivation is transformed according to the last rule in the derivation of the right premiss. There are four additional subcases with quantifier rules:

4.7. $L\forall$, and $\Delta = \forall x A, \Delta'$. The derivation

$$\frac{\Gamma \Rightarrow D \quad \dfrac{D, A(t/x), \forall x A, \Delta' \Rightarrow C}{D, \forall x A, \Delta' \Rightarrow C} \; L\forall}{\forall x A, \Gamma, \Delta' \Rightarrow C} \; Cut$$

is transformed into

$$\frac{\dfrac{\Gamma \Rightarrow D \quad D, A(t/x), \forall x A, \Delta' \Rightarrow C}{A(t/x), \forall x A, \Gamma, \Delta' \Rightarrow C} \; Cut}{\forall x A, \Gamma, \Delta' \Rightarrow C} \; L\forall$$

4.8. $L\exists$, and $\Delta = \exists x A, \Delta'$. We have

$$\frac{\Gamma \Rightarrow D \quad \dfrac{D, A(y/x), \Delta' \Rightarrow C}{D, \exists x A, \Delta' \Rightarrow C} \; L\exists}{\exists x A, \Gamma, \Delta' \Rightarrow C} \; Cut$$

Let z be a fresh variable. Using Lemma 4.1.2, we obtain the derivation

$$\frac{\dfrac{\Gamma \Rightarrow D \quad D, A(z/x), \Delta' \Rightarrow C}{A(z/x), \Gamma, \Delta' \Rightarrow C} \; Cut}{\exists x A, \Gamma, \Delta' \Rightarrow C} \; L\exists$$

4.9. $R\forall$ and $C = \forall x A$. We have the derivation

$$\frac{\Gamma \Rightarrow D \quad \dfrac{D, \Delta \Rightarrow A(y/x)}{D, \Delta \Rightarrow \forall x A} \; R\forall}{\Gamma, \Delta \Rightarrow \forall x A} \; Cut$$

By substituting a fresh variable z, we obtain by Lemma 4.1.2:

$$\frac{\dfrac{\Gamma \Rightarrow D \quad D, \Delta \Rightarrow A(z/x)}{\Gamma, \Delta \Rightarrow A(z/x)} \; Cut}{\Gamma, \Delta \Rightarrow \forall x A} \; R\forall$$

4.10. $R\exists$, and $C = \exists x A$. The derivation

$$\dfrac{\Gamma \Rightarrow D \quad \dfrac{D, \Delta \Rightarrow A(t/x)}{D, \Delta \Rightarrow \exists x A}\, R\exists}{\Gamma, \Delta \Rightarrow \exists x A}\, Cut$$

is transformed into

$$\dfrac{\dfrac{\Gamma \Rightarrow D \quad D, \Delta \Rightarrow A(t/x)}{\Gamma, \Delta \Rightarrow A(t/x)}\, Cut}{\Gamma, \Delta' \Rightarrow \exists x A}\, L\exists$$

5. If D is **principal in both premisses**, we have two additional subcases:

5.4. $\forall x A$ is principal in both premisses, and we have the derivation

$$\dfrac{\dfrac{\Gamma \Rightarrow A(y/x)}{\Gamma \Rightarrow \forall x A}\, R\forall \quad \dfrac{A(t/x), \forall x A, \Delta \Rightarrow C}{\forall x A, \Delta \Rightarrow C}\, L\forall}{\Gamma, \Delta \Rightarrow C}\, Cut$$

By Lemma 4.1.2, $\Gamma \Rightarrow A(t/x)$ is derivable with the same height as the premiss of the left branch $\Gamma \Rightarrow A(y/x)$, and we transform the derivation into

$$\dfrac{\Gamma \Rightarrow A(t/x) \quad \dfrac{\dfrac{\Gamma \Rightarrow \forall x A \quad A(t/x), \forall x A, \Delta \Rightarrow C}{A(t/x), \Gamma, \Delta \Rightarrow C}\, Cut}{\dfrac{\Gamma, \Gamma, \Delta \Rightarrow C}{\Gamma, \Delta \Rightarrow C}\, Ctr}\, Cut}{}$$

5.5. Finally, we have the case of $\exists x A$ principal in both premisses:

$$\dfrac{\dfrac{\Gamma \Rightarrow A(t/x)}{\Gamma \Rightarrow \exists x A}\, R\exists \quad \dfrac{A(y/x), \Delta \Rightarrow C}{\exists x A, \Delta \Rightarrow C}\, L\exists}{\Gamma, \Delta \Rightarrow C}\, Cut$$

Again, $A(t/x), \Delta \Rightarrow C$ is derivable with the same height as $A(y/x), \Delta \Rightarrow C$, and we transform the derivation into

$$\dfrac{\Gamma \Rightarrow A(t/x) \quad A(t/x), \Delta \Rightarrow C}{\Gamma, \Delta \Rightarrow C}\, Cut$$

QED.

(b) Admissibility of structural rules for G3c: The proofs extend those for the propositional calculus in Chapter 3. We indicate only the differences with the proofs for **G3i** and additions to those for **G3cp**.

Lemma 4.2.6: *Sequents of the form $C, \Gamma \Rightarrow \Delta, C$ are derivable in* **G3c**.

Proof: Similar to the proof of Lemma 4.2.1. QED.

Theorem 4.2.7: Height-preserving weakening for G3c.
If $\vdash_n \Gamma \Rightarrow \Delta$, then $\vdash_n D, \Gamma \Rightarrow \Delta$. If $\vdash_n \Gamma \Rightarrow \Delta$, then $\vdash_n \Gamma \Rightarrow \Delta, D$.

Proof: Similar to the proof of Theorem 4.2.2. QED.

In order to prove contraction admissible in **G3c**, we need the analogue of Lemma 4.2.3, concerning invertibility of $L\exists$, plus invertibility of $R\forall$. The former is used, as in the intuitionistic calculus, in the proof of admissibility of left contraction, the latter of right contraction.

Lemma 4.2.8: Height-preserving inversion of $L\exists$ and $R\forall$ for G3c.
 (i) If $\vdash_n \exists x A, \Gamma \Rightarrow \Delta$, then $\vdash_n A(y/x), \Gamma \Rightarrow \Delta$.
 (ii) If $\vdash_n \Gamma \Rightarrow \Delta, \forall x A$, then $\vdash_n \Gamma \Rightarrow \Delta, A(y/x)$.

Proof: The proof of (i) is similar to that of Lemma 4.2.3, by induction on n. The proof of (ii) is symmetric to the proof of (i). QED.

Theorem 4.2.9: Height-preserving contraction for G3c.
 (i) If $\vdash_n D, D, \Gamma \Rightarrow \Delta$, then $\vdash_n D, \Gamma \Rightarrow \Delta$.
 (ii) If $\vdash_n \Gamma \Rightarrow \Delta, D, D$, then $\vdash_n \Gamma \Rightarrow \Delta, D$.

Proof: The proof extends the proof for **G3cp** by considering the new cases arising from the addition of the quantifier rules, as in Theorem 4.2.4. The only essentially new case is the one for right contraction in which the contraction formula is $\forall x A$ and this is principal in the last rule used in the derivation. This case is taken care of by the inversion lemma for $R\forall$. QED.

Theorem 4.2.10: *The rule of cut*

$$\frac{\Gamma \Rightarrow \Delta, D \quad D, \Gamma' \Rightarrow \Delta'}{\Gamma, \Gamma' \Rightarrow \Delta, \Delta'}\ Cut$$

is admissible in **G3c**.

Proof: The proof is an extension of the proof for **G3cp**, similar to the proof for **G3i**. The only new case to be considered is the one in which the cut formula is not principal in the last rule used in the derivation of the left premiss, and this is a right rule with variable restrictions, i.e., $R\forall$. This situation is treated as the case in which the last rule used to derive the left premiss is $L\exists$: first, the free variable of the active formula is substituted by a fresh variable, by use of the substitution lemma, then cut is permuted to the premiss of the left premiss, and finally rule $R\forall$ is applied. QED.

4.3. APPLICATIONS OF CUT ELIMINATION

As applications of cut elimination, we conclude subformula properties and the existence property and underivability results for **G3i**.

(a) Subformula property and existence property for intuitionistic derivations: The notion of subformulas for predicate logic has to be independent of the particular choice of bound variables and substitution instances:

Definition 4.3.1: $A(t/x)$ *is a* subformula *of* $\forall x A$ *and* $\exists x A$ *for all terms* t.

By inspecting the respective rules, we obtain from the admissibility of cut, Theorem 4.2.5, the **subformula property** for **G3i** and **G3c**:

Corollary 4.3.2: *All formulas in the derivation of* $\Gamma \Rightarrow C$ *in* **G3i** (*of* $\Gamma \Rightarrow \Delta$ *in* **G3c**) *are subformulas of* Γ, C (*of* Γ, Δ).

As a consequence of the subformula property, we obtain underivability of the sequent $\Rightarrow \perp$ in the systems **G3i** and **G3c**; therefore we have:

Corollary 4.3.3: *The systems* **G3i** *and* **G3c** *are consistent.*

Since in a cut-free derivation of $\Rightarrow \exists x A$ in **G3i** the last rule must be $R\exists$, we obtain the **existence property** of intuitionistic predicate logic:

Corollary 4.3.4: *If* $\Rightarrow \exists x A$ *is derivable in* **G3i**, *then* $\Rightarrow A(t/x)$ *is derivable for some term* t.

In **G3c**, instead, $\Rightarrow \exists x A$ can be concluded from $\Rightarrow \exists x A, A(t/x)$, and there is no existence property. In case A is quantifier-free, a weaker result than Corollary 4.3.4 can be obtained: If $\Rightarrow \exists x A$ is derivable in **G3c**, there are terms $t_1, \ldots, t_n$ such that $\Rightarrow A(t_1/x) \lor \ldots \lor A(t_n/x)$ is derivable. By the subformula property, the derivation uses only propositional logic. The formula $A(t_1/x) \lor \ldots \lor A(t_n/x)$ is called the **Herbrand disjunction** of $\exists x A$. This result follows from a more general result to be given in Section 6.6.

(b) Underivability results for intuitionistic predicate logic: We show for intuitionistic logic that existence is not definable in terms of the universal quantifier, that Glivenko's theorem does not extend to predicate logic, and that there is no prenex normal form for formulas contrary to classical predicate logic.

Theorem 4.3.5: *The sequent* $\Rightarrow \sim\forall x \sim A \supset \exists x A$ *is not derivable in* **G3i**.

Proof: We show that a root-first proof search for a derivation of the sequent goes on forever. The last two steps in root-first order must be $R\supset, L\supset$ or $R\supset, R\exists$. In the second case the continuation can only be $L\supset$, so we have search trees beginning with

$$\frac{\frac{\sim\forall x \sim A \Rightarrow \forall x \sim A \quad \bot \Rightarrow \exists x A}{\sim\forall x \sim A \Rightarrow \exists x A}\ {}_{L\supset}}{\Rightarrow \sim\forall x \sim A \supset \exists x A}\ {}_{R\supset}$$

$$\frac{\frac{\frac{\sim\forall x \sim A \Rightarrow \forall x \sim A \quad \bot \Rightarrow A(t/x)}{\sim\forall x \sim A \Rightarrow A(t/x)}\ {}_{L\supset}}{\frac{\sim\forall x \sim A \Rightarrow \exists x A}{}}\ {}_{R\exists}}{\Rightarrow \sim\forall x \sim A \supset \exists x A}\ {}_{R\supset}$$

The second search succeeds only if the first one does, and the first one has two continuations, with $L\supset$ and $R\forall$. Continuation with $L\supset$ leads to a loop since it reproduces the conclusion in its first premiss. Therefore we continue with $R\forall$, which gives the premiss $\sim\forall x \sim A \Rightarrow \sim A(y/x)$. Continuation with $L\supset$ gives again a loop, and the same pattern repeats itself. What remains is the search tree:

$$\frac{\frac{\frac{\frac{\sim\forall x \sim A, A(y/x), A(z/x) \Rightarrow \bot}{\sim\forall x \sim A, A(y/x) \Rightarrow \sim A(z/x)}\ {}_{R\supset}}{\sim\forall x \sim A, A(y/x) \Rightarrow \forall x \sim A}\ {}_{R\forall} \quad \bot, A(y/x) \Rightarrow \bot}{\frac{\sim\forall x \sim A, A(y/x) \Rightarrow \bot}{\sim\forall x \sim A \Rightarrow \sim A(y/x)}\ {}_{R\supset}}\ {}_{L\supset}}{\frac{\sim\forall x \sim A \Rightarrow \forall x \sim A}{} \quad \bot \Rightarrow \exists x A}\ {}_{R\forall}}{\frac{\sim\forall x \sim A \Rightarrow \exists x A}{\Rightarrow \sim\forall x \sim A \supset \exists x A}\ {}_{R\supset}}\ {}_{L\supset}$$

Each time $L\supset$ is applied, its left premiss must be the conclusion of $R\forall$ since $L\supset$ would give a loop. The only remaining search tree never terminates but produces, by the variable restriction in rule $R\forall$, ever-longer sequents $\sim\forall x \sim A, A(y/x), A(z/x), \ldots \Rightarrow \bot$ to be derived. QED.

In Chapter 5, Theorem 5.4.9, we prove Glivenko's theorem that states that if a negative formula of propositional logic is derivable classically, it is also derivable intuitionistically. A corresponding result for predicate logic fails, as is shown through a sequent that is easily derived in **G3c** but underivable in **G3i**:

Theorem 4.3.6: $\Rightarrow \sim\sim\forall x(A\vee \sim A)$ *is not derivable in* **G3i**.

Proof: We prove underivability for an atom $P(x)$ by showing that looping caused by rule $L\supset$ and variable restrictions produce an infinite derivation. The last two steps must be

$$\frac{\frac{\sim(\forall x(P(x)\vee \sim P(x))) \Rightarrow \forall x(P(x)\vee \sim P(x)) \quad \bot \Rightarrow \bot}{\sim(\forall x(P(x)\vee \sim P(x))) \Rightarrow \bot}\ {}_{L\supset}}{\Rightarrow \sim\sim(\forall x(P(x)\vee \sim P(x)))}\ {}_{R\supset}$$

If $L\supset$ is used, its left premiss will reproduce the left topsequent. Therefore we continue by $R\forall$, which gives as the premiss $\sim(\forall x(P(x)\vee\sim P(x))) \Rightarrow P(y)\vee\sim P(y)$. As before, the next rule cannot be $L\supset$, so it is one of the $R\vee$ rules. The first $R\vee$ rule leaves the atom $P(x)$ in the succedent so from there the continuation would have to be from the antecedent, by $L\supset$, but this is forbidden by looping. Therefore only the second $R\vee$ rule remains, with $\sim P(x)$ in the succedent. Now continuation is possible through $R\supset$, and we have the search tree

$$
\dfrac{
\dfrac{
\dfrac{
\dfrac{
\dfrac{\vdots}{\sim(\forall x(P(x)\vee\sim P(x))),\,P(y)\overset{\cdot}{\Rightarrow}\bot}
}{\sim(\forall x(P(x)\vee\sim P(x)))\Rightarrow\sim P(y)}\,{}^{R\supset}
}{\sim(\forall x(P(x)\vee\sim P(x)))\Rightarrow P(y)\vee\sim P(y)}\,{}^{R\vee}
}{\sim(\forall x(P(x)\vee\sim P(x)))\Rightarrow\forall x(P(x)\vee\sim P(x))}\,{}^{R\forall}
\qquad \bot\Rightarrow\bot
}{
\dfrac{\sim(\forall x(P(x)\vee\sim P(x)))\Rightarrow\bot}{\Rightarrow\sim\sim(\forall x(P(x)\vee\sim P(x)))}\,{}^{R\supset}
}\,{}^{L\supset}
$$

Proof search now goes on exactly as from the second line from root, except for the addition of the atom $P(y)$ in the antecedent. When we arrive at applying $R\forall$ for the second time root-first, since the antecedent has y free, a variable z distinct from y has to be chosen, which leads in two more steps to the sequent $\sim(\forall x(P(x)\vee\sim P(x))),\,P(y),\,P(z)\Rightarrow\bot$. Continuing again as from the second line from root, but with also $P(z)$ added in the antecedent, proof search produces a third formula $P(v)$ in the antecedent, with $v\neq y,z$, with no end.   QED.

A formula is in **prenex normal form** if it has a string of quantifiers followed by a formula with only propositional connectives. In classical logic, all formulas can be brought to an equivalent prenex normal form, but in intuitionistic logic, this is not so:

Theorem 4.3.7: *The following sequents, with x not free in A, are not derivable in* **G3i***:*

(i) $\Rightarrow\forall x(A\vee B)\supset A\vee\forall xB$,

(ii) $\Rightarrow(A\supset\exists xB)\supset\exists x(A\supset B)$,

(iii) $\Rightarrow(\forall xB\supset A)\supset\exists x(B\supset A)$.

Proof: We show that the sequents are not derivable when A and B are atoms P and $Q(x)$.

For (i), assume that there is a derivation of $\Rightarrow\forall x(P\vee Q(x))\supset P\vee\forall xQ(x)$. The last step is $R\supset$, and therefore $\forall x(P\vee Q(x))\Rightarrow P\vee\forall xQ(x)$ is derivable.

Since there is no implication in this sequent, in any of its derivations only formulas of the forms $\forall x(P \vee Q(x))$, $P \vee Q(t)$, P, $Q(t)$ can appear in antecedents and formulas of the forms $P \vee \forall x\,Q(x)$, P, $\forall x\,Q(x)$, $Q(y)$ in succedents. Further, top-sequents can be only of the forms $P, \Gamma \Rightarrow P$ or $Q(y), \Gamma \Rightarrow Q(y)$. We show that every proof search leads to a branch that cannot have an axiom of these two forms as topsequent.

A sequent is a **nonaxiom** in the derivation of (i) if: 1. Whenever P is a sub-formula of the succedent, P is not in the antecedent, or 2. For any y, whenever $Q(y)$ is the succedent, $Q(y)$ is not in the antecedent.

We now define a branch such that all of its sequents are nonaxioms, from which underivability of sequent (i) follows. Note that the only branchings that can appear are due to rule $L\vee$, with principal formula $P \vee Q(t)$. If the succedent contains P as subformula, we choose the premiss with $Q(t)$, and if not, we choose the premiss with P. The proof that all sequents in the branch so defined are nonaxioms is by induction on length of the branch. There are four cases that depend on the prin-cipal formula of an inference:

1. $\forall x(P \vee Q(x))$ in antecedent: The active formula in the premiss is $P \vee Q(x)$ so that if the conclusion is a nonaxiom the premiss also is.

2. $P \vee Q(x)$ in antecedent: If the succedent has P as subformula the chosen branch has $Q(x)$ in the premiss and the property of being a nonaxiom is preserved. Else premiss with P is chosen and being a nonaxiom is preserved.

3. $P \vee \forall x\,Q(x)$ in succedent: Since the conclusion is a nonaxiom, P is not in the antecedent so having P or $\forall x\,Q(x)$ in the succedent of the premiss preserves being a nonaxiom.

4. $\forall x\,Q(x)$ in the succedent: If $Q(y)$ is in the antecedent, by the variable restriction in $R\forall$ the premiss contains in the succedent $Q(z)$ with $z \neq y$ so the property of being a nonaxiom is preserved.

For (ii), we attempt a proof search of the sequent for atoms P and $Q(x)$. The last step is $R\supset$, the one above it either $L\supset$ or $R\exists$. The former gives as the left premiss $P \supset \exists x\,Q(x) \Rightarrow P$, but this is underivable since only $L\supset$ applies and it gives a loop. So the next-to-last step is $R\exists$ and the premiss is $P \supset \exists x\,Q(x) \Rightarrow P \supset Q(t)$ for some term t. Again, $L\supset$ would produce a loop, and the only remaining search tree is

$$
\dfrac{
  \dfrac{
    P \supset \exists x\,Q(x), P \Rightarrow P \quad
    \dfrac{
      \dfrac{Q(y), P \Rightarrow Q(t)}{\exists x\,Q(x), P \Rightarrow Q(t)}\,{\scriptstyle L\exists}
    }{}
  }{
    \dfrac{
      \dfrac{P \supset \exists x\,Q(x), P \Rightarrow Q(t)}{P \supset \exists x\,Q(x) \Rightarrow P \supset Q(t)}\,{\scriptstyle R\supset}
    }{}\,{\scriptstyle L\supset}
  }{}
}{}
$$

$$
\dfrac{
  \dfrac{
    \dfrac{
      \dfrac{P \supset \exists x\,Q(x) \Rightarrow P \supset Q(t)}{P \supset \exists x\,Q(x) \Rightarrow \exists x(P \supset Q(x))}\,{\scriptstyle R\exists}
    }{\Rightarrow P \supset \exists x\,Q(x) \supset \exists x(P \supset Q(x))}\,{\scriptstyle R\supset}
  }{}
}{}
$$

The sequent $Q(y)$, $P \Rightarrow Q(t)$ is not an axiom since, by the variable restriction in $L\exists$, we must have $t \neq y$.

The proof of (iii) is similar to that of (ii).   QED.

Derivations of sequents containing only formulas in prenex normal form can be turned into derivations in which the propositional rules precede all the quantifier rules. In the system **G3c** this result, called the **midsequent theorem**, can be stated as follows:

Theorem 4.3.8: *If* $\Gamma \Rightarrow \Delta$ *is derivable in* **G3c** *and* Γ, Δ *have all formulas in prenex normal form, the derivation has a* midsequent $\Gamma' \Rightarrow \Delta'$ *such that all inferences up to the midsequent are propositional and all inferences after it quantificational.*

Proof: For each derivation and each instance of a quantifier rule $\mathcal{Q}$ in it, consider the number $n(\mathcal{Q})$ of applications of propositional rules that are below the quantifier rule in the derivation, and let n be the sum of the $n(\mathcal{Q})$ for all the applications of quantifier rules. We show by induction on n that every derivation can be transformed into a derivation in which n is zero. If $n = 0$, there is nothing to prove. If $n > 0$, we consider the downmost application of a quantifier rule with a propositional rule *Prop* immediately below it. There are several cases, all dealt with similarly, and we consider the case in which the quantifier rule is $L\exists$ and the propositional rule has one premiss. We have the following steps of derivation:

$$\frac{\dfrac{A(y/x), \Gamma'' \Rightarrow \Delta''}{\exists x A, \Gamma'' \Rightarrow \Delta''} L\exists}{\Theta \Rightarrow \Lambda} Prop$$

Since by hypothesis the endsequent of the derivation consists of prenex formulas only, by the subformula property all the sequents in the derivation consist of prenex formulas only. Therefore $\exists x A$ cannot be active in the propositional rule, for otherwise Θ would contain a formula of the form $B \circ \exists x A$ or $\exists x A \circ B$ that is not in prenex form. Thus $\Theta = \exists x A, \Gamma'''$ and the two steps of derivation can be permuted as follows:

$$\frac{\dfrac{A(y/x), \Gamma'' \Rightarrow \Delta''}{A(y/x), \Gamma''' \Rightarrow \Lambda} Prop}{\exists x A, \Gamma''' \Rightarrow \Lambda} L\exists$$

The variable restrictions are satisfied since the propositional rules do not alter the variable binding. By inductive hypothesis the derivation of $A(y/x)$, $\Gamma'' \Rightarrow \Delta''$ can be transformed into one that satisfies the midsequent theorem, and all the inferences below the steps considered are quantificational.   QED.

4.4. COMPLETENESS OF CLASSICAL PREDICATE LOGIC

We shall give a proof of completeness for **pure** classical predicate logic, where pure means that the language contains no functions or constants. A denumerable set of variables $x_1, x_2, \ldots$ ordered by the indices is needed in the proof. The notion of a valuation for formulas of classical predicate logic is defined as an extension of the definition for classical propositional logic:

Definition 4.4.1: *A valuation is a function v from formulas of predicate logic to the values 0 and 1, assumed to be given for all atoms,*

$$v(P^n(x_i, \ldots, x_j)) = 0 \ or \ v(P^n(x_i, \ldots, x_j)) = 1,$$

and extended inductively to all formulas,

$$v(\bot) = 0,$$
$$v(A \& B) = min(v(A), v(B)),$$
$$v(A \lor B) = max(v(A), v(B)),$$
$$v(A \supset B) = max(1 - v(A), v(B)),$$
$$v(\forall x \, A) = inf(v(A(x_i/x))),$$
$$v(\exists x \, A) = sup(v(A(x_i/x))).$$

The infimum is taken over the denumerable sequence of values $v(A(x_i/x))$ for $x_1, x_2, \ldots$, and similarly for the supremum. These two classical valuations are in general infinitary, and no method of actually computing values is assumed. As in Section 3.3, we extend valuations to contexts by taking conjunctions and disjunctions of their formulas and by setting $v \bigwedge (\Gamma) = min(v(C))$ for formulas C in Γ and $v \bigvee (\Gamma) = max(v(C))$ for formulas C in Γ.

Definition 4.4.2: *A sequent $\Gamma \Rightarrow \Delta$ is refutable if there is a valuation v such that $v \bigwedge (\Gamma) > v \bigvee (\Delta)$. A sequent $\Gamma \Rightarrow \Delta$ is valid if it is not refutable.*

A valuation showing refutability is called a refuting valuation. A sequent $\Gamma \Rightarrow \Delta$ is valid if for all valuations v, $v \bigwedge (\Gamma) \leqslant v \bigvee (\Delta)$. We now prove **soundness** of the sequent calculus **G3c** for classical predicate logic, continuing the proof of Theorem 3.3.5:

Theorem 4.4.3: *If a sequent $\Gamma \Rightarrow \Delta$ is derivable in **G3cp**, it is valid.*

Proof: Assume that $\Gamma \Rightarrow \Delta$ is derivable. We prove by induction on height of derivation that it is valid. The new cases to consider are when the last rule in the derivation is a quantifier rule. If the last rule is $L\forall$, suppose

$$min(v(A(y/x)), v(\forall x \, A), v(\bigwedge (\Gamma))) \leqslant v \bigvee (\Delta)$$

and show that

$$min(v(\forall x A), v(\bigwedge(\Gamma))) \leqslant v\bigvee(\Delta).$$

This follows by $v(\forall x A) \leqslant v(A(y/x))$. If the last rule is $R\forall$, we have, by inductive hypothesis,

$$v\bigwedge(\Gamma) \leqslant max(v\bigvee(\Delta), v(A(y/x))).$$

Since y does not occur in Γ, this implies

$$v\bigwedge(\Gamma) \leqslant inf_y(max(v\bigvee(\Delta), v(A(y/x)))),$$

and since y does not occur in Δ,

$$v\bigwedge(\Gamma) \leqslant max(v\bigvee(\Delta), inf_y(v(A(y/x)))),$$

that is,

$$v\bigwedge(\Gamma) \leqslant max(v\bigvee(\Delta), v(\forall x A)).$$

The cases of $L\exists$ and $R\exists$ are symmetric to $R\forall$ and $L\forall$, respectively.   QED.

The main idea of the **completeness proof** for the system **G3c** of classical predicate logic is the following: Given a sequent $\Gamma \Rightarrow \Delta$ we construct, by applying root-first the rules **G3c** in all possible ways, a tree, called a **reduction tree** for $\Gamma \Rightarrow \Delta$. If all branches reach the form of an axiom or conclusion of $L\bot$, the tree gives a proof of the given sequent. Otherwise, we prove by classical reasoning that the construction does not terminate. By König's lemma, a nonconstructive result for infinite trees recalled below, the tree has an infinite branch. Given such an infinite branch, we define a refuting valuation for the sequent.

We remark that the procedure that follows gives as a special case the completeness proof for the calculus **G3cp**; as in the propositional case the procedure of construction of the reduction tree reduces to a finite one. In the first-order case instead, we cannot know in general if the tree terminates or goes on forever, and no decision method is obtained.

The following property of trees will be necessary in the proof of Lemma 4.4.3. Its proof is nonconstructive, and we shall not give it.

Lemma 4.4.4: König's lemma. *Every infinite, finitely branching tree has an infinite branch.*

Theorem 4.4.5: *Any sequent $\Gamma \Rightarrow \Delta$ either has a proof in* **G3c** *or there is a refuting valuation for the sequent.*

Proof: The proof consists of two parts. In the first part we define for each sequent a reduction tree that gives a proof when finite. In the second part we show that an infinite tree gives a refuting valuation.

1. Construction of the reduction tree: We define for each sequent $\Gamma \Rightarrow \Delta$ a reduction tree having $\Gamma \Rightarrow \Delta$ as root and a sequent at each node. The tree is constructed inductively in **stages**, as follows:

Stage 0 has $\Gamma \Rightarrow \Delta$ at the root of the tree. Stage $n > 0$ has two cases:

Case I: If every topmost sequent is an axiom or conclusion of $L\bot$, the construction of the tree ends.

Case II: If not every topmost sequent is an axiom or conclusion of $L\bot$, we continue the construction of the tree by writing above those topmost sequents that are not axioms or conclusions of $L\bot$ other sequents, which we obtain by applying root-first the rules of **G3c** whenever possible, in a given order. When no rule is applicable the topmost sequent has distinct atoms in the antecedent and succedent and no $\bot$ in the antecedent, and the sequent is repeated. (Thus, for propositional logic, each branch terminates or starts repeating itself identically.)

For stages $n = 1, \ldots, 10$, the reduction is illustrated below. For $n = 11$ we repeat stage 1, for $n = 12$ stage 2, and so on for each n.

We start for $n = 1$ with $L\&$: For each topmost sequent of the form

$$B_1 \& C_1, \ldots, B_m \& C_m, \Gamma' \Rightarrow \Delta$$

where $B_1 \& C_1, \ldots, B_m \& C_m$ are all the formulas in the antecedent with conjunction as outermost logical connective, we write

$$B_1, C_1, \ldots, B_m, C_m, \Gamma' \Rightarrow \Delta$$

on top of it. This step corresponds to applying root first m times rule $L\&$.

For $n = 2$, we consider all the sequents of the form

$$\Gamma \Rightarrow B_1 \& C_1, \ldots, B_m \& C_m, \Delta'$$

where $B_1 \& C_1, \ldots, B_m \& C_m$ are all the formulas in the succedent with conjunction as the outermost logical connective, and write on top of them the 2^m sequents

$$\Gamma \Rightarrow D_1, \ldots, D_m, \Delta'$$

where D_i is either B_i or C_i (and all possible choices are taken). This is equivalent to applying $R\&$ root-first consecutively with principal formulas $B_1 \& C_1, \ldots, B_m \& C_m$.

For $n = 3$ and 4 we consider $L\vee$ and $R\vee$ and define the reductions symmetrically to the cases $n = 2$ and $n = 1$, respectively.

For $n = 5$, for each topmost sequent having the formulas $B_1 \supset C_1, \ldots, B_m \supset C_m$ with implication as the outermost logical connective in the antecedent, Γ' the other formulas, and succedent Δ, write on top of it the 2^m sequents

$$C_{i_1}, \ldots, C_{i_k}, \Gamma' \Rightarrow B_{j_{k+1}}, \ldots, B_{j_m}, \Delta$$

where $i_1, \ldots, i_k \in \{1, \ldots, m\}$ and $j_{k+1}, \ldots, j_m \in \{1, \ldots, m\} - \{i_1, \ldots, i_k\}$. This step (perhaps less transparent because of the double indexing) corresponds to the root-first application of rule $L\supset$ with principal formulas $B_1 \supset C_1, \ldots, B_m \supset C_m$.

For $n = 6$, we consider all the sequents having implications in the succedent, say $B_1 \supset C_1, \ldots, B_m \supset C_m$, and Δ' the other formulas, and write on top of them

$$B_1, \ldots, B_m, \Gamma \Rightarrow C_1, \ldots, C_m, \Delta'$$

that is, we apply root-first m times rule $R\supset$.

For $n = 7$, consider all the topsequents having universally quantified formulas $\forall x_1 B_1, \ldots, \forall x_m B_m$ in the antecedent. For $i = 1, \ldots, m$, let y_i be the first variable not yet used for a reduction of $\forall x_i B_i$, and write on top of these sequents the sequents

$$B_1(y_1/x_1), \ldots, B_m(y_m/x_m), \forall x_1 B_1, \ldots, \forall x_m B_m, \Gamma' \Rightarrow \Delta$$

that is, apply root-first rule $L\forall$ with principal formulas $\forall x_1 B_1, \ldots, \forall x_m B_m$. It is essential that the variable y_i be chosen starting from the beginning of the ordered set of variables and by excluding those variables that have already been used in a similar reduction for $\forall x_i B_i$, as the purpose of this step of the reduction is to obtain, sooner or later, any substitution instance of B_i.

For $n = 8$, let $\forall x_1 B_1, \ldots, \forall x_m B_m$ be the universally quantified formulas occurring in the succedent of a topsequent of the tree, and let Δ' be the other formulas. Let $z_1, \ldots, z_m$ be fresh variables, not yet used in the reduction tree, and write on top of each such sequent the sequents

$$\Gamma \Rightarrow B_1(z_1/x_1), \ldots, B_m(z_m/x_m), \Delta'$$

that is, apply root-first m times rule $R\forall$.

For $n = 9$ and 10 we consider $L\exists$ and $R\exists$ and define the reduction in a way symmetric to the cases $n = 8$ and $n = 7$, respectively.

For each n, for sequents that are neither axioms nor conclusions of $L\bot$, nor treatable by the above reductions, we write the sequent itself above them.

If the reduction tree is finite, all its leaves are axioms. We observe that the tree, read top-down from leaves to root, gives a proof of the endsequent $\Gamma \Rightarrow \Delta$.

2. Definition of the refuting valuation: If the reduction tree is infinite, by König's lemma it has an infinite branch. Let $\Gamma_0 \Rightarrow \Delta_0$ be the given sequent $\Gamma \Rightarrow \Delta$ and let

$$\Gamma_0 \Rightarrow \Delta_0, \ldots, \Gamma_i \Rightarrow \Delta_i, \ldots$$

be one such branch, and consider the sets of formulas

$$\Gamma \equiv \bigcup_{i \geqslant 0} \Gamma_i$$

$$\Delta \equiv \bigcup_{i \geqslant 0} \Delta_i$$

We define a valuation in which all formulas in Γ have value 1 and all formulas in Δ have value 0, thereby refuting the sequent $\Gamma \Rightarrow \Delta$.

Observe that, by definition of the reduction tree, Γ and Δ have no atomic formulas in common. For if P is in Γ and Δ, then for some i, j, P is in Γ_i and Δ_j, but then P is in Γ_k and Δ_k for $k \geqslant i, j$, contrary to the assumption that $\Gamma_k \Rightarrow \Delta_k$ is not an axiom. Consider the valuation v defined by setting $v(P) = 1$ for each atomic formula in Γ and $v(P) = 0$ for P in Δ.

We show by induction on the weight of formulas that $v(A) = 1$ if A is in Γ and $v(A) = 0$ if A is in Δ. Therefore v is a refuting valuation.

If A is $\bot$, A cannot be in Γ, for otherwise the infinite branch would contain a conclusion of $L\bot$, so A can be in only Δ and $v(A) = 0$ by the definition of a valuation.

If A is atomic, the claim holds by definition of the refuting valuation.

If $A = B \& C$ is in Γ, there exists i such that $A \in \Gamma_i$, and therefore B, C are in Γ_{i+k} for some $k \geqslant 0$. By inductive hypothesis, $v(B) = 1$ and $v(C) = 1$, so $v(B \& C) = 1$.

If $A = B \& C$ is in Δ, consider the step i in which the reduction for A applies. This gives a branching, and one of the two branches belongs to the infinite branch, so either B or C is in Δ, and therefore by inductive hypothesis $v(B) = 0$ or $v(C) = 0$, and therefore $v(B \& C) = 0$ is satisfied.

If $A = B \vee C$ is in Γ, we reason similarly to the case of $A = B \& C$ in Δ. If $A = B \vee C$ is in Δ, we argue as with $A = B \& C$ in Γ.

If $A = B \supset C$ is in Γ, then either $B \in \Delta$ or $C \in \Gamma$. By inductive hypothesis, in the former case $v(B) = 0$ and in the latter $v(C) = 1$, so in both cases $v(B \supset C) = 1$.

If $A = B \supset C$ is in Δ, then for some i, $B \in \Gamma_i$ and $C \in \Delta_i$, so by inductive hypothesis $v(B) = 1$ and $v(C) = 0$, so $v(B \supset C) = 0$.

If $A = \forall x B$ is in Γ, let i be the least index such that A occurs in Γ_i. Any substitution instance $B(y/x)$ occurs sooner or later in Γ_j for $j \geqslant i$, so by inductive hypothesis $v(B(y/x)) = 1$ for all y, and therefore $v(\forall x B) = 1$.

If $A = \forall x B$ is in Δ, consider the step at which the reduction applies to A. At this step we have, for some z and some j, $B(z/x) \in \Delta_j$, and therefore by inductive hypothesis $v(B(z/x)) = 0$, so $v(\forall x B) = 0$.

The cases of $A = \exists x B$ in Γ and $A = \exists x B$ in Δ are symmetric to the cases of $A = \forall x B$ in Δ and of $A = \forall x B$ in Γ, respectively.   QED.

By the theorem, we conclude:

Corollary 4.4.6: *If a sequent* $\Gamma \Rightarrow \Delta$ *is valid, it is derivable in* **G3c**.

NOTES TO CHAPTER 4

Our general elimination rule for the universal quantifier is presented in von Plato (1998). It follows the pattern of general elimination rules as determined by the inversion principle of Section 1.2.

The calculus **G3i** is a single succedent version of Dragalin's (1988) contraction-free intuitionistic calculus, first given in Troelstra and Schwichtenberg (1996). The calculus **G3c** is the standard contraction-free classical calculus. The syntactic proofs of underivability in Section 4.3 follow Kleene (1952).

Our completeness proof for **G3c** uses valuations as a continuation of the proof for propositional logic in Theorem 3.3.6. This is suggested, although not carried out in detail, in Ketonen (1941). His result is mentioned in the introduction to Szabo's 1969 edition of Gentzen's papers (p. 7). The construction of the refutation tree is carried out in Schütte (1956). In Takeuti (1987), whose exposition we closely follow, Schütte's method is applied to Gentzen's original calculus **LK**. For the use of König's lemma in the completeness proof, see Beth (1959, p. 195).

5

Variants of Sequent Calculi

In this chapter we present different formulations of sequent calculi. In Sections 5.1 and 5.2 we give calculi with independent contexts in two versions, one with explicit rules of weakening and contraction, the other with these rules built into the logical rules similarly to natural deduction. The proofs of cut elimination for these calculi are quite different from each other and from the earlier proofs for the **G3** calculi. For calculi of the second type in particular, with implicit weakening and contraction, cut elimination will be limited to cut formulas that are principal somewhere in the derivation of the right premiss of cut. All other cut formulas are shown to be subformulas of the conclusion already.

The structure of derivations in calculi with independent contexts is closely connected to the structure of derivations in natural deduction. The correspondence will be studied in Chapter 8.

In Section 5.3 we present an intuitionistic multisuccedent calculus and its basic properties. The calculus has a right disjunction rule that is invertible, which is very useful in proof search. The calculus is also used in the study of extensions of logical sequent calculi with mathematical axioms in Chapter 6.

We also present a single succedent calculus for classical propositional logic. Its main advantage compared with the multisuccedent calculus **G3cp** is that it has an operational interpretation and a straightforward translation to natural deduction. The calculus is obtained from the intuitionistic calculus **G3ip** by the addition of a sequent calculus rule corresponding to the law of excluded middle. The propositional part, from this point of view, amounts to an intuitionistic calculus for theories with decidable basic relations.

In the last section, we present a calculus for intuitionistic propositional logic, called **G4ip**, with the property that proof search terminates.

5.1. SEQUENT CALCULI WITH INDEPENDENT CONTEXTS

The motivation of the rules of sequent calculus from those of natural deduction in Section 1.3 produced rules with independent contexts: Two-premiss rules had contexts that were added together into the context of the conclusion. In later chapters we used shared contexts in order to obtain calculi that do not need the rule of contraction and for purposes of proof search.

We present two calculi with independent contexts, analogous to Gentzen's original calculi **LJ** and **LK** in that weakening and contraction are primitive rules, and establish their basic properties.

A long-standing complication in the proof of cut elimination is removed by a more detailed analysis of derivations: Gentzen in his 1934–35 proof of the "Hauptsatz" for sequent calculus, or cut elimination theorem, had to hide contraction into one of the cases. If the right premiss of cut is derived by contraction, the permutation of cut with contraction does not move the cut higher up in the derivation. The rule of **multicut** permits eliminating $m \geq 1$ occurrences $A, \ldots, A = A^m$ of the cut formula in the right premiss in one step:

$$\frac{\Gamma \Rightarrow A \quad A^m, \Delta \Rightarrow C}{\Gamma, \Delta \Rightarrow C} \, Cut^*$$

The reason for having to make recourse to this rule is the following: Consider the derivation

$$\frac{\Gamma \Rightarrow A \quad \dfrac{A, A, \Delta \Rightarrow C}{A, \Delta \Rightarrow C} \, Ctr}{\Gamma, \Delta \Rightarrow C} \, Cut$$

Permuting cut with contraction, we obtain

$$\frac{\Gamma \Rightarrow A \quad \dfrac{\Gamma \Rightarrow A \quad A, A, \Delta \Rightarrow C}{A, \Gamma, \Delta \Rightarrow C} \, Cut}{\dfrac{\Gamma, \Gamma, \Delta \Rightarrow C}{\Gamma, \Delta \Rightarrow C} \, Ctr} \, Cut$$

Here the second cut is on the same formula A and has a sum of heights of derivations of the premisses not less than the one in the first derivation. With multicut, instead, we transform the derivation into

$$\frac{\Gamma \Rightarrow A \quad A^2, \Delta \Rightarrow C}{\Gamma, \Delta \Rightarrow C} \, Cut^*$$

Here the height of derivation of the right premiss is diminished by one. A proof of multicut elimination can be given by induction on the length of the cut formula and a subinduction on cut-height. The proof consists of permuting multicut up with the rules used for deriving its premisses, until it reaches logical axioms the derivation started with, and disappears (see, for example, Takeuti 1987). A calculus with multicut is equivalent to a calculus with cut, in the sense that the same sequents are derivable. Ordinary cut is a special case of multicut, so that cut elimination follows from elimination of multicut.

We shall give proofs of cut elimination without multicut for an intuitionistic single succedent and a classical multisuccedent calculus. These can be considered standard calculi when contexts in rules with two premisses are treated as independent.

(a) Cut elimination for the intuitionistic calculus: In the proof of cut elimination without multicut, the problematic case of contraction is treated by a more global proof transformation by cases on the derivation of the premiss of contraction. The proof is given for a sequent calculus with independent contexts, **G0i**, with the following rules:

G0i

Logical axiom:

$A \Rightarrow A$

Logical rules:

$$\frac{A, B, \Gamma \Rightarrow C}{A \& B, \Gamma \Rightarrow C} L\& \qquad \frac{\Gamma \Rightarrow A \quad \Delta \Rightarrow B}{\Gamma, \Delta \Rightarrow A \& B} R\&$$

$$\frac{A, \Gamma \Rightarrow C \quad B, \Delta \Rightarrow C}{A \vee B, \Gamma, \Delta \Rightarrow C} L\vee \qquad \frac{\Gamma \Rightarrow A}{\Gamma \Rightarrow A \vee B} R\vee_1 \qquad \frac{\Gamma \Rightarrow B}{\Gamma \Rightarrow A \vee B} R\vee_2$$

$$\frac{\Gamma \Rightarrow A \quad B, \Delta \Rightarrow C}{A \supset B, \Gamma, \Delta \Rightarrow C} L\supset \qquad \frac{A, \Gamma \Rightarrow B}{\Gamma \Rightarrow A \supset B} R\supset$$

$$\frac{}{\bot \Rightarrow C} L\bot$$

$$\frac{A(t/x), \Gamma \Rightarrow C}{\forall x A, \Gamma \Rightarrow C} L\forall \qquad \frac{\Gamma \Rightarrow A(y/x)}{\Gamma \Rightarrow \forall x A} R\forall$$

$$\frac{A(y/x), \Gamma \Rightarrow C}{\exists x A, \Gamma \Rightarrow C} L\exists \qquad \frac{\Gamma \Rightarrow A(t/x)}{\Gamma \Rightarrow \exists x A} R\exists$$

Rules of weakening and contraction:

$$\frac{\Gamma \Rightarrow C}{A, \Gamma \Rightarrow C} Wk \qquad \frac{A, A, \Gamma \Rightarrow C}{A, \Gamma \Rightarrow C} Ctr$$

The variable restrictions in $R\forall$ and $L\exists$ are that y is not free in the conclusion. In the derivations below, Ctr^* will indicate repeated contractions.

To prove the admissibility of cut, formula length and cut-height are defined as before. The proof of cut elimination is organized as follows: We first consider cases in which one premiss is an axiom of form $A \Rightarrow A$ or the left premiss a conclusion of $L\bot$, then cases with either premiss obtained by weakening. Next we have cases in which the cut formula is principal in both premisses and cases in which the cut formula is principal in the right premiss only. Then we have the case that both premises are derived by a logical rule and the cut formula is not principal in either. The last cases concern contraction. Cut elimination proceeds by first eliminating cuts that are not preceded by other cuts. The following lemma will be used in the proof:

Lemma 5.1.1: *The following inversions hold in* **G0i**:
  (i) *If* $A\&B, \Gamma \Rightarrow C$ *is derivable, also* $A, B, \Gamma \Rightarrow C$ *is derivable.*
  (ii) *If* $A \vee B, \Gamma \Rightarrow C$ *is derivable, also* $A, \Gamma \Rightarrow C$ *and* $B, \Gamma \Rightarrow C$ *are derivable.*
  (iii) *If* $A \supset B, \Gamma \Rightarrow C$ *is derivable, also* $B, \Gamma \Rightarrow C$ *is derivable.*
  (vi) *If* $\Gamma \Rightarrow \forall x A$ *is derivable, also* $\Gamma \Rightarrow A(t/x)$ *is derivable.*
  (v) *If* $\exists x A, \Gamma \Rightarrow C$ *is derivable, also* $A(t/x), \Gamma \Rightarrow C$ *is derivable.*

Proof: In each case, trace up from the endsequent the occurrence of the formula in question. If at some stage the formula is principal in contraction, trace up from the premiss both occurrences. In this way, a number of **first occurrences** of the formula are located. (i) If a first occurrence of $A\&B$ is obtained by weakening, weaken with A and with B and continue as before after the weakenings until either a derivation of $A, B, \Gamma \Rightarrow C$ is reached or a step is found in which a contraction on $A\&B$ was done in the given derivation. In the latter case, the transformed derivation will have A, A, B, B in place of $A\&B, A\&B$, and a contraction on A and on B is done and the derivation continued as before. If a first occurrence $A\&B$ is obtained by an axiom $A\&B \Rightarrow A\&B$, the axiom is substituted by

$$\frac{A \Rightarrow A \quad B \Rightarrow B}{A, B \Rightarrow A\&B} R\&$$

and the derivation is continued as before. Otherwise, a first occurrence $A\&B$ is obtained by $L\&$, and deletion of this rule will give a derivation of $A, B, \Gamma \Rightarrow C$ as before. For (ii), weakening and axiom are treated similarly as in (i). Otherwise, the $L\vee$ rule introducing $A \vee B$ in the antecedent is

$$\frac{A, \Gamma'' \Rightarrow C' \quad B, \Gamma''' \Rightarrow C'}{A \vee B, \Gamma' \Rightarrow C'} L\vee$$

where $\Gamma' = \Gamma'', \Gamma'''$. Repeated weakening of the premisses gives $A, \Gamma' \Rightarrow C'$ and $B, \Gamma' \Rightarrow C'$, and continuing as before derivations of $A, \Gamma \Rightarrow C$ and $B, \Gamma \Rightarrow C$ are obtained. (iii) is proved similarly to (ii). (iv) If a first occurrence of $\forall x A$ is obtained by an axiom, replace it by $A(t/x) \Rightarrow A(t/x)$ and then apply $L\forall$. If it is obtained by $R\forall$ from $\Gamma' \Rightarrow A(y/x)$, y is not free in Γ' and, by arguments similar to the substitution Lemma 4.2.1 for **G3i**, a substitution of t for x gives a derivation of $\Gamma' \Rightarrow A(t/x)$. (v) If $\exists x A, \Gamma \Rightarrow C$ is obtained by $L\exists$, the premiss is $A(y/x), \Gamma \Rightarrow C$ and substitution gives $A(t/x), \Gamma \Rightarrow C$. QED.

The lemma gives inversions of the left rules, with shared contexts as in the inversions for the calculus **G3i**, except that the inversions are not height-preserving.

Theorem 5.1.2: *The rule of cut*

$$\frac{\Gamma \Rightarrow D \quad D, \Delta \Rightarrow C}{\Gamma, \Delta \Rightarrow C} \, Cut$$

is admissible in **G0i**.

Proof: The proof is by induction on length of cut formula with a subinduction on cut-height. It is sufficient to consider a derivation with just one cut. For all cases, a transformation is given that either reduces length of cut formula or reduces cut-height while leaving the cut formula unchanged. We proceed by analyzing the cut formula and show how to dispense with the cut or replace it by a cut on shorter formulas. In all the cases in which the cut formula cannot be reduced the derivations of the premisses of cut are analyzed. There are seven cases according to the form of the cut formula:

1. The cut formula D is $\bot$. Consider the left premiss of cut $\Gamma \Rightarrow \bot$. If it is an axiom, then the conclusion of cut is the right premiss. If it is a conclusion of $L\bot$, then $\bot$ is in Γ and thus the conclusion of cut is also a conclusion of $L\bot$. Else $\Gamma \Rightarrow \bot$ is obtained by a rule R with $\bot$ not principal in it. If R is a one-premiss rule the derivation ends with

$$\frac{\dfrac{\Gamma' \Rightarrow \bot}{\Gamma \Rightarrow \bot} \, R \quad \bot, \Delta \Rightarrow C}{\Gamma, \Delta \Rightarrow C} \, Cut$$

and the rule and cut are permuted into

$$\frac{\dfrac{\Gamma' \Rightarrow \bot \quad \bot, \Delta \Rightarrow C}{\Gamma', \Delta \Rightarrow C} \, Cut}{\Gamma, \Delta \Rightarrow C} \, R$$

with reduced cut-height. A similar conversion applies if R is a two-premiss rule.

2. The cut formula D is an atom P. If the left premiss $\Gamma \Rightarrow P$ of cut is an axiom, then the conclusion of cut is given by the right premiss $P, \Delta \Rightarrow C$. If $\Gamma \Rightarrow P$ is a conclusion of $L\bot$, the conclusion of cut is also a conclusion of $L\bot$. If $\Gamma \Rightarrow P$ is derived by a rule, P is not principal in it and cut is permuted with the rule as in case *1*.

We consider now the cases in which the cut formula is a compound formula. Observe that if the cut formula is not principal in the last rule used to derive the left premiss, cut can be permuted. If the left premiss is an axiom, the right premiss gives the conclusion of cut. If the left premiss is a conclusion of $L\perp$, the conclusion of cut is also a conclusion of $L\perp$. Thus those cases are left in which the cut formula is principal in the last rule used to derive the left premiss of cut.

3. The cut formula D is $A\&B$. The derivation

$$\frac{\dfrac{\Gamma \Rightarrow A \quad \Delta \Rightarrow B}{\Gamma, \Delta \Rightarrow A\&B}R\& \quad A\&B, \Theta \Rightarrow C}{\Gamma, \Delta, \Theta \Rightarrow C}Cut$$

is transformed by Lemma 5.1.1 into

$$\frac{\Gamma \Rightarrow A \quad \dfrac{\Delta \Rightarrow B \quad \dfrac{A\&B, \Theta \Rightarrow C}{A, B, \Theta \Rightarrow C}Inv}{\Delta, A, \Theta \Rightarrow C}Cut}{\Gamma, \Delta, \Theta \Rightarrow C}Cut$$

Note that cut-height can increase in the transformation, but the cut formula is reduced.

4. The cut formula D is $A \vee B$. The derivation

$$\frac{\dfrac{\Gamma \Rightarrow A}{\Gamma \Rightarrow A \vee B}R\vee_1 \quad A \vee B, \Delta \Rightarrow C}{\Gamma, \Delta \Rightarrow C}Cut$$

is transformed by Lemma 5.1.1 into

$$\frac{\Gamma \Rightarrow A \quad \dfrac{A \vee B, \Delta \Rightarrow C}{A, \Delta \Rightarrow C}Inv}{\Gamma, \Delta \Rightarrow C}Cut$$

and similarly if the second $R\vee$ rule was used. Length of cut formula is reduced.

We shall next analyze the case of cut formula $\exists x A$ and consider last the cases of $A \supset B$ and $\forall A$. Because of lack of invertibility, these last cases require a different analysis.

5. The cut formula D is $\exists x A$. The derivation

$$\frac{\dfrac{\Gamma \Rightarrow A(t/x)}{\Gamma \Rightarrow \exists x A}R\exists \quad \exists x A, \Delta \Rightarrow C}{\Gamma, \Delta \Rightarrow C}Cut$$

is transformed by Lemma 5.1.1 into

$$\frac{\Gamma \Rightarrow A(t/x) \quad \dfrac{\exists x A, \Delta \Rightarrow C}{A(t/x), \Delta \Rightarrow C}^{Inv}}{\Gamma, \Delta \Rightarrow C}_{Cut}$$

with length of cut formula reduced.

6. The cut formula D is $A \supset B$. There are three subcases.

6.1. The formula $A \supset B$ is not principal in the right premiss, and the last rule used to derive the right premiss is not a contraction on $A \supset B$. In all such cases cut is permuted with the last rule used to derive the right premiss of cut.

6.2. The formula $A \supset B$ is principal in the right premiss; thus the derivation is

$$\frac{\dfrac{A, \Gamma \Rightarrow B}{\Gamma \Rightarrow A \supset B}^{R\supset} \quad \dfrac{\Delta \Rightarrow A \quad B, \Theta \Rightarrow C}{A \supset B, \Delta, \Theta \Rightarrow C}^{L\supset}}{\Gamma, \Delta, \Theta \Rightarrow C}_{Cut}$$

and it is transformed into the derivation with two cuts on shorter formulas

$$\frac{\dfrac{\Delta \Rightarrow A \quad A, \Gamma \Rightarrow B}{\Gamma, \Delta \Rightarrow B}^{Cut} \quad B, \Theta \Rightarrow C}{\Gamma, \Delta, \Theta \Rightarrow C}_{Cut}$$

6.3. The right premiss of cut is derived by contraction on $A \supset B$. Since $L\supset$ is invertible with respect to only its right premiss, we analyze the derivation of the right premiss of cut. Tracing up this derivation until the rule applied is not a contraction on $A \supset B$, we find a sequent with n copies of formula $A \supset B$ in the antecedent:

If $A \supset B$ is not principal in the rule concluding this sequent, we permute down the rule through the $n - 1$ contractions until it concludes the right premiss of cut, or, if copies of $A \supset B$ come from two premisses, these are contracted and cut permuted up with two cuts of reduced cut-height as result.

If the rule concluding the sequent with n copies of $A \supset B$ in the antecedent is weakening with $A \supset B$, the weakening step is removed and one contraction less applied.

The remaining case is that one occurrence of $A \supset B$ in the sequent with n copies of $A \supset B$ in the antecedent is derived by $L\supset$; thus the step concluding the premiss of the uppermost contraction is

$$\frac{\Delta \Rightarrow A \quad B, \Theta' \Rightarrow C}{A \supset B, (A \supset B)^{n-1}, \Theta \Rightarrow C}^{L\supset}$$

Here $(A \supset B)^{n-1}, \Theta = \Delta, \Theta'$. The $n - 1$ copies of formula $A \supset B$ are divided in Δ and Θ' with $\Delta = (A \supset B)^k$, Λ and $\Theta' = (A \supset B)^l$, Θ'' and $k + l = n - 1$. Each formula in Λ and in Θ'' is also in Θ. The derivation can now be written, with Ctr^n standing for an $n - 1$ fold contraction, as

$$
\frac{
  \dfrac{A, \Gamma \Rightarrow B}{\Gamma \Rightarrow A \supset B} R\supset
  \qquad
  \dfrac{
    \dfrac{
      \dfrac{(A \supset B)^k, \Lambda \Rightarrow A \quad B, (A \supset B)^l, \Theta'' \Rightarrow C}{A \supset B, (A \supset B)^{n-1}, \Theta \Rightarrow C} L\supset
    }{A \supset B, \Theta \Rightarrow C} Ctr^n
  }{}
}{\Gamma, \Theta \Rightarrow C} Cut
$$

The transformed derivation, with a $k - 1$ fold contraction before the first cut, is

$$
\frac{
  \dfrac{
    \dfrac{
      \dfrac{A, \Gamma \Rightarrow B}{\Gamma \Rightarrow A \supset B} R\supset \quad \dfrac{(A \supset B)^k, \Lambda \Rightarrow A}{A \supset B, \Lambda \Rightarrow A} Ctr^k
    }{\Gamma, \Lambda \Rightarrow A} Cut \qquad A, \Gamma \Rightarrow B
  }{
    \dfrac{\Gamma^2, \Lambda \Rightarrow B}{} Cut \qquad B, \Theta \Rightarrow C
  }
}{\dfrac{\Gamma^2, \Lambda, \Theta \Rightarrow C}{\Gamma, \Theta \Rightarrow C} Ctr^*} Cut
$$

where $B, \Theta \Rightarrow C$ follows by Lemma 5.1.1 from the right premiss of cut. Since $k \leqslant n - 1$, the first cut has a reduced cut-height. Reduction of k goes on until $k = 0$. The other two cuts are on shorter formulas, and finally the contractions in the end are justified by the fact that each formula of Λ is a formula of Θ.

7. The cut formula is $\forall x A$. Since $R\forall$ is not invertible, the derivation of the right premiss of cut is analyzed.

7.1. The formula $\forall x A$ is not principal in the right premiss of cut. As for case *6.1.*

7.2. The formula $\forall x A$ is principal in the right premiss, and the derivation

$$
\frac{\dfrac{\Gamma \Rightarrow A(y/x)}{\Gamma \Rightarrow \forall x A} R\forall \quad \dfrac{A(t/x), \Delta \Rightarrow C}{\forall x A, \Delta \Rightarrow C} L\forall}{\Gamma, \Delta \Rightarrow C} Cut
$$

is transformed into

$$
\frac{\Gamma \Rightarrow A(t/x) \quad A(t/x), \Delta \Rightarrow C}{\Gamma, \Delta \Rightarrow C} Cut
$$

in which the left premiss is obtained by substitution and length of cut formula is reduced.

7.3. The right premiss of cut is derived by contraction on $\forall x A$. As in case *6.3,* we trace up the derivation of the right premiss of cut until a rule that is not a contraction on $\forall x A$ is encountered. If the rule is neither $L\forall$ nor weakening on $\forall x A$, we proceed by permuting the rule down with the contraction steps, thus reducing this case to case *7.1*. If the rule is weakening on $\forall x\, A$ the weakening and one contraction step are removed and cut-height is diminished. Else the derivation is

$$
\dfrac{
\dfrac{\Gamma \Rightarrow A(y/x)}{\Gamma \Rightarrow \forall x\, A}\, R\forall
\qquad
\dfrac{
\dfrac{
\dfrac{A(t/x), (\forall x\, A)^{n-1}, \Delta \Rightarrow C}{\forall x\, A, (\forall x\, A)^{n-1}, \Delta \Rightarrow C}\, L\forall
}{\forall x\, A, \Delta \Rightarrow C}\, Ctr^n
}{}
}{\Gamma, \Delta \Rightarrow C}\, Cut
$$

This is transformed into the derivation

$$
\dfrac{
\Gamma \Rightarrow A(t/x)
\qquad
\dfrac{
\dfrac{\Gamma \Rightarrow A(y/x)}{\Gamma \Rightarrow \forall x\, A}\, R\forall
\qquad
\dfrac{
\dfrac{A(t/x), (\forall x\, A)^{n-1}, \Delta \Rightarrow C}{A(t/x), \forall x\, A, \Delta \Rightarrow C}\, Ctr^{n-1}
}{}
}{
\dfrac{A(t/x), \Gamma, \Delta \Rightarrow C}{}
}\, Cut
}{
\dfrac{\Gamma, \Gamma, \Delta \Rightarrow C}{\Gamma, \Delta \Rightarrow C}\, Ctr^{*}
}\, Cut
$$

in which the premiss $\Gamma \Rightarrow A(t/x)$ is derivable by Lemma 5.1.1 and the first cut has a reduced cut-height and the second a reduced cut formula. QED.

(b) Cut elimination for the classical calculus: The rules for the calculus, designated **G0c**, are as follows:

G0c

Logical axiom:

$A \Rightarrow A$

Logical rules:

$$\dfrac{A, B, \Gamma \Rightarrow \Delta}{A\&B, \Gamma \Rightarrow \Delta}\, L\&
\qquad\qquad
\dfrac{\Gamma \Rightarrow \Delta, A \quad \Gamma' \Rightarrow \Delta', B}{\Gamma, \Gamma' \Rightarrow \Delta, \Delta', A\&B}\, R\&$$

$$\dfrac{A, \Gamma \Rightarrow \Delta \quad B, \Gamma' \Rightarrow \Delta'}{A \vee B, \Gamma, \Gamma' \Rightarrow \Delta, \Delta'}\, L\vee
\qquad
\dfrac{\Gamma \Rightarrow \Delta, A, B}{\Gamma \Rightarrow \Delta, A \vee B}\, R\vee$$

$$\dfrac{\Gamma \Rightarrow \Delta, A \quad B, \Gamma' \Rightarrow \Delta'}{A \supset B, \Gamma, \Gamma' \Rightarrow \Delta, \Delta'}\, L\supset
\qquad
\dfrac{A, \Gamma \Rightarrow \Delta, B}{\Gamma \Rightarrow \Delta, A \supset B}\, R\supset$$

$$\dfrac{}{\bot \Rightarrow C}\, L\bot$$

$$\dfrac{A(t/x), \Gamma \Rightarrow \Delta}{\forall x\, A, \Gamma \Rightarrow \Delta}\, L\forall
\qquad
\dfrac{\Gamma \Rightarrow \Delta, A(y/x)}{\Gamma \Rightarrow \Delta, \forall x\, A}\, R\forall$$

$$\dfrac{A(y/x), \Gamma \Rightarrow \Delta}{\exists x\, A, \Gamma \Rightarrow \Delta}\, L\exists
\qquad
\dfrac{\Gamma \Rightarrow \Delta, A(t/x)}{\Gamma \Rightarrow \Delta, \exists x\, A}\, R\exists$$

Rules of weakening:

$$\frac{\Gamma \Rightarrow \Delta}{A, \Gamma \Rightarrow \Delta}\,LW \qquad\qquad \frac{\Gamma \Rightarrow \Delta}{\Gamma \Rightarrow \Delta, A}\,RW$$

Rules of contraction:

$$\frac{A, A, \Gamma \Rightarrow \Delta}{A, \Gamma \Rightarrow \Delta}\,LC \qquad\qquad \frac{\Gamma \Rightarrow \Delta, A, A}{\Gamma \Rightarrow \Delta, A}\,RC$$

The restrictions in the $L\exists$ and $R\forall$ rules are that y must not occur free in the conclusion.

Lemma 5.1.3: *The following inversions hold in* **G0c**:

(i)   *If $A\&B, \Gamma \Rightarrow \Delta$ is derivable, also $A, B, \Gamma \Rightarrow \Delta$ is derivable.*

(ii) *If $\Gamma \Rightarrow \Delta, A\&B$ is derivable, also $\Gamma \Rightarrow \Delta, A$ and $\Gamma \Rightarrow \Delta, B$ are derivable.*

(iii) *If $A \vee B, \Gamma \Rightarrow \Delta$ is derivable, also $A, \Gamma \Rightarrow \Delta$ and $B, \Gamma \Rightarrow \Delta$ are derivable.*

(iv) *If $\Gamma \Rightarrow \Delta, A \vee B$ is derivable, also $\Gamma \Rightarrow \Delta, A, B$ is derivable.*

(v)   *If $A \supset B, \Gamma \Rightarrow \Delta$ is derivable, also $B, \Gamma \Rightarrow \Delta$ is derivable.*

(vi) *If $\Gamma \Rightarrow \Delta, A \supset B$ is derivable, also $A, \Gamma \Rightarrow \Delta, B$ is derivable.*

(vii) *If $\Gamma \Rightarrow \Delta, \forall x A$ is derivable, also $\Gamma \Rightarrow \Delta, A(t/x)$ is derivable.*

(viii) *If $\exists x A, \Gamma \Rightarrow \Delta$ is derivable, also $A(t/x), \Gamma \Rightarrow C$ is derivable.*

Proof: (i) Similar to that of Lemma 5.1.1. For (ii), if $A\&B$ is obtained by weakening, weaken first with A, then with B. If it is obtained as an axiom, conclude instead $A\&B \Rightarrow A$ from $A \Rightarrow A$ by weakening with B and $L\&$, and similarly for $A\&B \Rightarrow B$. If $A\&B$ is introduced by $R\&$, apply repeated weakening instead, dually to case (ii) of Lemma 5.1.1. (iii) and (iv) are dual to previous. (v) If $A \supset B$ in the antecedent is obtained by weakening, weaken with B on left instead. If $A \supset B$ is obtained by an axiom $A \supset B \Rightarrow A \supset B$, conclude $B \Rightarrow A \supset B$ from $B \Rightarrow B$ by left weakening with A followed by $R\supset$. If $A \supset B$ is obtained by $L\supset$, proof is similar to that of (ii). (vi) If $A \supset B$ in the succedent is obtained by right weakening, do left weakening with A and right weakening with B instead. If $A \supset B$ is obtained by axiom $A \supset B \Rightarrow A \supset B$, conclude $A, A \supset B \Rightarrow B$ from $A \Rightarrow A$ and $B \Rightarrow B$ by $L\supset$ instead. If $A \supset B$ is concluded by $R\supset$, delete the rule. Proofs for (vii) and (viii) are similar to those above and to those of (iv) and (v) of Lemma 5.1.1.   QED.

With independent contexts, rule $L\supset$ is invertible with respect to only its right premiss, in contrast to the previous context-sharing classical calculus **G3c** that has all rules invertible. We note that, as with **G0i**, the inversions of **G0c** are not height-preserving.

Theorem 5.1.4: *The rule of cut*

$$\frac{\Gamma \Rightarrow \Delta, D \quad D, \Gamma' \Rightarrow \Delta'}{\Gamma, \Gamma' \Rightarrow \Delta, \Delta'} \, Cut$$

is admissible in **G0c**.

Proof: The proof is by induction on length of cut formula and cut-height. All cases in which the cut formula is not a contraction formula are treated by the methods used in Theorem 5.1.2. We show the cases in which the cut formula has been derived by contraction in the right premiss, and the premiss of contraction is derived by another contraction on the cut formula, until the cut formula is principal. There are five cases:

1. The derivation is

$$\frac{\Gamma \Rightarrow \Delta, A\&B \quad \dfrac{(A\&B)^n, \Gamma' \Rightarrow \Delta'}{A\&B, \Gamma' \Rightarrow \Delta'} LC^n}{\Gamma, \Gamma' \Rightarrow \Delta, \Delta'} \, Cut$$

By Lemma 5.1.3, the sequents $\Gamma \Rightarrow \Delta, A$ and $\Gamma \Rightarrow \Delta, B$ and $A, B, \Gamma' \Rightarrow \Delta'$ are derivable. The derivation is transformed into one with cuts on shorter formulas, analogously to case *3.1* of Theorem 5.1.2.

2. The cut formula is $A \vee B$. Application of Lemma 5.1.3 followed by cuts on A or B gives the result, similarly to above.

3. The cut formula is $A \supset B$. The derivation is

$$\frac{\Gamma \Rightarrow \Delta, A \supset B \quad \dfrac{\dfrac{(A \supset B)^k, \Gamma'' \Rightarrow \Delta'', A \quad B, (A \supset B)^l, \Gamma''' \Rightarrow \Delta'''}{A \supset B, (A \supset B)^{n-1}, \Gamma' \Rightarrow \Delta'} L\supset}{A \supset B, \Gamma' \Rightarrow \Delta'} LC^n}{\Gamma, \Gamma' \Rightarrow \Delta, \Delta'} \, Cut$$

Here $(A \supset B)^{n-1}, \Gamma' = (A \supset B)^k, \Gamma'', (A \supset B)^l, \Gamma'''$ with $k + l = n - 1$ and $\Delta' = \Delta'', \Delta'''$. All formulas of Γ'' and Γ''' are formulas of Γ' and all formulas of Δ'' are formulas of Δ'.

By Lemma 5.1.3, the sequents $A, \Gamma \Rightarrow \Delta, B$ and $B, \Gamma' \Rightarrow \Delta'$ are derivable. We transform the above derivation into

$$\frac{\dfrac{\Gamma \Rightarrow \Delta, A \supset B \quad \dfrac{(A \supset B)^k, \Gamma'' \Rightarrow \Delta'', A}{A \supset B, \Gamma'' \Rightarrow \Delta'', A} LC^*}{\dfrac{\Gamma, \Gamma'' \Rightarrow \Delta, \Delta'', A}{\Gamma^2, \Gamma'' \Rightarrow \Delta^2, \Delta'', B} Cut \quad A, \Gamma \Rightarrow \Delta, B} \, Cut \quad B, \Gamma' \Rightarrow \Delta'}{\dfrac{\Gamma^2, \Gamma', \Gamma'' \Rightarrow \Delta^2, \Delta', \Delta''}{\Gamma, \Gamma' \Rightarrow \Delta, \Delta'} C^*} \, Cut$$

Since $k \leqslant n - 1$, the uppermost cut has a reduced cut-height, and the other two are on shorter formulas. The final left and right contractions C^* are allowed by the inclusion of formulas of Γ'' in Γ' and Δ'' in Δ'.

4. The cut formula is $\forall x A$. The proof uses Lemma 5.1.3 and is analogous to case 7.6 of Theorem 5.1.2.

5. The cut formula is $\exists x A$. As for case 4. QED.

5.2. SEQUENT CALCULI IN NATURAL DEDUCTION STYLE

We give a formulation of sequent calculi "in natural deduction style," with no explicit rules of weakening or contraction, guided by the following points (compare also the discussion of weakening and contraction in Section 1.3):

> *Discharge in natural deduction corresponds to the application of a sequent calculus rule that has an active formula in the antecedent of a premiss.*
>
> *A vacuous discharge corresponds to an active formula that has been obtained by weakening, and a multiple discharge to an active formula that has been obtained by contraction.*

In an intuitionistic sequent calculus, the rules in question are the left rules and the right implication rule. Ever since Gentzen, weakening and contraction have been made into steps independent of the application of these rules. Cut elimination is much more complicated than normalization, with numerous cases of permutation of cut that do not have any correspondence in the normalization process. Moreover, in sequent calculi, because of the mentioned independence, there can be formulas concluded by weakening or contraction that remain inactive through a whole derivation. These steps with unused weakenings and contractions do not contribute anything, and the formulas can either be pruned out (for unused weakening) or left multiplied (for unused contraction). The calculi we present avoid such steps with unused formulas altogether.

In the calculi we give, only those cuts need be eliminated in which the cut formula is principal in at least the right premiss of cut, or principal somewhere higher up in the derivation of the right premiss of cut, and the cut is moved up there in one step. For all other cases of cut, we prove that the cut formula is a subformula of the conclusion. Therefore the subformula property, Gentzen's original aim in the "Hauptsatz," can be concluded by eliminating only those cuts in which the cut formula is principal in the derivation leading to the right premiss. The proof of cut elimination uses induction on formula length and the height of derivation of the left premiss of cut.

(a) Cut elimination for the intuitionistic calculus: In the detour conversions of natural deduction, the multiset of assumptions can become changed, in that formulas become multiplied, where zero multiplicity (i.e., deletion) is also possible. The changed context is called a **multiset reduct** of the original one:

Definition 5.2.1: *If a multiset Δ is obtained from Γ by multiplying formulas in Γ, where zero multiplicity is also permitted, Δ is a* multiset reduct *of Γ.*

The relation of being a multiset reduct is reflexive and transitive. We also call a sequent a reduct of another if its antecedent is a multiset reduct. These reducts are generated by steps of cut elimination in the same way as assumptions are multiplied in the conversions to normal form in natural deduction.

The intuitionistic single succedent sequent calculus in natural deduction style is denoted by **GN**. As before, multiple occurrence of a formula is denoted by A^m.

GN

Logical axiom:

$$A \Rightarrow A$$

Logical rules:

$$\frac{A^m, B^n, \Gamma \Rightarrow C}{A \& B, \Gamma \Rightarrow C}\, L\& \qquad \frac{\Gamma \Rightarrow A \quad \Delta \Rightarrow B}{\Gamma, \Delta \Rightarrow A \& B}\, R\&$$

$$\frac{A^m, \Gamma \Rightarrow C \quad B^n, \Delta \Rightarrow C}{A \vee B, \Gamma, \Delta \Rightarrow C}\, L\vee \qquad \frac{\Gamma \Rightarrow A}{\Gamma \Rightarrow A \vee B}\, R\vee_1 \qquad \frac{\Gamma \Rightarrow B}{\Gamma \Rightarrow A \vee B}\, R\vee_2$$

$$\frac{\Gamma \Rightarrow A \quad B^m, \Delta \Rightarrow C}{A \supset B, \Gamma, \Delta \Rightarrow C}\, L\supset \qquad \frac{A^m, \Gamma \Rightarrow B}{\Gamma \Rightarrow A \supset B}\, R\supset$$

$$\frac{}{\bot \Rightarrow C}\, L\bot$$

$$\frac{A(t/x)^m, \Gamma \Rightarrow C}{\forall x A, \Gamma \Rightarrow C}\, L\forall \qquad \frac{\Gamma \Rightarrow A(y/x)}{\Gamma \Rightarrow \forall x A}\, R\forall$$

$$\frac{A(y/x)^m, \Gamma \Rightarrow C}{\exists x A, \Gamma \Rightarrow C}\, L\exists \qquad \frac{\Gamma \Rightarrow A(t/x)}{\Gamma \Rightarrow \exists x A}\, R\exists$$

The variable restrictions in $R\forall$ and $L\exists$ are that y is not free in the conclusion. Rules with exponents have instances for any $m, n \geqslant 0$. For example, from $L\&$ with $m = 1, n = 0$ we get the first of Gentzen's original left conjunction rules:

$$\frac{A, \Gamma \Rightarrow C}{A \& B, \Gamma \Rightarrow C}\, L\&$$

We say that formulas A and B with exponents are **used** in the rules of **GN**. Whenever $m = 0$ or $n = 0$ in an instance, there is a **vacuous use**, corresponding to weakening, and whenever $m > 1$ or $n > 1$, there is a **multiple use**, corresponding to contraction. Since in Gentzen's $L\&$ rules $m = 0$ or $n = 0$, they contain a hidden step of weakening.

The logical rules of the calculus are just like those of **G0i**, with the exponents in the rules added. There will be no structural rules. An example shows a derivation with structural rules in **G0i** and the corresponding derivation with an implicit treatment of weakening and contraction in **GN**:

$$\cfrac{\cfrac{\Gamma \Rightarrow C}{A, \Gamma \Rightarrow C}\,Wk \quad \cfrac{B, B, \Delta \Rightarrow C}{B, \Delta \Rightarrow C}\,Ctr}{A \vee B, \Gamma, \Delta \Rightarrow C}\,L\vee \qquad \cfrac{\Gamma \Rightarrow C \quad B, B, \Delta \Rightarrow C}{A \vee B, \Gamma, \Delta \Rightarrow C}\,L\vee$$

Many steps of cut elimination lead to a sequent the antecedent of which is a reduct of the antecedent of the original cut. In usual cut elimination procedures, once the cut has been permuted up, this original antecedent is restored by weakenings and contractions following the permuted cut. In our calculus, these are not explicitly available, but the restriction is not essential. The following proposition shows that the new antecedent can be left as it is:

Proposition 5.2.2: *If in the derivation of* $\Gamma \Rightarrow C$ *in* **GN**$+$Cut *the sequent* $\Delta \Rightarrow D$ *occurs and if the subderivation down to* $\Delta \Rightarrow D$ *is substituted by a derivation of* $\Delta^* \Rightarrow D$, *where* Δ^* *is a multiset reduct of* Δ, *then the derivation can be continued to conclude* $\Gamma^* \Rightarrow C$, *with* Γ^* *a multiset reduct of* Γ.

Proof: It is sufficient to consider an uppermost cut that we may assume to be the last step of the whole derivation. First consider the part before the cut, having only axioms and logical rules. Starting with the derivation of $\Delta^* \Rightarrow D$, the derivation is continued as with $\Delta \Rightarrow D$, save for the steps that use formulas. It is enough to consider such rules when one premiss is $\Delta^* \Rightarrow D$. If in the original derivation a formula from Δ was used that does not occur in Δ^*, a vacuous use is made, and similarly for formulas that occur multiplied with Δ^*, as compared with Δ, a multiple use is made.

It remains to show that the conclusion of cut can be replaced with a sequent having a multiset reduct as antecedent. Let the original cut concluding $\Gamma \Rightarrow C$ be

$$\cfrac{\Gamma_1 \Rightarrow A \quad A, \Gamma_2 \Rightarrow C}{\Gamma_1, \Gamma_2 \Rightarrow C}\,Cut$$

where $\Gamma_1, \Gamma_2 = \Gamma$, and let the reduced premisses be $\Gamma_1{}^* \Rightarrow A$ and $A^n, \Gamma_2{}^* \Rightarrow C$. If $n = 1$, a cut with the reduced premisses will give a conclusion with a multiset reduct of Γ as antecedent. If $n = 0$, the conclusion of cut is replaced by the right premiss. If $n > 1$, we make n cuts with left premiss $\Gamma_1^* \Rightarrow A$ in succession and the conclusion of the last cut has a multiset reduct of Γ as antecedent. QED.

The proposition shows two things: 1. It is enough to consider derivability in **GN** modulo multiset reducts. 2. It is enough to perform cut elimination modulo multiset reducts.

Definition 5.2.3: *A cut with premisses* $\Gamma \Rightarrow A$ *and* $A, \Delta \Rightarrow C$ *is* redundant *in the following cases:*

   (i) Γ *contains* A,
   (ii) Γ *or* Δ *contains* $\bot$,
   (iii) $A = C$,
   (iv) Δ *contains* C,
   (v) *The derivation of* $A, \Delta \Rightarrow C$ *contains a sequent with a multiple occurrence of* A.

Theorem 5.2.4: Elimination of redundant cuts. *Given a derivation of* $\Gamma \Rightarrow C$ *in* **GN+Cut** *there is a derivation with redundant cuts eliminated.*

Proof: In case (i) of redundant cut, if Γ contains A, then A, Δ is a multiset reduct of Γ, Δ and by Proposition 5.2.2, the cut is deleted and the derivation continued with $A, \Delta \Rightarrow C$. In case (ii), the conclusion has $\bot$ in the antecedent and the derivation begins with $\bot \Rightarrow C$. In case (iii), if $A = C$, the cut is deleted and the derivation continued with $\Gamma \Rightarrow C$. In case (iv), if Δ contains C, the derivation begins with $C \Rightarrow C$.

Case (v) of redundant cut can obtain in two ways: 1. It can happen that A has another occurrence in the context Δ of the right premiss and therefore also in the conclusion of cut. In this case, the antecedent A, Δ of the right premiss is a multiset reduct of Γ, Δ and, by Proposition 5.2.2, the cut can be deleted. 2. It can happen that there was a multiple occurrence of A in some antecedent in the derivation of the right premiss and all but one occurrence were active in earlier cuts or logical rules. In the former case, if the right premiss of a cut is $A, \Delta' \Rightarrow C'$ and Δ' contains another occurrence of A, the cut is deleted and the derivation continued from $\Delta' \Rightarrow C'$. In the latter case, using all occurrences of A will give a derivation of $\Delta \Rightarrow C$. Again, since Δ is a multiset reduct of the antecedent of conclusion of cut, the cut can be deleted and the derivation continued from $\Delta \Rightarrow C$. QED.

Redundant cuts (i)–(iv) have as one premiss a sequent from which an axiom or conclusion of $L\bot$ is obtainable as a multiset reduct. In particular, if one premiss already is an axiom or conclusion of $L\bot$, a special case of redundant cuts (i)–(iv) obtains.

Definition 5.2.5: *A cut is* hereditarily principal (hereditarily nonprincipal) *in a derivation if its cut formula is principal (is never principal) in some rule in the derivation of the right premiss of cut.*

Proposition 5.2.6: *The first occurrence of a hereditarily principal cut formula in a derivation without redundant cuts is unique.*

Proof: Assume there are at least two such occurrences. Then there is a sequent with a multiple occurrence of the cut formula and the derivation has a redundant cut as in case (v) of Definition 5.2.3. QED.

A principal cut is the special case of hereditarily principal cut, with the cut formula principal in the last rule deriving the right premiss. The idea of cut elimination is to consider only hereditarily principal cuts and to permute them up in one step to the first occurrence of a cut formula A hereditarily principal in the derivation of the right premiss.

It can happen that the instance of a rule concluding a hereditarily principal cut formula had vacuous or multiple uses of active formulas. These cuts are **hereditarily vacuous** and **hereditarily multiple**, respectively:

Definition 5.2.7: *If a hereditarily principal cut formula is concluded by rule $L\&$ and $m, n = 0$, by rule $L\vee$ and $m = 0$ or $n = 0$, by rule $L\supset$, $L\forall$, $L\exists$ and $m = 0$, the cut is* hereditarily vacuous. *If the formula is concluded by $L\&$ and $m > 1$ or $n > 1$, by $L\vee$ and $m, n > 1$, or by $L\supset$, $L\forall$, $L\exists$ and $m > 1$, a* hereditarily multiple cut *obtains.*

We now prove a cut elimination theorem for hereditarily principal cuts. The proof is by induction on the length of cut formula, with a subinduction on height of derivation of the left premiss of cut. Length is defined in the usual way, 0 for $\perp$, 1 for atoms, and sum of lengths of components plus 1 for proper connectives. Height of derivation is the greatest number of consecutive steps of inference in it. In the proof, multiplication of every formula occurrence in Γ to multiplicity m is written Γ^m.

Theorem 5.2.8: Elimination of hereditarily principal cuts. *Given a derivation of $\Gamma \Rightarrow C$ with cuts, there is a derivation of $\Gamma^* \Rightarrow C$ with no hereditarily principal cuts, with Γ^* a multiset reduct of Γ.*

Proof: First remove possible redundant cuts. Then consider the first hereditarily principal cut in the derivation; we may assume it to be the last step. If the cut formula is not principal in the left premiss, the cut is permuted in the derivation of the left premiss, with its height of derivation diminished.

There remain five cases with the cut formula principal in both premisses. In each case, if a step of cut elimination produces redundant cuts, these are at once eliminated.

1. Cut formula is $A \& B$. If $m > 0$ or $n > 0$ we have the derivation

$$\cfrac{\cfrac{\Gamma \Rightarrow A \quad \Delta \Rightarrow B}{\Gamma, \Delta \Rightarrow A \& B} R\& \qquad \cfrac{\cfrac{A^m, B^n, \Theta' \Rightarrow C'}{A \& B, \Theta' \Rightarrow C'} L\& }{\vdots}}{A \& B, \Theta \Rightarrow C}}{\Gamma, \Delta, \Theta \Rightarrow C} Cut$$

We make m cuts with $\Gamma \Rightarrow A$, starting with the premiss $A^m, B^n, \Theta' \Rightarrow C'$, and up to $B^n, \Gamma^m, \Theta' \Rightarrow C'$, then continue with n cuts with $\Delta \Rightarrow B$, up to the conclusion $\Gamma^m, \Delta^n, \Theta' \Rightarrow C'$. Now the derivation is continued as before from where $A \& B$ was principal, to conclude $\Gamma^m, \Delta^n, \Theta \Rightarrow C$, all cuts in the derivation being on shorter formulas than in the initial derivation.

If $m, n = 0$, we have a hereditarily vacuous cut, with $\Theta' \Rightarrow C'$ the premiss of rule $L\&$. It is not a special case of the previous since there is nothing to cut. Instead, the derivation is continued without rule $L\&$ until $\Theta \Rightarrow C$ is concluded.

2. Cut formula is $A \lor B$. With $A \lor B$ principal in the left premiss, assume that the rule is $R\lor_1$ with $m > 0$:

$$\cfrac{\cfrac{\Gamma \Rightarrow A}{\Gamma \Rightarrow A \lor B} R\lor_1 \qquad \cfrac{\cfrac{A^m, \Delta' \Rightarrow C' \quad B^n, \Theta' \Rightarrow C'}{A \lor B, \Delta', \Theta' \Rightarrow C'} L\lor }{\vdots}}{A \lor B, \Delta, \Theta \Rightarrow C}}{\Gamma, \Delta, \Theta \Rightarrow C} Cut$$

We make m cuts with $\Gamma \Rightarrow A$, starting with the premiss $A^m, \Delta' \Rightarrow C'$, obtaining $\Gamma^m, \Delta' \Rightarrow C'$. The derivation is continued as before from where $A \lor B$ was principal; where a formula from Θ' was used in the original derivation, there will be a vacuous use. The derivation ends with $\Gamma^m, \Lambda \Rightarrow C$ where Λ is a multiset reduct of the context Δ, Θ of the right premiss of the original cut. All cuts are on shorter formulas than the initial cut. If in the left premiss the rule was $R\lor_2$ and $n > 0$, the procedure is similar.

If $m = 0$, assuming still that the rule concluding the left premiss is $R\lor_1$, the cut is hereditarily vacuous, and, proceeding analogously to case *1*, we continue from the premiss $\Delta' \Rightarrow C'$ without cuts to a sequent $\Lambda \Rightarrow C$ where Λ is by Proposition 5.2.2 a reduct of Γ, Δ, Θ. The other cases of hereditarily vacuous cuts are handled similarly.

3. Cut formula is $A \supset B$. With $n > 0$, the derivation is

$$\cfrac{\cfrac{\Delta' \Rightarrow A \quad B^n, \Theta' \Rightarrow C'}{A \supset B, \Delta', \Theta' \Rightarrow C'} \, {}_{L\supset}}{}$$

$$\cfrac{\cfrac{A^m, \Gamma \Rightarrow B}{\Gamma \Rightarrow A \supset B} \, {}_{R\supset} \qquad \cfrac{\vdots}{A \supset B, \Delta, \Theta \Rightarrow C}}{\Gamma, \Delta, \Theta \Rightarrow C} \, {}_{Cut}$$

We first cut m times with $\Delta' \Rightarrow A$, starting with $A^m, \Gamma \Rightarrow B$, and obtain $\Gamma, \Delta'^m \Rightarrow B$, then cut with this n times, starting with $B^n, \Theta' \Rightarrow C'$, to obtain $\Gamma^n, \Delta'^{mn}, \Theta' \Rightarrow C'$. Continuing, if a formula from Δ' was used, there will be an mn-fold use. All cuts are on shorter formulas.

The case of $n = 0$ gives a hereditarily vacuous cut handled as in case *1*.

4. Cut formula is $\forall x A$. With $m > 0$ the derivation is

$$\cfrac{\cfrac{A(t/x)^m, \Delta' \Rightarrow C'}{\forall x A, \Delta' \Rightarrow C'} \, {}_{L\forall}}{}$$

$$\cfrac{\cfrac{\Gamma \Rightarrow A(y/x)}{\Gamma \Rightarrow \forall x A} \, {}_{R\forall} \qquad \cfrac{\vdots}{\forall x A, \Delta \Rightarrow C}}{\Gamma, \Delta \Rightarrow C} \, {}_{Cut}$$

In $\Gamma \Rightarrow A(y/x)$, substitution can be applied as in Lemma 4.1.2 to obtain a derivation of $\Gamma \Rightarrow A(t/x)$; then m cuts with $A(t/x)^m, \Delta' \Rightarrow C'$ give $\Gamma^m, \Delta' \Rightarrow C'$. The derivation is continued as before from where $\forall x A$ was principal.

If $m = 0$, the derivation is continued from $\Delta' \Rightarrow C'$, without $L\forall$, to $\Delta \Rightarrow C$.

5. Cut formula is $\exists x A$. With $m > 0$ the derivation is

$$\cfrac{\cfrac{A(y/x)^m, \Delta' \Rightarrow C'}{\exists x A, \Delta' \Rightarrow C'} \, {}_{L\exists}}{}$$

$$\cfrac{\cfrac{\Gamma \Rightarrow A(t/x)}{\Gamma \Rightarrow \exists x A} \, {}_{R\exists} \qquad \cfrac{\vdots}{\exists x A, \Delta \Rightarrow C}}{\Gamma, \Delta \Rightarrow C} \, {}_{Cut}$$

In $A(y/x)^m, \Delta' \Rightarrow C'$, substitution is applied to obtain a derivation of $A(t/x)^m, \Delta' \Rightarrow C'$. The rest of the proof is as in case *4*. QED.

Corollary 5.2.9: Subformula property. *If the derivation of $\Gamma \Rightarrow C$ has no hereditarily principal cuts, all formulas in the derivation are subformulas of Γ, C.*

Proof: Consider the uppermost hereditarily nonprincipal cut,

$$\frac{\Gamma' \Rightarrow A \quad A, \Delta' \Rightarrow C'}{\Gamma', \Delta' \Rightarrow C'} \; Cut$$

Since A is never active in the derivation of the right premiss, its first occurrence is in an axiom $A \Rightarrow A$. By the same, $A \Rightarrow A$ can be replaced by the derivation of the left premiss of cut, $\Gamma' \Rightarrow A$, and the derivation continued as before, until the sequent $\Gamma', \Delta' \Rightarrow C'$ is reached by the rule originally concluding the right premiss of cut. Therefore the succedent A is a subformula of $\Gamma', \Delta' \Rightarrow C'$. Repeating this for each nonhereditary cut formula in succession, we conclude that they all are subformulas of the endsequent.   QED.

Theorems 5.2.4 and 5.2.8 and the proof of Corollary 5.2.9 actually give an elimination procedure for all cuts:

Corollary 5.2.10: *Given a derivation of $\Gamma \Rightarrow C$ in* **GN**+Cut, *there is a derivation of $\Gamma^* \Rightarrow C$ in* **GN**, *with Γ^* a multiset reduct of Γ.*

There are sequents derivable in calculi with explicit weakening and contraction rules that have no derivation in the calculus **GN**, for example, $A \Rightarrow A\&A$. The last rule must be $R\&$, but its application in **GN** will give only $A, A \Rightarrow A\&A$. Even if the sequent $A \Rightarrow A\&A$ is not derivable, the sequent $\Rightarrow A \supset A\&A$ is, by a multiple use of A in rule $R\supset$.

The completeness of the calculus **GN** is easily proved, for example, by deriving any standard set of axioms of intuitionistic logic as sequents with empty antecedents and by noting that modus ponens in the form

$$\frac{\Rightarrow A \supset B \quad \Rightarrow A}{\Rightarrow B}$$

is admissible: Application of $L\supset$ to $A \Rightarrow A$ and $B \Rightarrow B$ gives $A \supset B, A \Rightarrow B$, and cuts with the premisses of modus ponens give $\Rightarrow B$. We also have completeness in another sense: Sequent calculi with weakening and contraction modify the derivability relation in an inessential way, for if $\Gamma \Rightarrow C$ is derivable in such calculi, obviously there is a derivation of $\Gamma^* \Rightarrow C$ in **GN**, with Γ^* a multiset reduct of Γ. In particular, if $\Rightarrow C$ is derivable in such calculi, it is derivable in **GN**.

(b) A multisuccedent calculus: We give a classical multisuccedent version of the calculus **GN**, called **GM**. We obtain it by writing the right rules in perfect symmetry to the left rules.

GM

Logical axiom:

$A \Rightarrow A$

Logical rules:

$$\dfrac{A^m, B^n, \Gamma \Rightarrow \Delta}{A\&B, \Gamma \Rightarrow \Delta} \, L\& \qquad\qquad \dfrac{\Gamma \Rightarrow \Delta, A^m \quad \Gamma' \Rightarrow \Delta', B^n}{\Gamma, \Gamma' \Rightarrow \Delta, \Delta', A\&B} \, R\&$$

$$\dfrac{A^m, \Gamma \Rightarrow \Delta \quad B^n, \Gamma' \Rightarrow \Delta'}{A \vee B, \Gamma, \Gamma' \Rightarrow \Delta, \Delta'} \, L\vee \qquad\qquad \dfrac{\Gamma \Rightarrow \Delta, A^m, B^n}{\Gamma \Rightarrow \Delta, A \vee B} \, R\vee$$

$$\dfrac{\Gamma \Rightarrow \Delta, A^m \quad B^n, \Gamma' \Rightarrow \Delta'}{A \supset B, \Gamma, \Gamma' \Rightarrow \Delta, \Delta'} \, L\supset \qquad\qquad \dfrac{A^m, \Gamma \Rightarrow \Delta, B^n}{\Gamma \Rightarrow \Delta, A \supset B} \, R\supset$$

$$\dfrac{}{\bot \Rightarrow \Delta} \, L\bot$$

$$\dfrac{A(t/x)^m, \Gamma \Rightarrow \Delta}{\forall x A, \Gamma \Rightarrow \Delta} \, L\forall \qquad\qquad \dfrac{\Gamma \Rightarrow \Delta, A(y/x)^m}{\Gamma \Rightarrow \Delta, \forall x A} \, R\forall$$

$$\dfrac{A(y/x)^m, \Gamma \Rightarrow \Delta}{\exists x A, \Gamma \Rightarrow \Delta} \, L\exists \qquad\qquad \dfrac{\Gamma \Rightarrow \Delta, A(t/x)^m}{\Gamma \Rightarrow \Delta, \exists x A} \, R\exists$$

Compared with the calculus **G0c**, **GM** has the exponents in the rules added and no explicit weakening or contraction.

Proposition 5.2.11: *If in the derivation of* $\Gamma \Rightarrow \Delta$ *a subderivation of* $\Theta \Rightarrow \Lambda$ *is substituted by a derivation of* $\Theta^* \Rightarrow \Lambda^*$ *with both contexts reducts of the original ones, then there is a derivation of* $\Gamma^* \Rightarrow \Delta^*$ *with contexts similarly reduced.*

Proof: Similar to proof of Proposition 5.2.2. QED.

A proof of elimination of hereditarily principal cuts and of the subformula property is obtained similarly to the results for the single succedent intuitionistic calculus:

Theorem 5.2.12: Elimination of hereditarily principal cuts. *Given a derivation of* $\Gamma \Rightarrow \Delta$ *with cuts, there is a derivation of* $\Gamma^* \Rightarrow \Delta^*$ *with no hereditarily principal cuts, with* Γ^* *and* Δ^* *multiset reducts of* Γ *and* Δ.

Theorem 5.2.13: Subformula property. *If the derivation of* $\Gamma \Rightarrow \Delta$ *has no hereditarily principal cuts, all formulas in the derivation are subformulas of* Γ, Δ.

Corollary 5.2.14: *Given a derivation of* $\Gamma \Rightarrow \Delta$ *in* **GM+Cut**, *there is a derivation of* $\Gamma^* \Rightarrow \Delta^*$ *in* **GM**, *with* Γ^* *and* Δ^* *multiset reducts of* Γ *and* Δ.

The calculus **GM** is complete for classical logic by the following derivation:

$$\frac{\dfrac{A \Rightarrow A}{\Rightarrow A, A \supset \bot} R\supset}{\dfrac{\Rightarrow A, A \supset \bot}{\Rightarrow A \vee \sim A} R\vee}$$

The instance of $R\supset$ has $m = 1, n = 0$, with Γ empty, $\Delta = A$, and $B = \bot$. More generally, we obtain the full versions of Gentzen's original left and right negation rules from $L\supset$ and $R\supset$ by suitable choices:

$$\frac{\Gamma \Rightarrow \Delta, A \quad \overline{\bot \Rightarrow}^{\,L\bot}}{A \supset \bot, \Gamma \Rightarrow \Delta} L\sim \qquad \frac{A, \Gamma \Rightarrow \Delta}{\Gamma \Rightarrow \Delta, A \supset \bot} R\sim$$

Gentzen had two $R\vee$ rules, dually to the two $L\&$ rules. We obtain them in **GM** by setting $m = 1, n = 0$ and $n = 1, m = 0$ in rule $R\vee$, respectively:

$$\frac{\Gamma \Rightarrow \Delta, A}{\Gamma \Rightarrow \Delta, A \vee B} R\vee \qquad \frac{\Gamma \Rightarrow \Delta, B}{\Gamma \Rightarrow \Delta, A \vee B} R\vee$$

Applications of Gentzen's rules contain, as for the single succedent version, a "hidden" weakening. It is quite conspicuous in derivations of $\Rightarrow A \vee \sim A$ in his classical calculus. The last step of the derivation can be only a contraction, doing away with what the hidden weakenings had brought into the derivation:

$$\frac{\dfrac{\dfrac{\dfrac{A \Rightarrow A}{\Rightarrow A, \sim A} R\sim}{\Rightarrow A \vee \sim A, \sim A} R\vee}{\Rightarrow A \vee \sim A, A \vee \sim A} R\vee}{\Rightarrow A \vee \sim A} Ctr$$

(c) Correspondence with earlier sequent calculi: By using weakening and contraction, derivations in **G0i** and **G0c** can be converted into derivations in **G3i** and **G3c** and the other way around. Derivations in **GN** can be translated into derivations in **G0i** by simple local transformations. Only rules that use assumptions are different, and vacuous uses are replaced by weakenings and multiple uses by contractions. The translation from **G0i** to **GN** consists in deleting the weakening and contraction steps and in adding the exponents to active formulas in rules that use formulas from antecedents. If in a derivation of $\Gamma \Rightarrow C$ in **G0i** there are no inactive weakening or contraction formulas, the translation produces a derivation of $\Gamma \Rightarrow C$, and if there are it produces a derivation of $\Gamma^* \Rightarrow C$, with Γ^* a multiset reduct of Γ. The relation between **G0c** and **GM** is analogous.

A direct translation from a derivation of $\Gamma \Rightarrow \Delta$ in **G3c** to a derivation in **GM** is obtained as follows: First trace all uppermost sequents of the forms $A, \Gamma' \Rightarrow \Delta', A$ and $\bot, \Gamma' \Rightarrow \Delta'$ and prune the subderivations above each. Next delete Γ' and Δ' from the former and Γ' from the latter form of sequents, and continue deleting all formulas descending from these contexts. Whenever there is a formula to be deleted that is used in a rule, a vacuous use is made in **GM**. The result is a derivation of $\Gamma^* \Rightarrow \Delta^*$ in **GM**, where Γ^* and Δ^* are multiset reducts of the original contexts. Similar remarks apply to the single succedent calculi.

5.3. AN INTUITIONISTIC MULTISUCCEDENT CALCULUS

The propositional part of the intuitionistic multisuccedent calculus **G3im** is the same as the classical calculus **G3cp**, except for the rules of implication. The left quantifier rules are the same as in **G3c**, but the right rules are different. This calculus is due to Dragalin (1988), who called it **GHPC** (for Gentzen-style Heyting predicate calculus). For the propositional part, we show only the two rules that are different from those of the classical calculus **G3c**:

G3im

Rules for implication:

$$\frac{A \supset B, \Gamma \Rightarrow A \quad B, \Gamma \Rightarrow \Delta}{A \supset B, \Gamma \Rightarrow \Delta} L\supset \qquad \frac{A, \Gamma \Rightarrow B}{\Gamma \Rightarrow \Delta, A \supset B} R\supset$$

Rules for quantifiers:

$$\frac{A(t/x), \forall x A, \Gamma \Rightarrow \Delta}{\forall x A, \Gamma \Rightarrow \Delta} L\forall \qquad \frac{\Gamma \Rightarrow A(y/x)}{\Gamma \Rightarrow \Delta, \forall x A} R\forall$$

$$\frac{A(y/x), \Gamma \Rightarrow \Delta}{\exists x A, \Gamma \Rightarrow \Delta} L\exists \qquad \frac{\Gamma \Rightarrow \Delta, \exists x A, A(t/x)}{\Gamma \Rightarrow \Delta, \exists x A} R\exists$$

The left implication rule has a repetition of the principal formula in the left premiss; further, its succedent is just A instead of Δ, A. With the former, the rules of the single succedent calculus **G3i** are directly special cases of the rules of **G3im**. The rule of right implication has only one formula in the succedent of its premiss, a feature discussed in the introduction to Chapter 3. However, it is essential that there be an arbitrary context in the succedent of the conclusion in order to guarantee admissibility of right weakening.

The universal quantifier behaves like implication (in a way made exact in type theory); thus there is a repetition of the principal formula in the premiss of the left rule and a restriction to one formula in the succedent of the premiss of the right rule. Variable restrictions are as in the classical calculus.

To start with the proof theory of the calculus **G3im**, we need a substitution lemma:

Lemma 5.3.1: Substitution lemma. *If* $\Gamma \Rightarrow \Delta$ *is derivable in* **G3im** *and if t is free for x in Γ, Δ, then* $\Gamma(t/x) \Rightarrow \Delta(t/x)$ *is derivable in* **G3im**, *with the same derivation height.*

Proof: The proof is analogous to the proof of the substitution lemma for the calculus **G3i**, Lemma 4.1.2. QED.

Theorem 5.3.2: Height-preserving weakening for G3im.

(i) *If* $\vdash_n \Gamma \Rightarrow \Delta$, *then* $\vdash_n D, \Gamma \Rightarrow \Delta$,

(ii) *If* $\vdash_n \Gamma \Rightarrow \Delta$, *then* $\vdash_n \Gamma \Rightarrow \Delta, D$.

Proof: For (i), proceed as in the proof for **G3i**, Theorems 2.3.4 and 4.2.2. For (ii) proceed similarly except in the cases in which the last rule is followed by a rule with a restriction on the succedent of its premiss, i.e., $R \supset$ or $R\forall$. In all such cases, weakening is absorbed into the downmost occurrence of the rule. QED.

All rules of the intuitionistic single succedent calculus **G3i** except $R\lor$ and $R\exists$ are special cases of rules of **G3im**, and $R\lor$ and $R\exists$ are easily shown admissible in **G3im** through the admissibility of weakening.

Lemma 5.3.3: *The sequent* $C, \Gamma \Rightarrow \Delta, C$ *is derivable in* **G3im**.

Proof: By right weakening from the corresponding result for **G3i**, Lemmas 2.3.3 and 4.2.1. QED.

Lemma 5.3.4: Height-preserving inversions. *Rules* $L\&$, $L\lor$, $R\lor$, $L\exists$, *and* $R\exists$ *are invertible and height-preserving in* **G3im**. *Rule* $L\supset$ *is invertible and height-preserving for the right premiss.*

Proof: By induction on the height of derivation. If the principal formula of the rule to be proved invertible is not principal in the last rule of the derivation and this last rule is $R \supset$ or $R\forall$, the inductive hypothesis cannot be applied because of the restriction in the succedent of the premiss. The conclusion is instead obtained by an instance of $R \supset$ or $R\forall$ with a matching context in the conclusion. For instance, we obtain invertibility of $R\lor$ in the case that the sequent $\Gamma \Rightarrow \Delta, A \lor B, C \supset D$ has been derived from $C, \Gamma \Rightarrow D$ by $R\supset$ by taking as the context Δ, A, B instead of $\Delta, A \lor B$. For the remaining cases, the proof goes through for the propositional part as in Lemma 2.3.5 and for invertibility of $L\exists$ as in Lemma 4.2.3. For $R\exists$, invertibility follows from height-preserving admissibility of right weakening. QED.

We remark that $R\forall$ also is invertible with height preserved, but this property is not needed below. In the variant of **G3im** with Δ, A as succedent in rule $L\supset$, invertibility obtains for its left premiss also.

Theorem 5.3.5: Height-preserving contraction for G3im.

(i) *If* $\vdash_n D, D, \Gamma \Rightarrow \Delta$, *then* $\vdash_n D, \Gamma \Rightarrow \Delta$,

(ii) *If* $\vdash_n \Gamma \Rightarrow \Delta, D, D$, *then* $\vdash_n \Gamma \Rightarrow \Delta, D$.

Proof: By induction on n. For (i), proceed, mutatis mutandis, as in the proof of Theorem 4.2.4. For (ii), if $\vdash_n \Gamma \Rightarrow \Delta, D$, D is an axiom, then also $\vdash_n \Gamma \Rightarrow \Delta, D$ is an axiom. If $n > 0$, we distinguish the cases in which D is principal and not principal in the last rule of the derivation. In the latter cases, we apply the inductive hypothesis to the premisses of the last rule and then the rule, except for $R\supset$ or $R\forall$. These latter are taken care of by an application of the same rule but with a suitably modified context. If D is principal and the last rule is $R\&$, $R\vee$, or $R\exists$, the conclusion is obtained by application of height-preserving invertibility to the premisses of the rule, the inductive hypothesis, and then the rule. If the last rule is $R\supset$ or $R\forall$, because of the restriction in the succedent, the formula D does not occur in the premiss and the conclusion is obtained by application of the rule with the context Δ in the conclusion. QED.

Theorem 5.3.6: *The rule of cut is admissible in* **G3im**.

Proof: The proof is similar to the proof of admissibility of cut for the calculus **G3c**, except for those cases involving rules with a restriction in the succedent of a premiss. Continuing the numbering in the proof of Theorem 3.2.3, we have the following cases to consider:

3. The cut formula is not principal in the left premiss: There are three new cases of last rule in the left premiss to consider:

3.3. The last rule in the left premiss is $L\supset$. The derivation

$$\cfrac{\cfrac{\Gamma'' \Rightarrow A \quad B, \Gamma'' \Rightarrow \Delta, C}{A \supset B, \Gamma'' \Rightarrow \Delta, C} {\scriptstyle L\supset} \quad C, \Gamma' \Rightarrow \Delta'}{A \supset B, \Gamma'', \Gamma' \Rightarrow \Delta, \Delta'} {\scriptstyle Cut}$$

is transformed into the derivation

$$\cfrac{\cfrac{\Gamma'' \Rightarrow A}{\Gamma'', \Gamma' \Rightarrow A} {\scriptstyle LW} \quad \cfrac{B, \Gamma'' \Rightarrow \Delta, C \quad C, \Gamma' \Rightarrow \Delta'}{B, \Gamma'', \Gamma' \Rightarrow \Delta, \Delta'} {\scriptstyle Cut}}{A \supset B, \Gamma'', \Gamma' \Rightarrow \Delta, \Delta'} {\scriptstyle L\supset}$$

with a cut of lower cut-height.

3.6. The last rule in the left premiss is $R\supset$, with $\Delta = \Delta''$, $A \supset B$. The derivation

$$\cfrac{\cfrac{\Gamma, A \Rightarrow B}{\Gamma \Rightarrow \Delta'', A \supset B, C} {\scriptstyle R\supset} \quad C, \Gamma' \Rightarrow \Delta'}{\Gamma, \Gamma' \Rightarrow \Delta'', A \supset B, \Delta'} {\scriptstyle Cut}$$

is transformed into the derivation

$$\frac{\dfrac{\Gamma, A \Rightarrow B}{\Gamma \Rightarrow \Delta'', A \supset B, \Delta'}{}^{R\supset}}{\Gamma, \Gamma' \Rightarrow \Delta'', A \supset B, \Delta'}{}^{LW}$$

in which no cut is used.

3.7. The last rule in the left premiss is $R\forall$. The derivation

$$\frac{\dfrac{\Gamma \Rightarrow B(y/x)}{\Gamma \Rightarrow \Delta'', \forall x B, C}{}^{R\forall} \quad C, \Gamma' \Rightarrow \Delta'}{\Gamma, \Gamma' \Rightarrow \Delta'', \forall x B, \Delta'}{}^{Cut}$$

is transformed into the derivation

$$\frac{\dfrac{\Gamma \Rightarrow B(y/x)}{\Gamma \Rightarrow \Delta'', \forall x B, \Delta'}{}^{R\supset}}{\Gamma, \Gamma' \Rightarrow \Delta'', \forall x B, \Delta'}{}^{LW}$$

in which, as in case *3.6*, no cut is used.

4. The cut formula is principal in the left premiss only and the right premiss is derived by a rule with a restriction in the succedent of the premiss, that is, $L\supset$, $R\supset$, or $R\forall$. In these cases, we cannot permute the cut up on the right, but consider instead the five subcases arising from the derivation of the left premiss:

4.i. The last rule is $L\&$, and the derivation

$$\frac{\dfrac{\Gamma \Rightarrow \Delta, A \quad \Gamma \Rightarrow \Delta, B}{\Gamma \Rightarrow \Delta, A\&B}{}^{L\supset} \quad A\&B, \Gamma' \Rightarrow \Delta'}{\Gamma, \Gamma' \Rightarrow \Delta, \Delta'}{}^{Cut}$$

is transformed into the derivation

$$\frac{\dfrac{\Gamma \Rightarrow \Delta, B \quad \dfrac{\Gamma \Rightarrow \Delta, A \quad \dfrac{A\&B, \Gamma' \Rightarrow \Delta'}{A, B, \Gamma' \Rightarrow \Delta'}{}^{Inv}}{B, \Gamma, \Gamma' \Rightarrow \Delta, \Delta'}{}^{Cut}}{\dfrac{\Gamma, \Gamma, \Gamma' \Rightarrow \Delta, \Delta, \Delta'}{\Gamma, \Gamma' \Rightarrow \Delta, \Delta'}{}^{LC,RC}}{}^{Cut}$$

4.ii. The last rule is $R\vee$, and the derivation

$$\frac{\dfrac{\Gamma \Rightarrow \Delta, A, B}{\Gamma \Rightarrow \Delta, A \vee B}{}^{L\supset} \quad A \vee B, \Gamma' \Rightarrow \Delta'}{\Gamma, \Gamma' \Rightarrow \Delta, \Delta'}{}^{Cut}$$

is transformed into

$$\cfrac{\Gamma \Rightarrow \Delta, A, B \qquad \cfrac{A \vee B, \Gamma' \Rightarrow \Delta'}{A, \Gamma' \Rightarrow \Delta'}\ Inv}{\cfrac{\Gamma, \Gamma' \Rightarrow \Delta, \Delta', B}{\cfrac{}{} } Cut} \qquad \cfrac{A \vee B, \Gamma' \Rightarrow \Delta'}{B, \Gamma' \Rightarrow \Delta'}\ Inv$$

$$\cfrac{\cfrac{\Gamma \Rightarrow \Delta, A, B \qquad \cfrac{A \vee B, \Gamma' \Rightarrow \Delta'}{A, \Gamma' \Rightarrow \Delta'} Inv}{\Gamma, \Gamma' \Rightarrow \Delta, \Delta', B} Cut \qquad \cfrac{A \vee B, \Gamma' \Rightarrow \Delta'}{B, \Gamma' \Rightarrow \Delta'} Inv}{\cfrac{\Gamma, \Gamma', \Gamma' \Rightarrow \Delta, \Delta', \Delta'}{\Gamma, \Gamma' \Rightarrow \Delta, \Delta'} LC,RC} Cut$$

4.iii. The last rule is $R\supset$ and the right premiss is derived by $L\supset$. The derivation

$$\cfrac{\cfrac{A, \Gamma \Rightarrow B}{\Gamma \Rightarrow \Delta, A \supset B}\ R\supset \qquad \cfrac{A \supset B, E \supset F, \Gamma'' \Rightarrow E \qquad F, A \supset B, \Gamma'' \Rightarrow \Delta'}{A \supset B, E \supset F, \Gamma'' \Rightarrow \Delta'}\ L\supset}{\Gamma, E \supset F, \Gamma'' \Rightarrow \Delta, \Delta'}\ Cut$$

is transformed into the derivation with two cuts of lower cut-heights:

$$\cfrac{\cfrac{\cfrac{A, \Gamma \Rightarrow B}{\Gamma \Rightarrow A \supset B}\ R\supset \quad A \supset B, E \supset F, \Gamma'' \Rightarrow E}{\Gamma, E \supset F, \Gamma'' \Rightarrow E}\ Cut \qquad \cfrac{\cfrac{A, \Gamma \Rightarrow B}{\Gamma \Rightarrow A \supset B}\ R\supset \quad A \supset B, F, \Gamma'' \Rightarrow \Delta'}{\cfrac{\Gamma, F, \Gamma'' \Rightarrow \Delta'}{\Gamma, E \supset F, \Gamma'' \Rightarrow \Delta'}\ L\supset}\ Cut}{\cfrac{\Gamma, E \supset F, \Gamma'' \Rightarrow \Delta'}{\Gamma, E \supset F, \Gamma'' \Rightarrow \Delta, \Delta'}\ RW}$$

If the right premiss is derived by $R\supset$ or $R\forall$ we proceed similarly: First we apply $R\supset$ to the first premiss to derive $\Gamma \Rightarrow A \supset B$, then cut with the premiss of the right premiss, and apply again $R\supset$ or $R\forall$.

4.iv. The last rule is $R\forall$. Instead of concluding $\Gamma \Rightarrow \Delta, \forall x B$, apply $R\forall$ with Δ empty to obtain $\Gamma \Rightarrow \forall x B$, then permute the cut up on the left premiss.

4.v. The last rule is $R\exists$, and the derivation

$$\cfrac{\cfrac{\Gamma \Rightarrow \Delta, \exists x B, B(t/x)}{\Gamma \Rightarrow \Delta, \exists x B}\ R\exists \qquad \exists x B, \Gamma' \Rightarrow \Delta'}{\Gamma, \Gamma' \Rightarrow \Delta, \Delta'}\ Cut$$

is transformed into

$$\cfrac{\cfrac{\Gamma \Rightarrow \Delta, \exists x B, B(t/x) \qquad \exists x B, \Gamma' \Rightarrow \Delta'}{\Gamma, \Gamma' \Rightarrow \Delta, B(t/x), \Delta'}\ Cut \qquad \cfrac{\exists x B, \Gamma' \Rightarrow \Delta'}{B(t/x), \Gamma' \Rightarrow \Delta'}\ Inv}{\cfrac{\Gamma, \Gamma', \Gamma' \Rightarrow \Delta, \Delta', \Delta'}{\Gamma, \Gamma' \Rightarrow \Delta, \Delta'}\ LC,RC}\ Cut$$

5. The cut formula is principal in both premisses.

5.3. The cut formula is $A \supset B$, and the derivation is

$$\dfrac{\dfrac{A, \Gamma \Rightarrow B}{\Gamma \Rightarrow \Delta, A \supset B} R\supset \quad \dfrac{A \supset B, \Gamma' \Rightarrow A \quad B, \Gamma' \Rightarrow \Delta'}{A \supset B, \Gamma' \Rightarrow \Delta'} L\supset}{\Gamma, \Gamma' \Rightarrow \Delta, \Delta'} Cut$$

This is transformed into

$$\dfrac{\dfrac{\dfrac{\Gamma \Rightarrow \Delta, A \supset B \quad A \supset B, \Gamma' \Rightarrow A}{\Gamma, \Gamma' \Rightarrow \Delta, A} Cut \quad A, \Gamma \Rightarrow B}{\dfrac{\Gamma, \Gamma, \Gamma' \Rightarrow \Delta, B}{\Gamma, \Gamma, \Gamma', \Gamma' \Rightarrow \Delta, \Delta'}Cut \quad B, \Gamma' \Rightarrow \Delta'} Cut}{\Gamma, \Gamma' \Rightarrow \Delta, \Delta'} LC,RC$$

with one cut of lesser cut-height and two on shorter formulas.

5.4. The cut formula is $\forall x\, B$. The derivation

$$\dfrac{\dfrac{\Gamma \Rightarrow B(y/x)}{\Gamma \Rightarrow \Delta, \forall x\, B} R\forall \quad \dfrac{\forall x\, B, B(t/x), \Gamma' \Rightarrow \Delta'}{\forall x\, B, \Gamma' \Rightarrow \Delta'} L\forall}{\Gamma, \Gamma' \Rightarrow \Delta, \Delta'} Cut$$

is transformed into

$$\dfrac{\dfrac{\dfrac{\Gamma \Rightarrow B(y/x)}{\Gamma \Rightarrow B(t/x)} Subst \quad \dfrac{\Gamma \Rightarrow \Delta, \forall x\, B \quad \forall x\, B, B(t/x), \Gamma' \Rightarrow \Delta'}{B(t/x), \Gamma, \Gamma' \Rightarrow \Delta, \Delta'} Cut}{\Gamma, \Gamma, \Gamma' \Rightarrow \Delta, \Delta'} Cut}{\Gamma, \Gamma' \Rightarrow \Delta, \Delta'} LC$$

QED.

Next we show that the single succedent and multisuccedent intuitionistic calculi are equivalent. In the proof, $\bigvee \Delta$ denotes the disjunction of formulas in Δ, with $\bigvee \Delta = \bot$ if Δ is empty. We also write iterated disjunctions as multiple disjunctions without parentheses:

Theorem 5.3.7: Equivalence of G3i and G3im. *The sequent* $\Gamma \Rightarrow \bigvee \Delta$ *is derivable in* **G3i** *if and only if* $\Gamma \Rightarrow \Delta$ *is derivable in* **G3im**.

Proof: Assume that $\Gamma \Rightarrow \bigvee \Delta$ is derivable in **G3i**. Since all rules in the derivation are also rules in **G3im** or admissible in **G3im**, $\Gamma \Rightarrow \bigvee \Delta$ is derivable in **G3im**. By invertibility of $R\vee$, also $\Gamma \Rightarrow \Delta$ is derivable.

The other direction is proved by induction on height of derivation of $\Gamma \Rightarrow \Delta$. If $\Gamma \Rightarrow \Delta$ is an axiom, Γ and Δ have a common atom P. Then $\Gamma \Rightarrow P$ is an axiom of **G3i** and $\Gamma \Rightarrow \bigvee \Delta$ follows by repeated application of $R\vee$. If $\Gamma \Rightarrow \Delta$ is a conclusion of $L\bot$, then $\bot$ is in Γ and $\Gamma \Rightarrow \bigvee \Delta$ also is a conclusion of $L\bot$.

Assume now that $\Gamma \Rightarrow \Delta$ has been derived in **G3im** and that Δ is nonempty and $\Delta = C_1, \ldots, C_n$. If the last rule is a left rule, we apply the inductive hypothesis to the premisses and then we apply the rule again. If the last rule is a right rule, we have five cases according to the form of principal formula; say it is C_n.

1. $C_n = A \& B$. The premisses are $\Gamma \Rightarrow C_1, \ldots, C_{n-1}, A$ and $\Gamma \Rightarrow C_1, \ldots, C_{n-1}, B$, and the inductive hypothesis gives that $\Gamma \Rightarrow C_1 \vee \ldots \vee C_{n-1} \vee A$ and $\Gamma \Rightarrow C_1 \vee \ldots \vee C_{n-1} \vee B$ are derivable in **G3i**. Now apply $R\&$, and a cut with the easily derivable sequent

$$(C_1 \vee \ldots \vee C_{n-1} \vee A) \& (C_1 \vee \ldots \vee C_{n-1} \vee B) \Rightarrow C_1 \vee \ldots \vee C_{n-1} \vee A \& B$$

gives $\Gamma \Rightarrow C_1 \vee \ldots \vee C_{n-1} \vee A \& B$.

2. $C_n = A \vee B$. The premiss is $\Gamma \Rightarrow C_1, \ldots, C_{n-1}, A, B$ and the inductive hypothesis gives that $\Gamma \Rightarrow C_1 \vee \ldots \vee C_{n-1} \vee A \vee B$ is derivable in **G3i**.

3. $C_n = A \supset B$. The premiss is $A, \Gamma \Rightarrow B$ and by the inductive hypothesis and $R\supset$, the sequent $\Gamma \Rightarrow A \supset B$ is derivable in **G3i**. Now apply $R\vee$ to derive $\Gamma \Rightarrow C_1 \vee \ldots \vee C_{n-1} \vee (A \supset B)$.

4. $C_n = \forall x A$. First apply the inductive hypothesis and $R\forall$ to the premiss and then $R\vee$.

5. $C_n = \exists x A$. The premiss is $\Gamma \Rightarrow C_1, \ldots, C_{n-1}, A(t/x), \exists x A$, so by the inductive hypothesis, we obtain that $\Gamma \Rightarrow C_1 \vee \ldots \vee C_{n-1} \vee A(t/x) \vee \exists x A$ is derivable in **G3i**. By derivability of $C_1 \vee \ldots \vee C_{n-1} \vee A(t/x) \vee \exists x A \Rightarrow C_1 \vee \ldots \vee C_{n-1} \vee \exists x A$ and admissibility of cut, we have that $\Gamma \Rightarrow C_1 \vee \ldots \vee C_{n-1} \vee \exists x A$ is derivable in **G3i**.

If Δ is empty, the sequent $\Gamma \Rightarrow$ can only be a conclusion of $L\perp$. But then $\Gamma \Rightarrow \bigvee \Delta$, which is the same as $\Gamma \Rightarrow \perp$, also is a conclusion of $L\perp$.   QED.

The result shows that the comma on the right in sequents of the calculus **G3im** behaves like intuitionistic disjunction.

5.4. A CLASSICAL SINGLE SUCCEDENT CALCULUS

We show that the addition of a rule of excluded middle for atomic formulas to **G3ip** gives a complete calculus for classical propositional logic:

Rule of excluded middle:

$$\frac{P, \Gamma \Rightarrow C \quad \sim P, \Gamma \Rightarrow C}{\Gamma \Rightarrow C} \; Gem\text{-}at$$

The structural rules, weakening, contraction, and cut, are admissible, and, further, the rule of excluded middle for arbitrary formulas is admissible. Thus we have the rule

$$\frac{A, \Gamma \Rightarrow C \quad \sim A, \Gamma \Rightarrow C}{\Gamma \Rightarrow C} \, Gem$$

An analogous rule for arbitrary formulas, for natural deduction in sequent calculus style, was considered already in Gentzen (1936, §5. 26). But the subformula property fails for the rule, and Gentzen used a rule corresponding to the law of double negation instead. When the rule of excluded middle is restricted to atoms, the subformula property becomes the following: All formulas in the derivation of $\Gamma \Rightarrow C$ are either subformulas of the endsequent or of negations of atoms (i.e., atoms, negations of atoms, or $\perp$). This principle already is sufficient for establishing many properties, but we also show that a simple transformation converts derivations into ones in which the rule of excluded middle is applied only to atoms that appear in the succedent of the conclusion so that we have a subformula property of the usual kind.

In the single succedent sequent calculus, all connectives are present and obey the rules of intuitionistic logic, excluded middle is applied to atoms only, but still derivations remain cut-free. A single succedent calculus is immediately trans-latable into natural deduction: The rule of excluded middle for atoms gives a generalization of the usual principle of indirect proof for atoms in natural deduc-tion, and we obtain, in Chapter 8, as a corollary to admissibility of structural rules and excluded middle for arbitrary formulas a fully normal form for derivations in full classical propositional logic.

The main reason for formulating a single succedent calculus is to extend the operational meaning of sequents to classical propositional logic. The calculus can equally well be seen as a system of intuitionistic proof theory of decidable relations. Examples of this point of view will be treated in the next chapter.

We cannot prove decidability of $\forall x A$ or $\exists x A$ from decidability of $A(t/x)$ for arbitrary t; the addition of quantifiers will not result in classical predicate logic, but in a logic with a classical propositional part and intuitionistic quantifiers, such as encountered in, say, Heyting arithmetic.

(a) Admissibility of structural rules: In proving that admissibility of structural rules in **G3ip** extends to **G3ip**+*Gem-at*, we need the following inversions:

Lemma 5.4.1: *The following inversions are admissible in* **G3ip**+Gem-at. *Each conclusion has a derivation of at most the same height as the premiss:*

$$\frac{A\&B, \Gamma \Rightarrow C}{A, B, \Gamma \Rightarrow C} \quad \frac{A \vee B, \Gamma \Rightarrow C}{A, \Gamma \Rightarrow C} \quad \frac{A \vee B, \Gamma \Rightarrow C}{B, \Gamma \Rightarrow C} \quad \frac{A \supset B, \Gamma \Rightarrow C}{B, \Gamma \Rightarrow C}$$

Proof: By induction on the height of the derivation of the premiss. If in the first inversion $A\&B, \Gamma \Rightarrow C$ was derived by *Gem-at*, we have the derivation

$$\frac{\dfrac{P, A\&B, \Gamma \Rightarrow C \quad \sim P, A\&B, \Gamma \Rightarrow C}{A\&B, \Gamma \Rightarrow C} \textit{Gem-at}}{A, B, \Gamma \Rightarrow C} \textit{Inv}$$

and this is transformed into the derivation

$$\frac{\dfrac{P, A\&B, \Gamma \Rightarrow C}{P, A, B, \Gamma \Rightarrow C} \textit{Ind} \quad \dfrac{\sim P, A\&B, \Gamma \Rightarrow C}{\sim P, A, B, \Gamma \Rightarrow C} \textit{Ind}}{A, B, \Gamma \Rightarrow C} \textit{Gem-at}$$

where *Ind* denotes the inductive step. All the other cases of the first rule go through as in the proof for **G3ip**, Lemma 2.3.5. The proofs for disjunction and implication are similar to those for conjunction.   QED.

Structural rules are proved admissible by induction on formula length and derivation height, extending the proof for **G3ip**, Theorem 2.3.4.

Theorem 5.4.2: Height-preserving weakening. *If $\vdash_n \Gamma \Rightarrow C$, then $\vdash_n A, \Gamma \Rightarrow C$.*

Proof: By adding the formula A to the antecedent of each sequent in the derivation of $\Gamma \Rightarrow C$, we obtain a derivation of $A, \Gamma \Rightarrow C$.   QED.

Theorem 5.4.3:   Height-preserving contraction.   *If $\vdash_n A, A, \Gamma \Rightarrow C$, then $\vdash_n A, \Gamma \Rightarrow C$.*

Proof: The proof is by induction on height of derivation of $A, A, \Gamma \Rightarrow C$. We consider only the case in which $A, A, \Gamma \Rightarrow C$ has been derived by *Gem-at* and the last step is

$$\frac{P, A, A, \Gamma \Rightarrow C \quad \sim P, A, A, \Gamma \Rightarrow C}{A, A, \Gamma \Rightarrow C} \textit{Gem-at}$$

We transform this into

$$\frac{\dfrac{P, A, A, \Gamma \Rightarrow C}{P, A, \Gamma \Rightarrow C} \textit{Ind} \quad \dfrac{\sim P, A, A, \Gamma \Rightarrow C}{\sim P, A, \Gamma \Rightarrow C} \textit{Ind}}{A, \Gamma \Rightarrow C} \textit{Gem-at}$$

For the rest, the proof goes through as that for **G3ip** in Theorem 2.4.1.   QED.

We note that contraction for atoms is not only admissible in **G3ip+**_Gem-at_, but is actually derivable by _Gem-at_:

$$\cfrac{P,P,\Gamma \Rightarrow C \quad \cfrac{\sim P,P,\Gamma \Rightarrow P \quad \perp,P,\Gamma \Rightarrow C}{\cfrac{\sim P,P,\Gamma \Rightarrow C}{P,\Gamma \Rightarrow C} \; Gem\text{-}at} L\supset}{}$$

Theorem 5.4.4: *The cut rule*

$$\frac{\Gamma \Rightarrow A \quad A,\Delta \Rightarrow C}{\Gamma,\Delta \Rightarrow C} \; Cut$$

is admissible in **G3ip+**Gem-at.

Proof: The proof is by induction on weight of the cut formula with subinduction on cut-height. If the first premiss in a cut was derived by _Gem-at_ we have

$$\frac{\cfrac{P,\Gamma \Rightarrow A \quad \sim P,\Gamma \Rightarrow A}{\Gamma \Rightarrow A}\;Gem\text{-}at \quad A,\Delta \Rightarrow C}{\Gamma,\Delta \Rightarrow C}\;Cut$$

and cut is permuted upward to cuts on the same formula but with lower cut-height,

$$\frac{\cfrac{P,\Gamma \Rightarrow A \quad A,\Delta \Rightarrow C}{P,\Gamma,\Delta \Rightarrow C}\;Cut \quad \cfrac{\sim P,\Gamma \Rightarrow A \quad A,\Delta \Rightarrow C}{\sim P,\Gamma,\Delta \Rightarrow C}\;Cut}{\Gamma,\Delta \Rightarrow C}\;Gem\text{-}at$$

and similarly if the second premiss was derived by _Gem-at_. For the remaining cases, the proof goes through as that for **G3ip** in Theorem 2.4.3. QED.

(b) Applications: We first show the completeness of the classical single suc-cedent calculus, then give a proof of Glivenko's theorem through a proof-transformation, and finally derive a strict subformula property showing that in the derivation of a sequent $\Gamma \Rightarrow C$, the rule of excluded middle can be restricted to atoms of C.

To prove the admissibility of excluded middle for arbitrary formulas and thereby the completeness of the calculus **G3ip+**_Gem-at_, we need the follow-ing inversion lemma for implication:

Lemma 5.4.5: *The inversion*

$$\frac{A \supset B,\Gamma \Rightarrow C}{\sim A,\Gamma \Rightarrow C}\;Inv$$

is admissible in **G3ip+**Gem-at.

Proof: A cut of $A \supset B, \Gamma \Rightarrow C$ with the easily derivable sequent $\sim A \Rightarrow A \supset B$ gives the result.   QED.

Theorem 5.4.6: *The rule of excluded middle for arbitrary formulas*

$$\frac{A, \Gamma \Rightarrow C \quad \sim A, \Gamma \Rightarrow C}{\Gamma \Rightarrow C} Gem$$

is admissible in **G3ip**+Gem-at.

Proof: We prove admissibility of excluded middle by induction on formula length. Structural rules were shown admissible and can be used in the proof. To narrow down, quite literally, the derivations, we shall indicate routine propositional derivations in **G3ip** by vertical dots.

For $A := \bot$, we derive the conclusion from the right premiss already:

$$\frac{\dfrac{\dfrac{\bot \Rightarrow \bot}{\Rightarrow \sim \bot} R\supset}{} L\bot \quad \sim\bot, \Gamma \Rightarrow C}{\Gamma \Rightarrow C} Cut$$

For $A := P$, excluded middle is rule *Gem-at*.

For $A := A \& B$, we have the derivation

$$\frac{\dfrac{A\&B, \Gamma \Rightarrow C}{A, B, \Gamma \Rightarrow C} Inv \quad \dfrac{\sim A, B \overset{\vdots}{\Rightarrow} \sim(A\&B) \quad \sim(A\&B), \Gamma \Rightarrow C}{\dfrac{\sim A, B, \Gamma \Rightarrow C}{B, \Gamma \Rightarrow C} Ind} Cut \quad \dfrac{\sim B \overset{\vdots}{\Rightarrow} \sim(A\&B) \quad \sim(A\&B), \Gamma \Rightarrow C}{\sim B, \Gamma \Rightarrow C} Ind}{\Gamma \Rightarrow C} C$$

For $A := A \lor B$, we have the derivation

$$\frac{\dfrac{A \lor B, \Gamma \Rightarrow C}{B, \Gamma \Rightarrow C} Inv \quad \dfrac{\dfrac{\dfrac{A \lor B, \Gamma \Rightarrow C}{A, \Gamma \Rightarrow C} Inv}{A, \sim B, \Gamma \Rightarrow C} Wk \quad \dfrac{\sim A, \sim B \overset{\vdots}{\Rightarrow} \sim(A \lor B) \quad \sim(A \lor B), \Gamma \Rightarrow C}{\sim A, \sim B, \Gamma \Rightarrow C} Cut}{\dfrac{\sim B, \Gamma \Rightarrow C}{} Ind}}{\Gamma \Rightarrow C} Ind$$

For $A := A \supset B$, we have the derivation

$$\frac{\dfrac{A \supset B, \Gamma \Rightarrow C}{B, \Gamma \Rightarrow C} Inv \quad \dfrac{\dfrac{A, \sim B \overset{\vdots}{\Rightarrow} \sim(A \supset B) \quad \sim(A \supset B), \Gamma \Rightarrow C}{A, \sim B, \Gamma \Rightarrow C} Cut \quad \dfrac{\dfrac{A \supset B, \Gamma \Rightarrow C}{\sim A, \Gamma \Rightarrow C} Inv}{\sim A, \sim B, \Gamma \Rightarrow C} Wk}{\dfrac{\sim B, \Gamma \Rightarrow C}{} Ind}}{\Gamma \Rightarrow C} Ind$$

QED.

From $A \Rightarrow A \lor \sim A$ and $\sim A \Rightarrow A \lor \sim A$, we now conclude by *Gem* that each instance of the scheme $\Rightarrow A \vee \sim A$ is derivable. With cut admissible, we have proved:

Corollary 5.4.7: *The calculus* **G3ip**+Gem-at *is complete for classical propositional logic.*

We show that the classical calculus **G3ip**+*Gem-at* is conservative over the intuitionistic calculus **G3ip** for sequents with a negation as succedent.

Lemma 5.4.8: *If* $\Gamma \Rightarrow C$ *is derivable in* **G3ip**+Gem-at, *applications of rule Gem-at can be permuted last, each concluding a sequent with succedent* C.

Proof: Commutation of the rule *Gem-at* with all the rules of **G3ip** is readily verified. Premisses in *Gem-at* have the same succedent formula as the conclusion. QED.

The following result is a sequent calculus formulation of **Glivenko's theorem** for propositional logic, proved here by an explicit transformation of a classical derivation of a negative formula into an intuitionistic one.

Theorem 5.4.9: *If* $\Gamma \Rightarrow \sim C$ *is derivable in* **G3ip**+Gem-at, *it is derivable in* **G3ip**.

Proof: Using Lemma 5.4.8, permute down the applications of *Gem-at*, and let the first one of them be

$$\frac{P, \Delta \Rightarrow \sim C \quad \sim P, \Delta \Rightarrow \sim C}{\Delta \Rightarrow \sim C} \; Gem\text{-}at$$

The premisses are derivable in **G3ip**, and we have, by using invertibility of $R\supset$ and admissibility of contraction and cut in **G3ip**, the derivation

$$\frac{\dfrac{\dfrac{\dfrac{P, \Delta \Rightarrow \sim C}{P, C, \Delta \Rightarrow \perp} \, Inv}{C, \Delta \Rightarrow \sim P} \, R\supset \quad \dfrac{\dfrac{\sim P, \Delta \Rightarrow \sim C}{\sim P, C, \Delta \Rightarrow \perp} \, Inv}{} \, Cut}{\dfrac{C, \Delta, C, \Delta \Rightarrow \perp}{C, \Delta \Rightarrow \perp} \, Ctr^*}}{\Delta \Rightarrow \sim C} \, R\supset$$

Repeating the proof transformation, we obtain a derivation of $\Gamma \Rightarrow \sim C$ in **G3ip**. QED.

The admissibility of cut permits a structural analysis of proofs for the sequent calculus **G3ip**+*Gem-at*. This is based on the following:

Theorem 5.4.10: *All formulas in the derivation of* $\Gamma \Rightarrow C$ *in* **G3ip**+Gem-at *are either subformulas of the endsequent or of negations of atoms.*

Proof: Inspection of the rules in a cut-free derivation shows that only the rule *Gem-at* can make formulas disappear in a derivation, and these are atoms or negations of atoms. QED.

For the usual logical calculi, consistency is a trivial corollary to cut elimination, but here the argument is not altogether simple:

Corollary 5.4.11: G3ip+Gem-at *is consistent.*

Proof: Assume that $\Rightarrow \perp$ is derivable in **G3ip**+*Gem-at*. The only rule that can have an empty antecedent in the conclusion is *Gem-at*, and therefore the last step in the derivation is

$$\frac{P \Rightarrow \perp \quad \sim P \Rightarrow \perp}{\Rightarrow \perp} \ Gem\text{-}at$$

The left premiss can have been derived only by *Gem-at*, but this would lead to an infinite derivation. Therefore there is no derivation of $\Rightarrow \perp$. QED.

Many other standard results for logical sequent calculi extend to the calculus **G3ip**+*Gem-at*.

An application of the rule of excluded middle to an atom not appearing in the conclusion should have nothing to do with the derivation, for if such an atom were effectively used, it would be a subformula of the conclusion.

Theorem 5.4.12: *If* $\Gamma \Rightarrow C$ *is derivable in* **G3ip**+Gem-at, *it has a derivation with the rule Gem-at restricted to atoms of* C.

Proof: Permute down applications of *Gem-at* with atoms P that are not subformulas of C. Let the first such step be

$$\frac{P, \Delta \Rightarrow C \quad \sim P, \Delta \Rightarrow C}{\Delta \Rightarrow C} \ Gem\text{-}at$$

We transform this derivation into

$$\frac{C, \Delta \Rightarrow C \quad \dfrac{\dfrac{\dfrac{\dfrac{P, \Delta \Rightarrow C}{\sim C, P, \Delta \Rightarrow C}Wk \quad \dfrac{}{\perp, P, \Delta \Rightarrow \perp}L\perp}{\sim C, P, \Delta \Rightarrow \perp}L\supset}{\sim C, \Delta \Rightarrow \sim P}R\supset \quad \sim P, \Delta \Rightarrow C}{\dfrac{\sim C, \Delta, \Delta \Rightarrow C}{\sim C, \Delta \Rightarrow C}Ctr^*}Cut}{\Delta \Rightarrow C}Gem}$$

By Theorem 5.4.6, application of the rule of excluded middle to C converts to atoms of C. The proof transformation is repeated until $\Gamma \Rightarrow C$ is concluded, with *Gem-at* restricted to atoms of C. QED.

The proof gives an effective transformation into a derivation with *Gem-at* applied to atoms of C only. A root-first proof search for a sequent $\Gamma \Rightarrow C$ can begin by a splitting into $P, \Gamma \Rightarrow C$ and $\sim P, \Gamma \Rightarrow C$ for atoms P of C. For example, to derive **Peirce's law** $\Rightarrow ((A \supset B) \supset A) \supset A$ it is enough to derive $(A \supset B) \supset A \Rightarrow A$ and then to apply $R\supset$. By the theorem, Peirce's law is derivable with *Gem* applied to A only. In writing the derivation, we leave out repetitions of the principal formula in the left premiss of $L\supset$ as these are not needed and the rule without repetition is admissible:

$$\cfrac{A,(A \supset B) \supset A \Rightarrow A \qquad \cfrac{\cfrac{\cfrac{A \Rightarrow A \quad \bot, A \Rightarrow B}{\sim A, A \Rightarrow B}L\supset}{\sim A \Rightarrow A \supset B}R\supset \qquad A, \sim A \Rightarrow A}{\sim A, (A \supset B) \supset A \Rightarrow A}L\supset}{\cfrac{\cfrac{(A \supset B) \supset A \Rightarrow A}{\Rightarrow ((A \supset B) \supset A) \supset A}R\supset}{}}Gem$$

(c) Quantifier rules: The addition of quantifiers to the single succedent classical calculus gives a calculus with a classical propositional part and intuitionistic quantifiers, **G3i**+*Gem-at*. Proofs of the basic results for systems with quantifiers are mostly similar to previous proofs.

Lemma 5.4.13: *The rule of weakening is admissible and height-preserving in* G3i+Gem-at.

Proof: Similar to that of Theorem 5.4.2. QED.

Lemma 5.4.14: *The inversions of Lemmas 5.4.1 and 5.4.5 and the following two inversions are admissible in* **G3i**+Gem-at.

$$\cfrac{\forall x A, \Gamma \Rightarrow C}{A(t/x), \forall x A, \Gamma \Rightarrow C}Inv \qquad \cfrac{\exists x A, \Gamma \Rightarrow C}{A(y/x), \Gamma \Rightarrow C}Inv$$

Proof: Inversion for $L\forall$ follows by the admissibility of weakening. For $L\exists$, the proof is by induction on the height of derivation of the premiss. QED.

Theorem 5.4.15: *The rules of contraction and cut are admissible in* **G3i**+ Gem-at.

Proof: By induction, analogously to Theorems 5.4.3 and 5.4.4. QED.

The most natural interpretation of the calculus **G3i**+*Gem-at* is that it is a cut-free intuitionistic system for theories that have decidable atomic formulas. Illustrations of this point of view will be given in Section 6.6. Note that the proof of admissibility of excluded middle for arbitrary formulas for the propositional part cannot be extended to quantified formulas.

5.5. A TERMINATING INTUITIONISTIC CALCULUS

In Section 2.5(c), we gave proofs of underivability in an intuitionistic calculus for propositional logic. In these proofs, it was shown that each possible derivation tree of a given sequent either begins with at least one sequent of the form $P_1, \ldots, P_n \Rightarrow Q$, where $P_i \neq Q$ for all i, or else produces a loop, a subderivation that repeats itself to infinity. The latter is produced by the repetition of the principal formula $A \supset B$ of rule $L \supset$.

It was discovered in Hudelmaier (1992) and in Dyckhoff (1992) that the left implication rule of **G3ip** can be refined into four different rules, according to the form of the antecedent A of the principal formula, to the effect that proof search terminates. The refinement gives a calculus, designated as **G4ip** (not to be confused with the calculus **G4** in Kleene's book of 1967, p. 306), with left implication rules corresponding to the cases $A = P$, $A = C\&D$, $A = C \vee D$, $A = C \supset D$, respectively:

Left implication rules of G4ip:

$$\frac{P, B, \Gamma \Rightarrow E}{P, P \supset B, \Gamma \Rightarrow E} L0\supset \qquad \frac{C \supset (D \supset B), \Gamma \Rightarrow E}{C\&D \supset B, \Gamma \Rightarrow E} L\&\supset$$

$$\frac{C \supset B, D \supset B, \Gamma \Rightarrow E}{C \vee D \supset B, \Gamma \Rightarrow E} L\vee\supset \qquad \frac{C, D \supset B, \Gamma \Rightarrow D \quad B, \Gamma \Rightarrow E}{(C \supset D) \supset B, \Gamma \Rightarrow E} L\supset\supset$$

The rules for conjunction and disjunction and the right implication rule are identical to the rules of **G3ip**. The first three left implication rules of **G4ip** are based on the intuitionistically provable equivalences

$$P\&B \supset\subset P\&(P \supset B),$$
$$(C \supset (D \supset B)) \supset\subset (C\&D \supset B),$$
$$(C \supset B)\&(D \supset B) \supset\subset (C \vee D \supset B).$$

The fourth rule is not intuitive, but it can be justified as follows: From the left premiss $C, D \supset B, \Gamma \Rightarrow D$ we obtain by $R\supset$ the sequent $D \supset B, \Gamma \Rightarrow C \supset D$. A cut with the derivable sequent $(C \supset D) \supset B \Rightarrow D \supset B$ gives $(C \supset D) \supset B, \Gamma \Rightarrow C \supset D$. Now the conclusion of $L\supset\supset$ follows by $L\supset$ only:

$$\frac{(C \supset D) \supset B, \Gamma \Rightarrow C \supset D \quad B, \Gamma \Rightarrow E}{(C \supset D) \supset B, \Gamma \Rightarrow E} L\supset$$

Rules $L0\supset$, $L\&\supset$, and $L\vee\supset$ are invertible with height of derivation preserved. Similarly to rule $L\supset$ of **G3ip**, rule $L\supset\supset$ is invertible with respect to its second premiss only.

The structural proof theory of **G4ip** is an example of the subtle organization of many details in order to obtain admissibility of structural rules in a direct and purely syntactic way. We shall not give full details that would be too long to be included here, but just some of the leading ideas. Proofs of all results can be found in Dyckhoff and Negri (2000). To start with, the naive definition of formula weight as corresponding to formula length will be changed so that the active formulas in rules have a weight that is strictly less than the weight of the principal formula. Following Dyckhoff (1992), we set

$$w(\bot) = 0,$$
$$w(P) = 1 \text{ for atoms } P,$$
$$w(A \supset B) = w(A) + w(B) + 1,$$
$$w(A \& B) = w(A) + w(B) + 2,$$
$$w(A \lor B) = w(A) + w(B) + 3.$$

Other choices are also possible. The rules of **G4ip** are routinely shown admissible in **G3ip**, since structural rules can be used, as in the justification of $L\supset\supset$ above. In the other direction, one has to prove first the admissibility of the left implication rule without repetition of the principal formula in the antecedent of the left premiss. By induction on the height of derivation, we easily prove

Proposition 5.5.1: *The rule of weakening,*

$$\frac{\Gamma \Rightarrow C}{A, \Gamma \Rightarrow C} Wk$$

is admissible and height-preserving in **G4ip**.

Lemma 5.5.2: *The sequent* $A, \Gamma \Rightarrow A$ *is derivable in* **G4ip** *for any formula A.*

Proof: By induction on $w(A)$. If A is an atom or $\bot$, then $A, \Gamma \Rightarrow A$ is an axiom or conclusion of $L\bot$. Else A is a compound formula. For conjunction and disjunction, the claim follows from its validity for the components, obtained by the inductive hypothesis. For implication, we have to analyze the structure of the antecedent. If $A = \bot \supset B$, the sequent $\bot \supset B, \Gamma \Rightarrow \bot \supset B$ follows from the axiom $\bot, \bot \supset B, \Gamma \Rightarrow B$ by $R\supset$. For antecedents of the form $P, C\&D, C \lor D$, and $C \supset D$, application of the inductive hypothesis to lighter formulas combined with the rules for $L\supset$ of **G4ip** and corresponding inversions gives the conclusion. (For details, see Dyckhoff and Negri 2000.)   QED.

The proof of admissibility of contraction is not a routine matter: The essential step is given by a lemma in which duplication of a formula in the conclusion is shown admissible. The lemma shows that if the sequent $(C \supset D) \supset B, \Gamma \Rightarrow E$ is derivable, also $C, D \supset B, D \supset B, \Gamma \Rightarrow E$ is derivable. The effect of this lemma is to reduce contraction to lighter formulas in the problematic case of an implication, the antecedent of which is also an implication.

Theorem 5.5.3: *The rule of contraction*

$$\frac{A, A, \Gamma \Rightarrow E}{A, \Gamma \Rightarrow E}\ Ctr$$

is admissible in **G4ip**.

Proof: See Dyckhoff and Negri (2000).   QED.

The next step is to prove

Lemma 5.5.4: *The rule*

$$\frac{\Gamma \Rightarrow A \quad B, \Gamma \Rightarrow E}{A \supset B, \Gamma \Rightarrow E}$$

is admissible in **G4ip**.

Proof: See Dyckhoff and Negri (2000).   QED.

Theorem 5.5.5: *The rule*

$$\frac{A \supset B, \Gamma \Rightarrow A \quad B, \Gamma \Rightarrow E}{A \supset B, \Gamma \Rightarrow E}\ L\supset$$

is admissible in **G4ip**.

Proof: From the right premiss, by admissibility of weakening, we obtain $B, A \supset B, \Gamma \Rightarrow E$, and from the first premiss, by Lemma 5.5.4, we obtain $A \supset B, A \supset B, \Gamma \Rightarrow E$. The conclusion follows by admissibility of contraction. QED.

The theorem shows that the calculi **G3ip** and **G4ip** are equivalent. By admissibility of cut in **G3ip**, we can conclude closure with respect to cut for **G4ip**, but the stronger result of direct cut elimination is also provable. The proof, in Dyckhoff and Negri (2000), is quite long and involved.

Proofs of admissibility of structural rules can also be given to a multisuccedent version of **G4ip** and to corresponding systems with quantifier rules. They can also be given to extensions of these calculi with nonlogical rules, as shown in Dyckhoff and Negri (2001).

NOTES TO CHAPTER 5

The proofs of cut elimination for **G0i** and **G0c** come from von Plato (2001). The rules of the calculus **GN** were first found in Negri (2000), in connection with studies on linear logic. The results of Section 5.2 come from Negri and von Plato (2001). The multisuccedent intuitionistic calculus is presented in Dragalin (1988, Russian original 1979). Dragalin's proof is given in outline only, and few readers seem to have worked their way through it. Our detailed proof of cut elimination for this calculus follows mainly Dyckhoff (1997), who in turn refers to correspondence with Dragalin on the details of the proof. The classical single succedent calculus is due to von Plato (1998a). The terminating intuitionistic calculus was discovered by Hudelmaier (1992) and Dyckhoff (1992), or actually rediscovered, for related ideas were already presented by Vorob'ev in the early 1950s (see Vorob'ev 1970). The direct proof of admissibility of structural rules for **G4ip** and its extensions was found by Dyckhoff and Negri (2000, 2001).

6

Structural Proof Analysis of Axiomatic Theories

In this chapter, we give a method of adding axioms to sequent calculus, in the form of nonlogical rules of inference. When formulated in a suitable way, cut elimination will not be lost by such addition. By the conversion of axioms into rules, it becomes possible to prove properties of systems by induction on the height of derivations.

The method of extension by nonlogical rules works uniformly for systems based on classical logic. For constructive systems, there will be some special forms of axioms, notably $(P \supset Q) \supset R$, that cannot be treated through cut-free rules.

In the conversion of axiom systems into systems with nonlogical rules, the multisuccedent calculi **G3im** and **G3c** are most useful. All structural rules will be admissible in extensions of these calculi, which has profound consequences for the structure of derivations. The first application is a cut-free system of predicate logic with equality. In earlier systems, cut was reduced to cuts on atomic formulas in instances of the equality axioms, but by the method of this chapter, there will be no cuts anywhere. Other applications of the structural proof analysis of mathematical theories include elementary theories of equality and apartness, order and lattices, and elementary geometry.

6.1. FROM AXIOMS TO RULES

When classical logic is used, all free-variable axioms (purely universal axioms) can be turned into rules of inference that permit cut elimination. The constructive case is more complicated, and we shall deal with it first.

(a) The representation of axioms as rules: We shall be using the intuitionistic multisuccedent sequent calculus **G3ipm** of Section 5.3. In adding nonlogical rules representing axioms, we follow

Principle 6.1.1: *In nonlogical rules, the premises and conclusion are sequents that have atoms as active and principal formulas in the antecedent and an arbitrary context in the succedent.*

126

The most general scheme corresponding to this principle, with shared contexts, is

$$\frac{Q_1, \Gamma \Rightarrow \Delta \quad \cdots \quad Q_n, \Gamma \Rightarrow \Delta}{P_1, \ldots, P_m, \Gamma, \Rightarrow \Delta} Reg$$

where Γ, Δ are arbitrary multisets, $P_1, \ldots, P_m, Q_1, \ldots, Q_n$ are fixed atoms, and the number of premisses n can be zero.

Once we have shown structural rules to be admissible, we can conclude that a rule admitting several atoms in the antecedents of the premisses reduces to as many rules with one atom; for example, the rule

$$\frac{Q_1, Q_2, \Gamma \Rightarrow \Delta \quad R, \Gamma \Rightarrow \Delta}{P, \Gamma \Rightarrow \Delta}$$

reduces to the two rules

$$\frac{Q_1, \Gamma \Rightarrow \Delta \quad R, \Gamma \Rightarrow \Delta}{P, \Gamma \Rightarrow \Delta} \qquad \frac{Q_2, \Gamma \Rightarrow \Delta \quad R, \Gamma \Rightarrow \Delta}{P, \Gamma \Rightarrow \Delta}$$

The second and third rule follow from the first by weakening of the left premiss. In the other direction, weakening $R, \Gamma \Rightarrow \Delta$ to $R, Q_2, \Gamma \Rightarrow \Delta$, we obtain the conclusion $P, Q_2, \Gamma \Rightarrow \Delta$ from $Q_1, Q_2, \Gamma \Rightarrow \Delta$ by the second rule, and weakening again $R, \Gamma \Rightarrow \Delta$ to $R, P, \Gamma \Rightarrow \Delta$, we obtain by the third rule $P, P, \Gamma \Rightarrow \Delta$, which contracts to $P, \Gamma \Rightarrow \Delta$. This argument generalizes, so we do not need to consider premisses with several atoms.

The full rule-scheme corresponds to the formula $P_1 \& \ldots \& P_m \supset Q_1 \vee \ldots \vee Q_n$. In order to see what forms of axioms the rule-scheme covers, we write out a few cases, together with their corresponding axiomatic statements in Hilbert-style calculus. Omitting the contexts, the rules for axioms of the forms $Q \& R$, $Q \vee R$, and $P \supset Q$ are

$$\frac{Q \Rightarrow \Delta}{\Rightarrow \Delta}, \frac{R \Rightarrow \Delta}{\Rightarrow \Delta} \qquad \frac{Q \Rightarrow \Delta \quad R \Rightarrow \Delta}{\Rightarrow \Delta} \qquad \frac{Q \Rightarrow \Delta}{P \Rightarrow \Delta}$$

The rules for axioms of the forms Q, $\sim P$ and $\sim (P_1 \& P_2)$ are

$$\frac{Q \Rightarrow \Delta}{\Rightarrow \Delta} \qquad \overline{P \Rightarrow \Delta} \qquad \overline{P_1, P_2 \Rightarrow \Delta}$$

We recall the definition of regular sequents and their trace formulas from Section 3.1: A sequent is regular if it is of the form

$$P_1, \ldots, P_m \Rightarrow Q_1, \ldots, Q_n, \bot, \ldots, \bot$$

where the number of occurrences of $\bot$, m, and n can be 0, and $P_i \neq Q_j$ for all i, j. Regular sequents are grouped into four types, each with a corresponding trace formula:

1. $P_1 \& \ldots \& P_m \supset Q_1 \vee \ldots \vee Q_n$ if $m > 0, n > 0$,
2. $Q_1 \vee \ldots \vee Q_n$ if $m = 0, n > 0$,
3. $\sim(P_1 \& \ldots \& P_m)$ if $m > 0, n = 0$,
4. $\perp$ if $m = 0, n = 0$.

Regular sequents are precisely the sequents that correspond to rules (Latin "regulae") following our rule-scheme. In terms of the rule-scheme, the formation of trace formulas corresponds to the deletion of all but one of several identical premisses in a rule when any of the Q_j are identical and to the contraction of repetitions in the antecedent of the conclusion when any of the P_i are identical.

Given a sequent $\Rightarrow A$, we can perform a root-first decomposition by means of the rules of **G3ipm**. If the decomposition terminates, we reach leaves that are either axioms or conclusions of $L\perp$ or regular sequents. Among such leaves, we distinguish those that are reached from $\Rightarrow A$ by "invertible paths," ones that never pass via a noninvertible rule of **G3ipm**:

Definition 6.1.2: *In a terminating decomposition of a sequent $\Rightarrow A$ in* **G3ipm**, *if a topsequent is reached without passing through the left premiss of $L\supset$ or via an instance of $R\supset$ with a nonempty context Δ in its conclusion, it is an* invertible *leaf, and in the contrary case it is a* noninvertible *leaf.*

We now define the class of **regular formulas**:

Definition 6.1.3: *A formula A is* regular *if it has a decomposition that leads to invertible leaves that are logical axioms, conclusions of $L\perp$, or regular sequents and noninvertible leaves that are logical axioms or conclusions of $L\perp$.*

We observe that the invertible leaves in a decomposition of $\Rightarrow A$ are independent of the order of decomposition chosen, since any two rules among $L\&$, $R\&$, $L\vee$, $R\vee$, and $R\supset$ with empty right context Δ commute with each other, and each of them commutes with the right premiss of $L\supset$. This uniqueness justifies:

Definition 6.1.4: *For a regular formula A, its* regular decomposition *is the set* $\{A_1, \ldots, A_k\}$, *where the A_i are the formula traces of the regular sequents among the invertible leaves of A. The* regular normal form *of a regular formula A is* $A_1 \& \ldots \& A_k$.

Note that the regular decomposition of a regular formula A is unique, and A is equivalent to its regular normal form. Thus regular formulas are those that permit a constructive version of a conjunctive normal form, one in which each conjunct is an implication of form $P_1 \& \ldots \& P_m \supset Q_1 \vee \ldots \vee Q_n$, instead of the classically equivalent disjunctive form $\sim P_1 \vee \ldots \vee \sim P_m \vee Q_1 \vee \ldots \vee Q_n$. The class of formulas constructively equivalent to usual conjunctive normal form is strictly smaller than the class of formulas having regular normal form. The following proposition shows some closure properties of the latter class of formulas:

Proposition 6.1.5:

(i)   *If A has no $\supset$, then A is regular.*
(ii) *If A, B are regular, then A&B is regular.*
(iii) *If A has no $\supset$ and B is regular, then $A \supset B$ is regular.*

Proof: (i) By invertibility of the rules for & and $\vee$. (ii) Obvious. (iii) Starting with $R\supset$, a decomposition of $\Rightarrow A \supset B$ has invertible leaves of the form $P_1, \ldots, P_m, \Gamma \Rightarrow \Delta$, where $P_1, \ldots, P_m$ are atoms (from the decomposition of A) and $\Gamma \Rightarrow \Delta$ is either a logical axiom or a regular sequent. Thus also $P_1, \ldots, P_m, \Gamma \Rightarrow \Delta$ is either a logical axiom or a regular sequent.   QED.

From the two cases of noninvertible rules we see that typical formulas that need not be regular are disjunctions that contain an implication and implications that contain an implication in the antecedent. But sometimes even these are regular, such as the formula $(P \supset Q) \supset (P \supset R)$.

In the next section we show that the class of regular formulas consists of formulas the corresponding rules of which commute with the cut rule. The reason for adopting Principle 6.1.1 will then be clear.

(b) Extension of classical systems with nonlogical rules: For the extension of classical systems, we use the classical multisuccedent sequent calculus **G3c** in which all structural rules are built in. All propositional rules of **G3c** are invertible, but instead of analyzing regularity of formulas through decomposability as in Section 3.1, we can use the existence of conjunctive normal form in classical propositional logic: Each formula is equivalent to a conjunction of disjunctions of atoms and negations of atoms. Each conjunct can be converted into the classically equivalent form $P_1 \& \ldots \& P_m \supset Q_1 \vee \ldots \vee Q_n$ which is representable as a rule of inference. As special cases we can have $m = 0$ or $n = 0$ as in the four types of trace formulas. We therefore have

Proposition 6.1.6: *All classical quantifier-free axioms can be represented by formulas in regular normal form.*

Thus, to every classical quantifier-free theory, there is a corresponding sequent calculus with structural rules admissible.

(c) Conversion of axiom systems into systems with rules: Conversion of a Hilbert-style axiomatic system into a Gentzen-style sequent system proceeds, after quantifier-elimination, by first finding the regular decomposition of each axiom and then converting each conjunct into a corresponding rule following Principle 6.1.1. Right contraction is unproblematic because of the arbitrary context Δ in the succedents of the rule scheme. In order to handle left contraction, we have to augment this scheme. So assume that we have a derivation of $A, A, \Gamma \Rightarrow \Delta$, and assume that the last rule is nonlogical. Then the derivation of $A, A, \Gamma \Rightarrow \Delta$ can be of three different forms. First, neither occurrence of A is principal in the rule; second, one is principal; third, both are principal. The first case is handled by a straightforward induction, and the second case by the method, familiar from the work of Kleene and exemplified by the $L\supset$ rule of **G3ip**, of repeating the principal formulas of the conclusion in the premisses. Thus the general rule-scheme becomes

$$\frac{Q_1, P_1, \ldots, P_m, \Gamma \Rightarrow \Delta \quad \ldots \quad Q_n, P_1, \ldots, P_m, \Gamma \Rightarrow \Delta}{P_1, \ldots, P_m, \Gamma \Rightarrow \Delta} Reg$$

Here $P_1, \ldots, P_m$ in the conclusion are principal in the rule, and $P_1, \ldots, P_m$ and $Q_1, \ldots, Q_n$ in the premisses are active in the rule. Repetitions in the premisses will make left contractions commute with rules following the scheme. For the remaining case, with both occurrences of formula A principal in the last rule, consider the situation with a Hilbert-style axiomatization. We have some axiom, say $\sim(a < b \ \& \ b < a)$ in the theory of strict linear order, and substitution of b with a produces $\sim(a < a \ \& \ a < a)$ that we routinely abbreviate to $\sim a < a$, irreflexivity of strict linear order. This is in fact a contraction. For systems with rules, the case in which a substitution produces two identical formulas that are both principal in a nonlogical rule, is taken care of by the

Closure condition 6.1.7: *Given a system with nonlogical rules, if it has a rule where a substitution instance in the atoms produces a rule of the form*

$$\frac{Q_1, P_1, \ldots, P_{m-2}, P, P, \Gamma \Rightarrow \Delta \quad \ldots \quad Q_n, P_1, \ldots, P_{m-2}, P, P, \Gamma \Rightarrow \Delta}{P_1, \ldots, P_{m-2}, P, P, \Gamma \Rightarrow \Delta} Reg$$

then it also has to contain the rule

$$\frac{Q_1, P_1, \ldots, P_{m-2}, P, \Gamma \Rightarrow \Delta \quad \ldots \quad Q_n, P_1, \ldots, P_{m-2}, P, \Gamma \Rightarrow \Delta}{P_1, \ldots, P_{m-2}, P, \Gamma \Rightarrow \Delta} Reg$$

The condition is unproblematic, since the number of rules to be added to a given system of nonlogical rules is bounded. Often the closure condition is superfluous; For example, the rule expressing irreflexivity in the constructive theory of strict linear order is derivable from the other rules, as will be shown in Section 6.6.

6.2. ADMISSIBILITY OF STRUCTURAL RULES

In this section we shall prove the admissibility of the structural rules of weakening, contraction, and cut for extensions of logical systems with nonlogical rules of inference. We shall deal in detail with constructive systems and just note that the proofs go through for classical systems with inessential modifications.

We shall denote by **G3im*** any extension of the system **G3im** with rules following our general rule-scheme and satisfying the closure condition. Starting from the proof of admissibility of structural rules for **G3im** in Section 5.1, we then prove admissibility of the structural rules for **G3im***.

Theorem 6.2.1: *The rules of weakening*

$$\frac{\Gamma \Rightarrow \Delta}{A, \Gamma \Rightarrow \Delta}\,LW \qquad \frac{\Gamma \Rightarrow \Delta}{\Gamma \Rightarrow \Delta, A}\,RW$$

are admissible and height-preserving in **G3im***.

Proof: For left weakening, since the axioms and all the rules have an arbitrary context in the antecedent, adding the weakening formula to the antecedent of each sequent will give a derivation of $A, \Gamma \Rightarrow \Delta$. For right weakening, adding the weakening formula to the succedents of all sequents that are not followed by an instance of rules $R\supset$ or $R\forall$ will give a derivation of $\Gamma \Rightarrow \Delta, A$. QED.

The proof of admissibility of the contraction rules and the cut rule for **G3im** requires the use of inversion lemmas. We observe that all the inversion lemmas of Section 5.1, holding for **G3im**, hold for **G3im*** as well. This is achieved by having only atomic formulas as principal in nonlogical rules, a property guaranteed by the restriction given in Principle 6.1.1.

Theorem 6.2.2: *The rules of contraction*

$$\frac{A, A, \Gamma \Rightarrow \Delta}{A, \Gamma \Rightarrow \Delta}\,LC \qquad \frac{\Gamma \Rightarrow \Delta, A, A}{\Gamma \Rightarrow \Delta, A}\,RC$$

are admissible and height-preserving in **G3im***.

Proof: For left contraction, the proof is by induction on the height of the derivation of the premiss. If it is an axiom or conclusion of $L\perp$, the conclusion also is.

If A is not principal in the last rule (either logical or nonlogical), apply inductive hypothesis to the premisses and then apply the rule.

If A is principal and the last rule is logical, for $L\&$ and $L\vee$ apply height-preserving invertibility, inductive hypothesis, and then the rule. For $L\supset$ apply inductive hypothesis to the left premiss, invertibility and inductive hypothesis to the right premiss, and then apply the rule. If the last rule is $L\forall$, apply the inductive hypothesis to its premiss, and $L\forall$. If the last rule is $L\exists$, apply height-preserving invertibility of $L\exists$, the inductive hypothesis, and $L\exists$.

If the last rule is nonlogical, A is an atomic formula P and there are two cases. In the first case, one occurrence of A belongs to the context, another is principal in the rule, say, $A = P_m(= P)$. The derivation ends with

$$\frac{Q_1, P_1, \ldots, P_{m-1}, P, P, \Gamma' \Rightarrow \Delta \quad \ldots \quad Q_n, P_1, \ldots, P_{m-1}, P, P, \Gamma' \Rightarrow \Delta}{P_1, \ldots, P_{m-1}, P, P, \Gamma' \Rightarrow \Delta} \, Reg$$

and we obtain

$$\frac{\dfrac{Q_1, P_1, \ldots, P_{m-1}, P, P, \Gamma' \Rightarrow \Delta}{Q_1, P_1, \ldots, P_{m-1}, P, \Gamma' \Rightarrow \Delta} \, Ind \quad \ldots \quad \dfrac{Q_n, P_1, \ldots, P_{m-1}, P, P, \Gamma' \Rightarrow \Delta}{Q_n, P_1, \ldots, P_{m-1}, P, \Gamma' \Rightarrow \Delta} \, Ind}{P_1, \ldots, P_{m-1}, P, \Gamma' \Rightarrow \Delta} \, Reg$$

In the second case both occurrences of A are principal in the rule, say, $A = P_{m-1} = P_m = P$; thus the derivation ends with

$$\frac{Q_1, P_1, \ldots, P_{m-2}, P, P, \Gamma' \Rightarrow \Delta \quad \ldots \quad Q_n, P_1, \ldots, P_{m-2}, P, P, \Gamma' \Rightarrow \Delta}{P_1, \ldots, P_{m-2}, P, P, \Gamma' \Rightarrow \Delta} \, Reg$$

and we obtain

$$\frac{\dfrac{Q_1, P_1, \ldots, P_{m-2}, P, P, \Gamma' \Rightarrow \Delta}{Q_1, P_1, \ldots, P_{m-2}, P, \Gamma' \Rightarrow \Delta} \, Ind \quad \ldots \quad \dfrac{Q_n, P_1, \ldots, P_{m-2}, P, P, \Gamma' \Rightarrow \Delta}{Q_n, P_1, \ldots, P_{m-2}, P, \Gamma' \Rightarrow \Delta} \, Ind}{P_1, \ldots, P_{m-2}, P, \Gamma' \Rightarrow \Delta} \, Reg$$

with the last rule given by Closure Condition 6.1.7.

The proof of admissibility of right contraction in **G3im*** does not present any additional difficulty with respect to the proof of admissibility in **G3im** since in nonlogical rules the succedent in both the premisses and the conclusion is an arbitrary multiset Δ. So in the case in which the last rule in a derivation of $\Gamma \Rightarrow \Delta, A, A$ is a nonlogical rule, we simply proceed by applying the inductive hypothesis to the premisses and then applying the rule. QED.

Theorem 6.2.3: *The cut rule*

$$\frac{\Gamma \Rightarrow \Delta, A \quad A, \Gamma' \Rightarrow \Delta'}{\Gamma, \Gamma' \Rightarrow \Delta, \Delta'} \, Cut$$

is admissible in **G3im***.

Proof: The proof is by induction on the length of A with subinduction on the sum of the heights of the derivations of $\Gamma \Rightarrow \Delta, A$ and $A, \Gamma' \Rightarrow \Delta'$. We consider here in detail only the cases arising from the addition of nonlogical rules. The other cases are treated in the corresponding proof for the intuitionistic multisuccedent calculus **G3im**, Theorem 5.3.6.

1. If the left premiss is a **nonlogical axiom** (zero-premiss nonlogical rule), then also the conclusion is, since nonlogical axioms have an arbitrary succedent.

2. If the right premiss is a nonlogical axiom with A not principal in it, the conclusion is a nonlogical axiom for the same reason as in case *1.*

3. If the right premiss is a nonlogical axiom with A principal in it, A is atomic and we consider the left premiss. The case that it is a nonlogical axiom is covered by case *1.* If it is a logical axiom with A not principal, the conclusion is a logical axiom; else Γ contains the atom A and the conclusion follows from the right premiss by weakening. In the remaining cases we consider the last rule in the derivation of $\Gamma \Rightarrow \Delta, A$. Since A is atomic, A is not principal in the rule. Let us consider the case of a nonlogical rule (the others being dealt with similarly, except $R{\supset}$ and $R\forall$, which are covered in case *4*). We transform the derivation, where $\mathbf{P}_m$ stands for $P_1, \ldots, P_m$,

$$\frac{\dfrac{Q_1, \mathbf{P}_m, \Gamma'' \Rightarrow \Delta, A \ \ldots \ Q_n, \mathbf{P}_m, \Gamma'' \Rightarrow \Delta, A}{\mathbf{P}_m, \Gamma'' \Rightarrow \Delta, A}Reg \quad A, \Gamma' \Rightarrow \Delta'}{\mathbf{P}_m, \Gamma', \Gamma'' \Rightarrow \Delta, \Delta'}Cut$$

into

$$\frac{\dfrac{Q_1, \mathbf{P}_m, \Gamma'' \Rightarrow \Delta, A \quad A, \Gamma' \Rightarrow \Delta'}{Q_1, \mathbf{P}_m, \Gamma', \Gamma'' \Rightarrow \Delta, \Delta'}Cut \ \ldots \ \dfrac{Q_n, \mathbf{P}_m, \Gamma'' \Rightarrow \Delta, A \quad A, \Gamma' \Rightarrow \Delta'}{Q_n, \mathbf{P}_m, \Gamma', \Gamma'' \Rightarrow \Delta, \Delta'}Cut}{\mathbf{P}_m, \Gamma', \Gamma'' \Rightarrow \Delta, \Delta'}Reg$$

where the cut has been replaced by n cuts with left premiss with derivation of lower height.

Let us now consider the cases in which neither premiss is an axiom.

4. A is not principal in the left premiss. These are dealt with as above, with cut permuted upward to the premisses of the last rule used in the derivation of the left premiss (with suitable variable renaming in order to match the variable restrictions in the cases of quantifier rules), except for $R{\supset}$ and $R\forall$. By the intuitionistic restriction in this rule, A does not appear in the premiss, and the conclusion is obtained without cut by $R{\supset}$ or $R\forall$ and weakening.

5. A is principal in the left premiss only. Then A has to be a compound formula. Therefore, if the last rule of the right premiss is a nonlogical rule, A cannot be principal in the rule, because only atomic formulas are principal in nonlogical rules. In this case cut is permuted to the premisses of the right premiss. If the right rule is a logical one with A not principal in it, the usual reductions are applied.

6. A is principal in both premisses. This case can involve only logical rules and is dealt with as in the usual proof for pure logic. QED.

The conversions used in the proof of admissibility of cut show why it is necessary to formulate the nonlogical rules so that they have an arbitrary context in the succedent, both in the premises and in the conclusion. Besides, as already observed, active and principal formulas have to be atomic and appear in the antecedent.

Theorem 6.2.4: *The rules of weakening, contraction, and cut are admissible in* **G3c***.

Proof: The proof is an extension of the results for the purely logical calculus in Sections 3.2 and 4.2. The new cases are analogous to the intuitionistic case. QED.

6.3. Four approaches to extension by axioms

We found in Section 1.4 that the addition of axioms A into sequent calculus in the form of sequents $\Rightarrow A$, by which derivations can start, will lead to failure of cut elimination. Another way of adding axioms, used by Gentzen (1938, sec. 1.4) already, is to add "mathematical basic sequents" which are (substitution instances of) sequents

$$P_1, \ldots, P_m \Rightarrow Q_1, \ldots, Q_n.$$

Here P_i, Q_j are atomic formulas (typically containing free parameters) or $\bot$. By Gentzen's "Hauptsatz," the use of the cut rule can be pushed into such basic sequents. A third way of adding axioms, first found in Gentzen's consistency proof of elementary arithmetic (1934–35, sec. IV.3), is to treat axioms as a context Γ and to relativize all theorems into Γ, thus proving results of the form $\Gamma \Rightarrow C$. Now the sequent calculus derivations have no nonlogical premises and cut elimination applies. A fourth way of adding axioms is the one of this chapter.

We shall specify formally the four different ways of extending logical sequent systems by axioms and then establish their equivalence. Below, let $\mathcal{D}$ be a finite set of regular formulas. We define sequent systems of four kinds:

Definition 6.3.1:

(a) *An A-system for $\mathcal{D}$ is a sequent system with* axioms **G3ipm+LW+RW+LC+RC+Cut+$A\mathcal{D}$**, *where $A\mathcal{D}$ is the set of sequents $\Rightarrow D$ obtained from elements D in $\mathcal{D}$. In derivations of a sequent $\Gamma \Rightarrow \Delta$ in an A-system, sequents from $A\mathcal{D}$ may appear as premises. Derivability is denoted by $\vdash_{A\mathcal{D}} \Gamma \Rightarrow \Delta$.*

(b) *A B-system for $\mathcal{D}$ is a sequent system with* basic sequents, **G3ipm+LW+RW+LC+RC+Cut+$B\mathcal{D}$**, *where $B\mathcal{D}$ is the set of regular sequents 1–3 of Definition 6.1.2 that correspond to elements of $\mathcal{D}$. In derivations of a sequent $\Gamma \Rightarrow \Delta$ in a*

B-system, sequents from $B\mathcal{D}$ *may appear as premisses. Derivability is denoted by* $\vdash_{BD} \Gamma \Rightarrow \Delta$.

(c) *A C-system for* $\mathcal{D}$ *is a sequent system with a* context. *In derivations of a sequent* $\Gamma \Rightarrow \Delta$ *in a C-system, instances of formulas in* $C\mathcal{D}$ *are permitted in the antecedent. Derivability is denoted by* $\vdash_{CD} \Gamma \Rightarrow \Delta$. *(We can also write it as derivability in* **G3ipm**, *that is, as* $\vdash_{G3} \Gamma, \Theta \Rightarrow \Delta$, *where* Θ *is the multiset of instances of formulas in* $\mathcal{D}$ *used in the derivation.)*

(d) *An R-system for* $\mathcal{D}$ *is a sequent system with* rules, **G3ipm**+$R\mathcal{D}$, *where* $R\mathcal{D}$ *is the set of rules of inference given by the regular decomposition of the formulas in* $\mathcal{D}$. *In derivations of a sequent* $\Gamma \Rightarrow \Delta$ *in an R-system, rules from* $R\mathcal{D}$ *are permitted. Derivability is denoted by* $\vdash_{RD} \Gamma \Rightarrow \Delta$.

Theorem 6.3.2: $\vdash_{AD} \Gamma \Rightarrow \Delta$ *iff* $\vdash_{BD} \Gamma \Rightarrow \Delta$ *iff* $\vdash_{CD} \Gamma \Rightarrow \Delta$ *iff* $\vdash_{RD} \Gamma \Rightarrow \Delta$.

Proof: Axioms and basic sequents are interderivable by cuts, so *A*- and *B*-systems are equivalent. We show equivalence of *R*-systems with *A*-systems and *C*-systems. If a regular formula has to be considered, we take it to be the **split formula** $P \supset Q \vee R$, as other formulas convertible to rules are special cases or inessential generalizations of it.

1. Equivalence of R- and A-systems: The rule

$$\frac{Q, P \Rightarrow \Delta \quad R, P \Rightarrow \Delta}{P \Rightarrow \Delta}\; Split$$

can be derived in the *A*-system with axiom $\Rightarrow P \supset Q \vee R$ by means of cuts and contractions:

$$\cfrac{\Rightarrow P \supset Q \vee R \quad \cfrac{\cfrac{\cfrac{P \supset Q \vee R, P \Rightarrow P \quad Q \vee R, P \Rightarrow Q \vee R}{P \supset Q \vee R, P \Rightarrow Q \vee R}L\supset \quad \cfrac{Q, P \Rightarrow \Delta \quad R, P \Rightarrow \Delta}{Q \vee R, P \Rightarrow \Delta}L\vee}{P \supset Q \vee R, P, P \Rightarrow \Delta}Cut}{\cfrac{P, P \Rightarrow \Delta}{P \Rightarrow \Delta}LC}}{}Cut$$

In the other direction, $\Rightarrow P \supset Q \vee R$ is provable in the *R*-system with *Split*.

$$\cfrac{\cfrac{\cfrac{Q, P \Rightarrow Q, R}{Q, P \Rightarrow Q \vee R}R\vee \quad \cfrac{R, P \Rightarrow Q, R}{R, P \Rightarrow Q \vee R}R\vee}{P \Rightarrow Q \vee R}Split}{\Rightarrow P \supset Q \vee R}R\supset$$

2. Equivalence of C- and R-systems: Assume that $\Gamma \Rightarrow \Delta$ was derived in the *R*-system with *Split*, and show that $\Gamma \Rightarrow \Delta$ can be derived in the *C*-system with

$P \supset Q \vee R$. We assume that *Split* is the last rule in the derivation and therefore $\Gamma = P, \Gamma'$. By induction, $\vdash_{CD} Q, P, \Gamma' \Rightarrow \Delta$ and $\vdash_{CD} R, P, \Gamma' \Rightarrow \Delta$; thus there are instances $A_1, \ldots, A_m$ and $A'_1, \ldots, A'_n$ of the schemes in CD such that

$$\vdash_{G3} Q, P, \Gamma', A_1, \ldots, A_m \Rightarrow \Delta \quad \text{and} \quad \vdash_{G3} R, P, \Gamma', A'_1, \ldots, A'_n \Rightarrow \Delta$$

Structural rules can be used, and we have, in **G3ipm**, a derivation starting with weakening of the A_i and A'_j into a common context $A''_1, \ldots, A''_k$ of instances from CD:

$$\cfrac{P \supset Q \vee R, P, \Gamma', A''_1, \ldots, A''_k \Rightarrow P \quad \cfrac{\cfrac{Q, P, \Gamma', A_1, \ldots, A_m \Rightarrow \Delta}{Q, P, \Gamma', A''_1, \ldots, A''_k \Rightarrow \Delta}LW \quad \cfrac{R, P, \Gamma', A'_1, \ldots, A'_n \Rightarrow \Delta}{R, P, \Gamma', A''_1, \ldots, A''_k \Rightarrow \Delta}LW}{Q \vee R, P, \Gamma', A''_1, \ldots, A''_k \Rightarrow \Delta}L\vee}{P \supset Q \vee R, P, \Gamma', A''_1, \ldots, A''_k \Rightarrow \Delta}L\supset$$

Since the split formula and the $A''_1, \ldots, A''_k$ are in CD, we have shown $\vdash_{CD} \Gamma \Rightarrow \Delta$.

In the other direction, assume $\vdash_{CD} \Gamma \Rightarrow \Delta$. Suppose for simplicity that only one axiom occurs in the context, i.e., that $\vdash_{G3} P \supset Q \vee R, \Gamma \Rightarrow \Delta$. We have the derivation in **G3ipm**+RD+Cut:

$$\cfrac{\cfrac{\cfrac{\cfrac{Q, P \Rightarrow Q, R}{Q, P \Rightarrow Q \vee R}R\vee \quad \cfrac{R, P \Rightarrow Q, R}{R, P \Rightarrow Q \vee R}R\vee}{P \Rightarrow Q \vee R}Split}{\Rightarrow P \supset Q \vee R}R\supset \quad P \supset Q \vee R, \Gamma \Rightarrow \Delta}{\Gamma \Rightarrow \Delta}Cut$$

By admissibility of cut in **G3ipm***, the conclusion follows. QED.

Derivations in A- and B-systems can have premisses, and therefore cut must be assumed, whereas C- and R-systems are cut-free. The strength of R-systems is that they permit proofs by induction on rules used in a derivation. This leads to some surprisingly simple, purely syntactic proofs of properties of elementary axiom systems.

6.4. PROPERTIES OF CUT-FREE DERIVATIONS

The properties of sequent systems representing axiomatic systems are based on the subformula property for systems with nonlogical rules:

Theorem 6.4.1: *If $\Gamma \Rightarrow \Delta$ is derivable in **G3im*** or **G3c***, then all formulas in the derivation are either subformulas of the endsequent or atomic formulas.*

Proof: Only nonlogical rules can make formulas disappear in a derivation, and all such formulas are atomic. QED.

The subformula property is weaker than that for purely logical systems, but sufficient for structural proof analysis. Some general consequences are obtained: Consider a theory having as axioms a finite set $\mathcal{D}$ of regular formulas. Define $\mathcal{D}$ to be **inconsistent** if $\Rightarrow \perp$ is derivable in the corresponding extension and **consistent** if it is not inconsistent. For a theory $\mathcal{D}$, inconsistency surfaces with the axioms through regular decomposition, with no consideration of the logical rules:

Theorem 6.4.2: *Let $\mathcal{D}$ be inconsistent. Then*

   (i)   *All rules in the derivation of $\Rightarrow \perp$ are nonlogical.*
   (ii) *All sequents in the derivation have $\perp$ as succedent.*
   (iii) *Each branch in the derivation begins with a nonlogical rule of the form*

$$\overline{P_1, \ldots, P_m \Rightarrow \perp}$$

   (iv) *The last step in the derivation is a rule of form*

$$\frac{Q_1 \Rightarrow \perp \quad \cdots \quad Q_n \Rightarrow \perp}{\Rightarrow \perp}$$

Proof: (i) By Theorem 6.4.1, no logical constants except $\perp$ can occur in the derivation. (ii) If the conclusion of a nonlogical rule has Δ as succedent, the premisses of the rule also have. Since the endsequent is $\Rightarrow \perp$, (ii) follows. (iii) By (ii) and by $\perp$ not being atomic, no derivation begins with $P, \Gamma \Rightarrow P$. Since only atoms can disappear from antecedents in a nonlogical rule, no derivation begins with $\perp, \Gamma \Rightarrow \perp$. This leaves only zero-premiss nonlogical rules. (iv) By observing that the endsequent has an empty antecedent. QED.

It follows that if an axiom system is inconsistent, its formula traces contain negations and atoms or disjunctions. Therefore, if there are neither atoms nor disjunctions, the axiom system is consistent, and similarly if there are no negations.

By our method, the logical structure in axioms as they are usually expressed is converted into combinatorial properties of derivation trees and completely separated from steps of logical inference. This is especially clear in the classical quantifier-free case, in which theorems to be proved can be converted into a finite number of regular sequents $\Gamma \Rightarrow \Delta$. By the subformula property, derivations of these sequents use only the nonlogical rules and axioms of the corresponding sequent calculus, with the succedent remaining the same throughout all derivations. It becomes possible to use proof theory for syntactic proofs of mutual independence of axiom systems, as follows. Let the axiom to be proved independent be expressed by the logic-free sequent $\Gamma \Rightarrow \Delta$. When the rule corresponding to the axiom is left out of the system of nonlogical rules, underivability of $\Gamma \Rightarrow \Delta$ is usually very easily seen. Examples will be given in the last section of this chapter.

6.5. PREDICATE LOGIC WITH EQUALITY

Axiomatic presentations of predicate logic with equality assume a primitive relation $a = b$ with the axiom of **reflexivity**, $a = a$, and the **replacement scheme**, $a = b \,\&\, A(a/x) \supset A(b/x)$. In sequent calculus, the standard way of treating equality is to add regular sequents with which derivations can start (as in Troelstra and Schwichtenberg 1996, p. 98). These sequents are of the form $\Rightarrow a = a$ and $a = b, P(a/x) \Rightarrow P(b/x)$, with P atomic, and Gentzen's "extended Hauptsatz" says that cuts can be reduced to cuts on these equality axioms. For example, symmetry of equality is derived by letting P be $x = a$. Then the second axiom gives $a = b, a = a \Rightarrow b = a$, and a cut with the first axiom $\Rightarrow a = a$ gives $a = b \Rightarrow b = a$. But there is no cut-free derivation of symmetry. Note also that, in this approach, the rules of weakening and contraction must be assumed, and only then can cuts be reduced to cuts on axioms. (Weakening could be made admissible by letting arbitrary contexts appear on both sides of the regular sequents, but not contraction.)

By our method, cuts on equality axioms are avoided. We first restrict the replacement scheme to atomic predicates $P, Q, R, \ldots$, and then convert the axioms into rules:

$$\frac{a = a, \Gamma \Rightarrow \Delta}{\Gamma \Rightarrow \Delta} \, Ref \qquad \frac{P(b/x), a = b, P(a/x), \Gamma \Rightarrow \Delta}{a = b, P(a/x), \Gamma \Rightarrow \Delta} \, Repl$$

There is a separate replacement rule for each predicate P, and $a = b, P(a/x)$ are repeated in the premiss to obtain admissibility of contraction. By the restriction to atomic predicates, both forms of rules follow the rule-scheme. A case of duplication is produced in the conclusion of the replacement rule in case P is $x = b$. The replacement rule concludes $a = b, a = b, \Gamma \Rightarrow \Delta$ from the premiss $b = b, a = b, a = b, \Gamma \Rightarrow \Delta$. We note that the rule in which both duplications are contracted is an instance of the reflexivity rule so that the closure condition is satisfied. Intuitionistic and classical **predicate logic with equality** is obtained by adding to **G3im** and **G3c**, respectively, rules *Ref* and *Repl*.

Theorem 6.5.1: *The rules of weakening, contraction, and cut are admissible in predicate logic with equality.*

Next we have to show the replacement rule admissible for arbitrary predicates.

Lemma 6.5.2: *The replacement axiom $a = b, A(a/x) \Rightarrow A(b/x)$ is derivable for arbitrary A.*

Proof: The proof is by induction on length of A. If $A = \perp$, the sequent follows by $L\perp$, and if A is an atom, it follows from the replacement rule. If $A = B \,\&\, C$

or $A = B \vee C$, we apply inductive hypothesis to B and C and then left and right rules. If $A = B \supset C$, we have the derivation

$$\cfrac{\cfrac{\cfrac{\cfrac{b = a, B(b/x) \Rightarrow B(a/x)}{\cfrac{b = a, a = b, a = a, B(b/x) \Rightarrow B(a/x)}{\cfrac{a = b, a = a, B(b/x) \Rightarrow B(a/x)}{a = b, B(b/x) \Rightarrow B(a/x)} Ref} Repl} W,W}{a = b, B(a/x) \supset C(a/x), B(b/x) \Rightarrow B(a/x)} W \quad \cfrac{a = b, C(a/x) \Rightarrow C(b/x)}{a = b, C(a/x), B(b/x) \Rightarrow C(b/x)} W}{\cfrac{a = b, B(a/x) \supset C(a/x), B(b/x) \Rightarrow C(b/x)}{a = b, B(a/x) \supset C(a/x) \Rightarrow B(b/x) \supset C(b/x)} R\supset} L\supset}$$

If $A = \forall y B$, the sequent $a = b, \forall y B(a/x) \Rightarrow \forall y B(b/x)$ is derived from $a = b, B(a/x) \Rightarrow B(b/x)$ by applying first $L\forall$ and then $R\forall$. Finally, the sequent $a = b, \exists y B(a/x) \Rightarrow \exists y B(b/x)$ is derived by applying first $R\exists$ and then $L\exists$. QED.

Theorem 6.5.3: *The replacement rule*

$$\frac{A(b/x), a = b, A(a/x), \Gamma \Rightarrow \Delta}{a = b, A(a/x), \Gamma \Rightarrow \Delta} Repl$$

is admissible for arbitrary predicates A.

Proof: By Lemma 6.5.2, $a = b, A(a/x) \Rightarrow A(b/x)$ is derivable. A cut with the premiss of the replacement rule and contractions lead to $a = b, A(a/x), \Gamma \Rightarrow \Delta$. Therefore, by admissibility of contraction and cut in the calculus of predicate logic with equality, admissibility of the replacement rule follows. QED.

Our cut- and contraction-free calculus is equivalent to the usual calculi: the sequents $\Rightarrow a = a$ and $a = b, P(a/x) \Rightarrow P(b/x)$ follow at once from the reflexivity rule and the replacement rule. In the other direction, the two rules are easily derived from $\Rightarrow a = a$ and $a = b, P(a/x) \Rightarrow P(b/x)$ by cut and contraction. The formulation of equality axioms as rules has the advantage of permitting proofs by induction on height of derivation. The conservativity of predicate logic with equality over predicate logic illustrates such proofs. In a cut-free derivation of a sequent $\Gamma \Rightarrow \Delta$ that contains no equalities, the last nonlogical rule must be *Ref*. To prove the conservativity, we show that instances of this rule can be eliminated from the derivation. Above we noticed that the rule of replacement has an instance with a duplication, but that the closure condition is satisfied since the instance in which both duplications are contracted is an instance of reflexivity. For the proof of conservativity, the closure condition will be satisfied by the addition of the contracted instance of *Repl* as a rule *Repl**:

$$\frac{b = b, a = b, \Gamma \Rightarrow \Delta}{a = b, \Gamma \Rightarrow \Delta} Repl^*$$

Lemma 6.5.4: *If* $\Gamma \Rightarrow \Delta$ *has no equalities and is derivable in* **G3c**+Ref+Repl+ Repl*, *no sequents in its derivation have equalities in the succedent.*

Proof: Assume that there is an equality in a succedent. Only a logical rule can move it, but then it is a subformula of the endsequent. QED.

Lemma 6.5.5: *If* $\Gamma \Rightarrow \Delta$ *has no equalities and is derivable in* **G3c**+Ref+Repl+ Repl*, *it is derivable in* **G3c**+Repl+Repl*.

Proof: It is enough to show that a topmost instance of *Ref* can be eliminated from a given derivation. The proof is by induction on the height of derivation of a topmost instance:

$$\frac{a = a, \Gamma' \Rightarrow \Delta'}{\Gamma' \Rightarrow \Delta'} Ref$$

If the premiss is an axiom the conclusion also is, since by Lemma 6.5.4 the succedent Δ' contains no equality, and the same if it is a conclusion of $L\bot$. If the premiss has been concluded by a one-premiss logical rule R, we have

$$\frac{\dfrac{a = a, \Gamma'' \Rightarrow \Delta''}{a = a, \Gamma' \Rightarrow \Delta'} R}{\Gamma' \Rightarrow \Delta'} Ref$$

and this is transformed into

$$\frac{\dfrac{a = a, \Gamma'' \Rightarrow \Delta''}{\Gamma'' \Rightarrow \Delta''} Ref}{\Gamma' \Rightarrow \Delta'} R$$

There is by the inductive hypothesis a derivation of $\Gamma'' \Rightarrow \Delta''$ without rule *Ref*. If a two-premiss logical rule has been applied, the case is similar.

If the premiss has been concluded by *Repl*, there are two cases, according to whether $a = a$ is or is not principal. In the latter case the derivation is, with $\Gamma' = P(b/x), \Gamma''$,

$$\frac{\dfrac{P(c/x), a = a, b = c, P(b/x), \Gamma'' \Rightarrow \Delta'}{a = a, b = c, P(b/x), \Gamma'' \Rightarrow \Delta'} Repl}{b = c, P(b/x), \Gamma'' \Rightarrow \Delta'} Ref$$

By permuting the two rules, the inductive hypothesis can be applied. If $a = a$ is principal, the derivation is, with $\Gamma' = P(a/x), \Gamma''$,

$$\frac{\dfrac{P(a/x), a = a, P(a/x), \Gamma'' \Rightarrow \Delta'}{a = a, P(a/x), \Gamma'' \Rightarrow \Delta'} Repl}{P(a/x), \Gamma'' \Rightarrow \Delta'} Ref$$

By height-preserving contraction, there is a derivation of $a = a, P(a/x), \Gamma'' \Rightarrow \Delta'$

so that the premiss of *Ref* is obtained by a derivation with lower height. The inductive hypothesis applies, giving a derivation of $\Gamma' \Rightarrow \Delta'$ without rule *Ref*.

Last, if the premiss of *Ref* has been concluded by *Repl**, with $a = a$ not principal, the derivation is

$$\frac{\dfrac{c = c, a = a, b = c, \Gamma' \Rightarrow \Delta'}{a = a, b = c, \Gamma' \Rightarrow \Delta'}\ Repl^*}{b = c, \Gamma'' \Rightarrow \Delta'}\ Ref$$

The rules are permuted and the inductive hypothesis applied. If $a = a$ is principal, the derivation is

$$\frac{\dfrac{a = a, a = a, \Gamma' \Rightarrow \Delta'}{a = a, \Gamma' \Rightarrow \Delta'}\ Repl^*}{\Gamma' \Rightarrow \Delta'}\ Ref$$

and we apply height-preserving contraction and the inductive hypothesis.   QED.

Theorem 6.5.6: *If* $\Gamma \Rightarrow \Delta$ *is derivable in* **G3c**+Ref+Repl+Repl* *and if* Γ, Δ *contain no equality, then* $\Gamma \Rightarrow \Delta$ *is derivable in* **G3c**.

Proof: By Lemma 6.5.5, there is a derivation without rule *Ref*. Since the end-sequent has no equality, *Repl* and *Repl** cannot have been used in this derivation.   QED.

Note that if cuts on atoms had not been eliminated, the proof would not go through. Also, if the closure condition were satisfied by considering the contracted rule to be an instance of *Ref*, elimination of contraction could introduce new instances of *Ref* above the *Ref* to be eliminated in Lemma 6.5.5.

6.6. APPLICATION TO AXIOMATIC SYSTEMS

All classical systems permitting quantifier-elimination, and most intuitionistic ones, can be converted into systems of cut-free nonlogical rules of inference. In the previous section, we gave the first application, predicate logic with equality. In Section 5.4, we showed how to turn the logical axiom of excluded middle for atomic formulas into a sequent calculus rule. Also the calculus **G3ip**+*Gem-at* can be seen as an intuitionistic calculus to which a rule corresponding to the decidability of atomic formulas has been added, and, from this point of view, it is more natural to consider the law of excluded middle as a nonlogical rather than a logical axiom.

We shall first give, as a general result for theories with purely universal axioms, a version of **Herbrand's theorem**. Then specific examples from elementary intu-itionistic axiomatics are given: Theories of equality, apartness, and order, as well as algebraic theories with operations, such as lattices and Heyting algebras, are representable as cut-free intuitionistic systems. On the other hand, the intuitionistic theory of negative equality does not admit of a good structural proof theory under the present approach: This theory has a primitive relation $a \neq b$, and the two axioms $\sim a \neq a$ and $\sim a \neq c$ & $\sim b \neq c \supset \sim a \neq b$ expressing reflexivity and transitivity of negative equality.

As a further application of the methods of this chapter, we give a structural proof theory of classical plane affine geometry, with a proof of the independence of Euclid's fifth postulate obtained by proof-theoretical means. Another application of the fact that logical rules can be dispensed with is proof search. We can start root-first from a logic-free sequent $\Gamma \Rightarrow \Delta$ to be derived: The succedent will be the same throughout in derivations with nonlogical rules, and in typical cases very few nonlogical rules match the sequent to be derived.

(a) Herbrand's theorem for universal theories: Let **T** be a theory with a finite number of purely universal axioms and classical logic. We turn the theory **T** into a system of nonlogical rules by first removing the quantifiers from each axiom, then converting the remaining part into nonlogical rules. The resulting system will be denoted by **G3cT**.

Theorem 6.6.1: Herbrand's theorem. *If the sequent* $\Rightarrow \forall x \exists y_1 \ldots \exists y_k A$, *with A quantifier-free, is derivable in* **G3cT**, *then there are terms* t_{i_j} *with* $i \leqslant n$, $j \leqslant k$ *such that*

$$\bigvee_{i=1}^{n} A(t_{i_1}/y_1, \ldots, t_{i_k}/y_k)$$

is derivable in **G3cT**.

Proof: Suppose, to narrow things down, that $k = 1$. Then the derivation of $\Rightarrow \forall x \exists y A$ ends with

$$\frac{\dfrac{\Rightarrow A(z/x, t_1/y), \exists y A(z/x)}{\Rightarrow \exists y A(z/x)} \, R\exists}{\Rightarrow \forall x \exists y A} \, R\forall$$

If the derivation continues, root-first, with a propositional inference, the next premiss is $\Gamma_1 \Rightarrow \Delta_1, \exists y A(z/x)$, where Γ_1, Δ_1 consist of subformulas of $A(z/x, t_1/y)$. (For the sake of simplicity, only a one-premiss rule is considered.) Otherwise $R\exists$ was applied, and the premiss is

$$\Rightarrow A(z/x, t_1/y), A(z/x, t_2/y), \exists y A(z/x)$$

The derivation can continue up from the second alternative in the same way, producing possible derivations in which $R\exists$ is applied and instances of the formula

$\exists y A(z/x)$ multiplied, but since the derivation cannot grow indefinitely, at some stage a conclusion must come from an inference that is not $R\exists$.

Every sequent in the derivation is of the form

$$\Gamma \Rightarrow \Delta, A(z/x, t_m/y), \ldots, A(z/x, t_{m+l}/y), \exists y A(z/x)$$

where Γ, Δ consist of subformulas of $A(z/x, t_i/y)$, with $i < m$. In particular, the formula $\exists y A(z/x)$ can occur in only the succedent. Consider the topsequents of the derivation. If they are axioms or conclusions of $L\bot$, they remain so after deletion of the formula $\exists y A(z/x)$. If they are conclusions of zero-premiss nonlogical rules, they remain so after the deletion since the right context in these rules is arbitrary. After deletion, every topsequent in the derivation is of the form

$$\Gamma \Rightarrow \Delta, A(z/x, t_m/y), \ldots, A(z/x, t_{m+l}/y)$$

Making the propositional and nonlogical inferences as before, but without the formula $\exists y A(z/x)$ in the succedent, produces a derivation of

$$\Rightarrow A(z/x, t_1/y), \ldots, A(z/x, t_{m-1}/y), A(z/x, t_m/, \ldots, A(z/x, t_n/y)$$

and repeated application of rule $R\lor$ now leads to the conclusion.   QED.

In the end of Section 4.3(a) we anticipated a simple form of Herbrand's theorem for classical predicate logic as a result that corresponds to the existence property of intuitionistic predicate logic: Dropping the universal theory from Theorem 6.6.1, we have no nonlogical rules to consider and we obtain

Corollary 6.6.2: *If $\Rightarrow \exists x A$ is derivable in* **G3c**, *there are terms $t_1, \ldots, t_n$ such that $\Rightarrow A(t_1/x) \lor \ldots \lor A(t_n/x)$ is derivable.*

(b) Theories of equality and apartness: The axioms of an apartness relation were introduced in Section 2.1. We shall turn first the equality axioms and then the apartness axioms into systems of cut-free rules.

1. The theory of **equality** has one basic relation $a = b$ that obeys the following axioms:

EQ1. $a = a,$
EQ2. $a = b \,\&\, a = c \supset b = c.$

Symmetry of equality follows by substitution of a for c in EQ2. Note that the formulation is slightly different from the transitivity of equality as given in Section 2.1, where we had $a = c \,\&\, b = c \supset a = b$. The change is dictated by the form of the replacement axiom of Section 6.5: Now transitivity is directly an instance of the replacement axiom, with A equal to $x = c$.

Addition of the rules

$$\frac{a = a, \Gamma \Rightarrow \Delta}{\Gamma \Rightarrow \Delta} \, Ref \qquad \frac{b = c, a = b, a = c, \Gamma \Rightarrow \Delta}{a = b, a = c, \Gamma \Rightarrow \Delta} \, Trans$$

where $a = b, a = c$ are repeated in the premiss of rule *Trans*, gives a calculus **G3im**+*Ref*+*Trans* the rules of which follow the rule-scheme. As noted in Section 6.5, a duplication in *Trans* is produced if b is identical to c, but the corresponding contracted rule is an instance of rule *Ref*. The closure condition is satisfied and the structural rules admissible.

2. The theory of **decidable equality** is given by the above axioms EQ1 and EQ2 and

DEQ. $a = b \vee {\sim} a = b.$

The corresponding rule is an instance of a multisuccedent version of the scheme *Gem-at*:

$$\frac{a = b, \Gamma \Rightarrow \Delta \quad {\sim} a = b, \Gamma \Rightarrow \Delta}{\Gamma \Rightarrow \Delta} \, Deq$$

Admissibility of structural rules for this rule is proved similarly to the single succedent version in Section 5.4. For the language of equality, we have **G3im**+*Gem-at* = **G3im**+*Deq*, a cut-free calculus. Proof of admissibility of structural rules is modular for the rules *Ref, Trans,* and *Deq,* and it follows that the intuitionistic theory of decidable equality, which is the same as the classical theory of equality, is cut-free.

3. The theory of **apartness** has the basic relation $a \neq b$ (a and b are apart, a and b are distinct), with the axioms

AP1. ${\sim} a \neq a,$
AP2. $a \neq b \supset a \neq c \vee b \neq c.$

The rules are

$$\frac{}{a \neq a, \Gamma \Rightarrow \Delta} \, Irref \qquad \frac{a \neq c, a \neq b, \Gamma \Rightarrow \Delta \quad b \neq c, a \neq b, \Gamma \Rightarrow \Delta}{a \neq b, \Gamma \Rightarrow \Delta} \, Split$$

The first, premissless rule represents ${\sim} a \neq a$ by licensing any inference from $a \neq a$; the second has repetition of $a \neq b$ in the premisses. Both rules follow the rule-scheme; the closure condition does not arise because there is only one principal formula, and therefore structural rules are admissible in **G3im**+*Irref*+*Split*.

4. **Decidability of apartness** is expressed by the axiom

DAP. $a \neq b \vee {\sim} a \neq b,$

and the corresponding rule is

$$\frac{a \neq b, \Gamma \Rightarrow \Delta \quad \sim a \neq b, \Gamma \Rightarrow \Delta}{\Gamma \Rightarrow \Delta} Dap$$

As before, it follows that the calculus **G3im**+*Irref*+*Split*+*Dap* is cut-free.

5. The intuitionistic theory of **negative equality** is obtained from the axioms of apartness, with the second axiom replaced by its constructively weaker contraposition:

NEQ1. $\sim a \neq a$,
NEQ2. $\sim a \neq c \ \& \sim b \neq c \supset \sim a \neq b$.

It is not possible to extend **G3im** into a cut-free theory of negative equality by the present methods. If a classical calculus such as **G3c** or **G3i**+*Gem-at* is used, a cut-free system is obtained since NEQ2 becomes equivalent to AP2.

The elementary theories in *1–4* can also be given in a single succedent formulation based on extension of the calculus **G3i**, as in Negri (1999). As a consequence of the admissibility of structural rules in such extensions, we have the following result for the theory of apartness:

Corollary 6.6.3: Disjunction property for the theory of apartness. *If* $\Rightarrow A \vee B$ *is derivable in the single succedent calculus for the theory of apartness, either* $\Rightarrow A$ *or* $\Rightarrow B$ *is derivable.*

Proof: Consider the last rule in the derivation. The rules for apartness cannot conclude a sequent with an empty antecedent, and therefore the last rule must be rule $R\vee$ of **G3i**. QED.

Let us compare the result to the treatment of axiom systems as a context, the third of the approaches described in Section 6.3. Each derivation uses a finite number of instances of the universal closures of the two axioms of apartness, say, Γ. The assumption becomes that $\Gamma \Rightarrow A \vee B$ is derivable in **G3i**. Whenever Γ contains an instance of the "split" axiom, it has a formula with a disjunction in the consequent of an implication. Therefore Γ does not consist of Harrop formulas only (Definition 2.5.3), so that Corollary 6.6.3 gives a proper extension of the disjunction property under hypotheses that are Harrop formulas, Theorem 2.5.4.

(c) Theories of order: We first consider a constructive version of linear order and, next, partial order. The latter is then extended in 6.6(d) by the addition of lattice operations and their axioms.

1. **Constructive linear order:** We have a set with a strict order relation with the two axioms:

LO1. $\sim(a < b \,\&\, b < a)$,
LO2. $a < b \supset a < c \lor c < b$.

Contraposition of the second axiom expresses transitivity of weak linear order. Two rules, denoted by *Asym* and *Split*, are uniquely determined from the axioms. Both rules follow the rule scheme, and the first one has an instance with a duplication, produced when a and b are identical:

$$\frac{}{a < a, a < a, \Gamma \Rightarrow \Delta}\, {}^{Asym}$$

The contracted sequent $a < a, \Gamma \Rightarrow \Delta$ is derived by

$$\frac{\dfrac{}{a < a, a < a, \Gamma \Rightarrow \Delta}\, {}^{Asym} \qquad \dfrac{}{a < a, a < a, \Gamma \Rightarrow \Delta}\, {}^{Asym}}{a < a, \Gamma \Rightarrow \Delta}\, {}^{Split}$$

We observe that the contracted rule is only admissible, rather than being a rule of the system. This makes no difference unless height-preserving admissibility of contraction is required. It is not needed for admissibility of cut.

2. **Partial order:** We have a set with an order relation satisfying the two axioms

PO1. $a \leqslant a$,
PO2. $a \leqslant b \,\&\, b \leqslant c \supset a \leqslant c$.

Equality is defined by $a = b \equiv a \leqslant b \,\&\, b \leqslant a$. It follows that equality is an equivalence relation. Further, since equality is defined in terms of partial order, the principle of substitution of equals for the latter is provable. The axioms of partial order determine by the rule-scheme two rules, the one corresponding to transitivity producing a duplication in case $a = b$ and $b = c$. The rule in which both the premiss and conclusion are contracted is an instance of the rule corresponding to reflexivity, and therefore the structural rules are admissible. The rules corresponding to the two axioms are denoted by *Ref* and *Trans*:

$$\frac{a \leqslant a, \Gamma \Rightarrow \Delta}{\Gamma \Rightarrow \Delta}\, {}^{Ref} \qquad \frac{a \leqslant c, a \leqslant b, b \leqslant c, \Gamma \Rightarrow \Delta}{a \leqslant b, b \leqslant c, \Gamma \Rightarrow \Delta}\, {}^{Trans}$$

Derivations of a regular sequent $\Gamma \Rightarrow \Delta$ in the theory of partial order begin with logical axioms, followed by applications of the above rules. As is seen from the rules, these derivations have the following peculiar form: They are all linear and each step consists in the deletion of one atom from the antecedent. If classical logic is used, by invertibility of all its rules, every derivation consists of derivations of regular sequents followed by application of logical rules only.

3. **Nondegenerate partial order:** We add to the axioms of partial order two constant 0, 1 satisfying the axiom of **nondegeneracy** $\sim 1 \leqslant 0$. The corresponding rule has zero premisses:

$$\frac{}{1 \leqslant 0, \Gamma \Rightarrow \Delta}\; Nondeg$$

Partial order is conservative over nondegenerate partial order:

Theorem 6.6.4: *If* $\Gamma \Rightarrow \Delta$ *is derivable in the classical theory of nondegenerate partial order and* Γ, Δ *are quantifier-free and do not contain* 0, 1, *then* $\Gamma \Rightarrow \Delta$ *is derivable in the theory of partial order.*

Proof: We can assume $\Gamma \Rightarrow \Delta$ to be a regular sequent. We prove that if a derivation of $\Gamma \Rightarrow \Delta$ contains atoms with 0 or 1 the atoms are instances of reflexivity, of the form $0 \leqslant 0$ or $1 \leqslant 1$. So suppose the derivation contains an atom with 0 or 1 and not of the above form. Its downmost occurrence can only disappear by an application of rule *Trans*

$$\frac{a \leqslant c, a \leqslant b, b \leqslant c, \Gamma \Rightarrow \Delta}{a \leqslant b, b \leqslant c, \Gamma \Rightarrow \Delta}\; Trans$$

where $a \leqslant c$ contains 0 or 1 and is not an instance of reflexivity. If $a \equiv 0$, i.e., a is syntactically equal to 0, then $a \leqslant b$ in the conclusion must be an instance of reflexivity and we have $b \equiv 0$, therefore also $c \equiv 0$. But then $a \leqslant c$ is an instance of reflexivity contrary to assumption. The same conclusion follows if $a \equiv 1$ or $c \equiv 0$ or $c \equiv 1$.

By the above, the derivation does not contain instances of $1 \leqslant 0$ and therefore no instances of rule *Nondeg*.   QED.

If intuitionistic logic is used, the result follows whenever $\Gamma \Rightarrow \Delta$ is a regular sequent.

(d) Lattice theory: We add to partial order the two lattice constructions and their axioms:

Lattice operations and axioms:

$a \wedge b$ *the meet of a and b,*	$a \vee b$ *the join of a and b,*
$a \wedge b \leqslant a$ *(Mtl)*,	$a \leqslant a \vee b$ *(Jnl)*,
$a \wedge b \leqslant b$ *(Mtr)*,	$b \leqslant a \vee b$ *(Jnr)*,
$c \leqslant a \;\&\; c \leqslant b \supset c \leqslant a \wedge b$ *(Unimt)*,	$a \leqslant c \;\&\; b \leqslant c \supset a \vee b \leqslant c$ *(Unijn)*.

All of the axioms follow the rule-scheme, and we shall use the above identifiers as names of the nonlogical rules of lattice theory:

$$\frac{a \wedge b \leqslant a, \Gamma \Rightarrow \Delta}{\Gamma \Rightarrow \Delta} Mtl \qquad \frac{a \leqslant a \vee b, \Gamma \Rightarrow \Delta}{\Gamma \Rightarrow \Delta} Jnl$$

$$\frac{a \wedge b \leqslant b, \Gamma \Rightarrow \Delta}{\Gamma \Rightarrow \Delta} Mtr \qquad \frac{b \leqslant a \vee b, \Gamma \Rightarrow \Delta}{\Gamma \Rightarrow \Delta} Jnr$$

$$\frac{c \leqslant a \wedge b, c \leqslant a, c \leqslant b, \Gamma \Rightarrow \Delta}{c \leqslant a, c \leqslant b, \Gamma \Rightarrow \Delta} Unimt \qquad \frac{a \vee b \leqslant c, a \leqslant c, b \leqslant c, \Gamma \Rightarrow \Delta}{a \leqslant c, b \leqslant c, \Gamma \Rightarrow \Delta} Unijn$$

The uniqueness rules for the meet and join constructions can have instances with a duplication in the premiss and conclusion:

$$\frac{c \leqslant a \wedge a, c \leqslant a, c \leqslant a, \Gamma \Rightarrow \Delta}{c \leqslant a, c \leqslant a, \Gamma \Rightarrow \Delta} Unimt$$

and similarly for join. The rule in which $c \leqslant a$ is contracted in both the premiss and conclusion can be added to the system to meet the closure condition. If height-preserving contraction is not required, the contracted rule can be proved admissible: Using admissibility of left weakening, admissibility of the rule obtained from *Unimt* is proved as follows, starting with the contracted premiss $c \leqslant a \wedge a, c \leqslant a, \Gamma \Rightarrow \Delta$:

$$\frac{\dfrac{\dfrac{\dfrac{\dfrac{c \leqslant a \wedge a, c \leqslant a, \Gamma \Rightarrow \Delta}{c \leqslant a \wedge a, c \leqslant a, a \leqslant a \wedge a, \Gamma \Rightarrow \Delta} LW}{c \leqslant a, a \leqslant a \wedge a, \Gamma \Rightarrow \Delta} Trans}{c \leqslant a, a \leqslant a \wedge a, a \leqslant a, a \leqslant a, \Gamma \Rightarrow \Delta} LW,LW}{c \leqslant a, a \leqslant a, a \leqslant a, \Gamma \Rightarrow \Delta} Unimt}{c \leqslant a, \Gamma \Rightarrow \Delta} Ref,Ref$$

All structural rules are admissible in the proof-theoretical formulation of lattice theory. The underivability of $\Rightarrow \perp$ follows, by Theorem 6.4.2, from the fact that no axiom of lattice theory is a negation,

As a consequence of having an equality relation defined through partial order, substitution of equals in the meet and join operations,

$$b = c \supset a \wedge b = a \wedge c \qquad b = c \supset a \vee b = a \vee c$$

need not be postulated but can instead be derived. For example, we have $a \wedge b \leqslant a$ by *Mtl* and $a \wedge b \leqslant c$ by *Mtl*, $b \leqslant c$ and *Trans*, so $a \wedge b \leqslant a \wedge c$ follows by *Unimt*.

Lattice theory is conservative over partial order:

Theorem 6.6.5: *If $\Gamma \Rightarrow \Delta$ is derivable in classical lattice theory and Γ, Δ are quantifier-free and do not contain lattice operations, then $\Gamma \Rightarrow \Delta$ is derivable in the theory of partial order.*

Proof: We can assume that $\Gamma \Rightarrow \Delta$ is a regular sequent. The topsequent is a logical axiom of the form $a \leqslant c, \Gamma' \Rightarrow \Delta', a \leqslant c$ where $\Delta', a \leqslant c = \Delta$ and a, c contain no lattice operations. We can also assume that the first step removes $a \leqslant c$ from the antecedent; if not, the steps and removed atoms before $a \leqslant c$ can be deleted.

If the first rule is *Ref*, then $a \equiv c$ and $a \leqslant c, \Gamma \Rightarrow \Delta$ is a logical axiom from which the conclusion follows by *Ref*. Else the first rule must be *Trans* with the step

$$\frac{a \leqslant c, a \leqslant b, b \leqslant c, \Gamma' \Rightarrow \Delta}{a \leqslant b, b \leqslant c, \Gamma' \Rightarrow \Delta} \; Trans$$

The atoms $a \leqslant b, b \leqslant c$ are **activated** in an instance of rule *Trans* by the **removed atom** $a \leqslant c$. They form a **chain** of two atoms $a \leqslant b, b \leqslant c$ in the topsequent. We may assume $a \leqslant b$ or $b \leqslant c$ to be the removed atom in the next step, for otherwise the step and its removed atom can be deleted. If $a \leqslant b$ is removed by *Trans*, two atoms $a \leqslant d, d \leqslant b$ are activated by $a \leqslant b$ and similarly if $b \leqslant c$ is removed by *Trans*. Among the atoms activated so far there is a chain of three atoms $a \leqslant d, d \leqslant b, b \leqslant c$ each of which is in the topsequent. Starting with $a \leqslant c$, we form the transitive closure of atoms activated in instances of *Trans*. Each such instance will substitute one atom in the chain by two, until we come to the last instance of *Trans* with the chain $a \leqslant b_0, b_0 \leqslant b_1, \ldots, b_n \leqslant c$ in the topsequent. If an atom is not in the chain and is removed by a rule other than *Trans*, the atom and rule are deleted. Thus, we only have to show that atoms in the chain with lattice operations can be removed, and let $b_k \leqslant b_{k+1}$ be the first such atom. (If there are none, there is nothing to prove.) If it is removed by *Ref*, we have $b_k \equiv b_{k+1}$ and consider the pair of atoms $b_{k-1} \leqslant b_k, b_k \leqslant b_{k+2}$, and so on, until both atoms must be removed by lattice rules. Similarly, let b_l with $l > k$ be the first of the b_i that does not contain lattice operations. (If there are none, consider the last term c in the chain.)

   We claim that in the chain $a \leqslant b_0, b_0 \leqslant b_1, \ldots, b_n \leqslant c$ there is a contiguous pair of atoms that are removed by rules *Unimt, Mt* or *Jn, Unijn*: Start with $b_{k-1} \leqslant b_k$. If the outermost lattice operation of b_k is $\wedge$, the atom $b_{k-1} \leqslant b_k$ has to be removed by *Unimt*, for b_{k-1} does not contain lattice operations. Then $b_k \leqslant b_{k+1}$ must be removed by *Jn, Unimt* or *Mt*. In the last case we are done, else we continue along the chain, analyzing $b_{k+1} \leqslant b_{k+2}$. If the first case had occurred, $b_{k+1} \leqslant b_{k+2}$ is removed by *Unijn, Jn*, or *Unimt*; if the second, it is removed by *Jn, Unimt*, or *Mt*. In the last case we have the conclusion. In the other cases, we continue the case analysis until we have that $b_{l-2} \leqslant b_{l-1}$ is removed by *Jn* or *Unimt*. But then $b_{l-1} \leqslant b_l$ is removed by *Unijn* or *Mt*, respectively, since b_l does not contain lattice operations.

We prove the result in a similar fashion if the outermost lattice operation of b_k is $\vee$.

Let two contiguous atoms $b \leqslant d \wedge e$ and $d \wedge e \leqslant d$ be removed by *Unimt, Mt*. For *Unimt* to be applicable, the topsequent has to contain the atoms $b \leqslant d$ and $b \leqslant e$. Then replace the two atoms $b \leqslant d \wedge e$ and $d \wedge e \leqslant d$ with the single atom $b \leqslant d$, and continue the derivation as before except for deleting the instances of *Trans* where the two atoms were active and the two steps *Unimt, Mt*. In this way the number of atoms containing lattice operations is decreased. If there are two contiguous atoms that are removed by *Jn, Unijn*, let them be $b \leqslant b \vee d$, $b \vee d \leqslant e$. Then replace them with the atom $b \leqslant e$ that is found in the topsequent and delete the steps where the two atoms were active. Again, this proof transformation decreases the number of atoms containing lattice operations.   QED.

If $\Gamma \Rightarrow \Delta$ is a regular sequent, the result applies also in the intuitionistic theory.

(e) Affine geometry: We have two sets of basic objects, points denoted by $a, b, c, \ldots$ and lines denoted by $l, m, n, \ldots$. In order to eliminate all logical structure from the nonlogical rules, we use a somewhat unusual set of basic concepts, written as follows:

$a \neq b$, a and b are *distinct* points,
$l \neq m$, l and m are *distinct* lines,
$l \nparallel m$, l and m are *convergent* lines,
$A(a, l)$,   point a is *outside* line l.

The usual concepts of equal points, equal lines, parallel lines, and incidence of a point with a line, are obtained as negations from the above. These are written as $a = b$, $l = m$, $l \parallel m$, and $I(a, l)$, respectively. The axioms, with names added, are as follows:

I. Axioms for apartness relations:

$\sim a \neq a$ *(Irref)*, $a \neq b \supset a \neq c \vee b \neq c$ *(Split)*,
$\sim l \neq l$ *(Irref)*, $l \neq m \supset l \neq n \vee m \neq n$ *(Split)*,
$\sim l \nparallel l$ *(Irref)*, $l \nparallel m \supset l \nparallel n \vee m \nparallel n$ *(Split)*.

These three basic relations are apartness relations, and their negations are equivalence relations.

Next we have three constructions, two of which have **conditions**: the **connecting line** $ln(a, b)$ that can be formed if $a \neq b$ has been proved, the **intersection point** $pt(l, m)$ where similarly $l \nparallel m$ is required to be proved, and the **parallel line** $par(l, a)$ that can be applied without any conditions uniformly in l and a.

Constructed objects obey incidence and parallelism properties expressed by the next group of axioms:

II. Axioms of incidence and parallelism:

$a \neq b \supset I(a, ln(a, b))$ (*Inc*), $a \neq b \supset I(b, ln(a, b))$ (*Inc*),

$l \nparallel m \supset I(pt(l, m), l)$ (*Inc*), $l \nparallel m \supset I(pt(l, m), m)$ (*Inc*),

$I(a, par(l, a))$ (*Inc*),

$l \parallel par(l, a)$ (*Par*).

Uniqueness of connecting lines, intersection points, and parallel lines is guaranteed by the following axioms:

III. Uniqueness axioms:

$a \neq b \ \& \ l \neq m \supset A(a, l) \vee A(b, l) \vee A(a, m) \vee A(b, m)$ (*Uni*),

$l \neq m \supset A(a, l) \vee A(a, m) \vee l \nparallel m$ (*Unipar*).

The contrapositions of these two principles express usual uniqueness properties. Last, we have the substitution axioms:

IV. Substitution axioms:

$A(a, l) \supset a \neq b \vee A(b, l)$ (*Subst*),

$A(a, l) \supset l \neq m \vee A(a, m)$ (*Subst*),

$l \nparallel m \supset l \neq n \vee m \nparallel n$ (*Subst*),

Again, the contrapositions of these three axioms give the usual substitution principles.

The above axiom system is equivalent to standard systems, such as Artin's (1957) axioms. These state the existence and uniqueness of connecting lines and parallel lines, and existence and properties of intersection points are obtained through a defined notion of parallels. As is typical in such an informal discourse, the principles corresponding to our groups I and IV are left implicit. There is a further axiom stating the existence of at least three noncollinear points, but as explained in von Plato (1995), we do not use such existential axioms, say $(\exists x : Pt)(\exists y : Pt)x \neq y$ and $(\forall x : Ln)(\exists y : Pt)A(y, x)$. We achieve the same effect by systematically considering only geometric situations containing the assumptions $a : Pt, b : Pt, a \neq b, c : Pt, A(c, ln(a, b))$.

An axiom such as $a \neq b \supset I(a, ln(a, b))$ hides a structure going beyond first-order logic. Contrary to appearance, it does **not** consist of two independent formulas $a \neq b$ and $I(a, ln(a, b))$ and a connective, for the latter is a well-formed formula only if $a \neq b$ has been proved. (For a detailed explanation of this structure, **dependent typing**, see Section 3 of Appendix B.) As an example of conditions for well-formed formulas, from our axioms a "triangle axiom"

$$A(c, ln(a, b)) \supset A(b, ln(c, a))$$

can be derived, but the conditions $a \neq b$ and $c \neq a$ are required for this to be well-formed. Here we can actually prove more, the lemma

$$a \neq b \,\&\, A(c, ln(a, b)) \supset c \neq a$$

Assume for this $a \neq b$ and $A(c, ln(a, b))$. By the first substitution axiom, $A(c, ln(a, b))$ gives $c \neq a \vee A(a, ln(a, b))$. By incidence axioms, $I(a, ln(a, b))$, so that $c \neq a$ follows. By the second substitution axiom, $A(c, ln(a, b))$ gives $ln(a, b) \neq ln(c, a) \vee A(c, ln(c, a))$, so that $ln(a, b) \neq ln(c, a)$ follows. By the uniqueness axiom, $a \neq b$ and $ln(a, b) \neq ln(c, a)$ give $A(a, ln(a, b)) \vee A(b, ln(a, b)) \vee A(a, ln(c, a)) \vee A(b, ln(c, a))$, so the incidence axioms lead to the conclusion $A(b, ln(c, a))$.

Examples of conditions can be found in mathematics whenever first-order logic is insufficient. A familiar case is field theory, where results involving inverses $x^{-1}, y^{-1}, \ldots$ can be expressed only after the conditions $x \neq 0$, $y \neq 0$, $\ldots$ have been established.

In a more formal treatment of conditions, they can be made into progressive contexts in the sense of type theory (see Martin-Löf 1984 and von Plato 1995). Such contexts can be arbitrarily complex, even if the formulas in them should all be atomic. For example, the formula $ln(pt(l, m), a) \neq l$ presupposes that $pt(l, m) \neq a$ which in turn presupposes that $l \nparallel m$.

The reason for having basic concepts different from the traditional ones is not only that the "apartness" style concepts suit a constructive axiomatization; there is a reason for the choice of these concepts in classical theories also, namely, if the conditions $a \neq b$ and $l \nparallel m$ were defined as $a = b \supset \bot$ and $l \parallel m \supset \bot$, the natural logic-free expression of the incidence axioms would be lost.

All of the axioms of plane affine geometry can be converted into nonlogical rules, moreover, closure condition 6.1.7 will not lead to any new rules. We conclude that the structural rules are admissible in the rule system for plane affine geometry.

We first derive a form of Euclid's fifth postulate from the geometrical rules: Given a point a outside a line l, no point is incident with both l and the parallel to l through point a. Axiomatically, we may express this by the formula

$$A(a, l) \supset \sim(I(b, l) \,\&\, I(b, par(l, a)))$$

The sequent

$$A(a, l) \Rightarrow A(b, l), A(b, par(l, a))$$

is classically equivalent to the previous one and expresses the same principle as a logic-free multisuccedent sequent. To derive this sequent, we note that, by admissibility of structural rules, all rules in its derivation are nonlogical, and therefore the succedent is always the same, $A(b, l)$, $A(b, par(l, a))$. Further, no conditions will appear. With these prescriptions, root-first proof search is very nearly deterministic. Inspecting the sequent to be derived, we find that the last step has to be a substitution rule in which the second premiss is immediately derived. In order to fit the derivations in, the principal formulas are not repeated in the premisses, and the second formula in the succedent is abbreviated by $A = A(b, par(l, a))$:

$$\frac{l \neq par(l, a) \Rightarrow A(b, l), A \quad \overline{A(a, par(l, a)) \Rightarrow A(b, l), A}^{\,Inc}}{A(a, l) \Rightarrow A(b, l), A} \, Subst$$

The first premiss can be derived by the uniqueness of parallels, and now the rest is obvious:

$$\frac{A(b, l) \Rightarrow A(b, l), A \quad A \Rightarrow A(b, l), A \quad \overline{l \nparallel par(l, a) \Rightarrow A(b, l), A}^{\,Par}}{l \neq par(l, a) \Rightarrow A(b, l), A} \, Unipar$$

We shall show that when the rule of uniqueness of parallels is left out, the sequent

$$A(a, l) \Rightarrow A(b, l), A(b, par(l, a))$$

is not derivable by the rules of affine geometry. We know already that if there is such a derivation, it must end with one of the two first substitution rules. If it is the first rule, we have

$$\frac{a \neq b \Rightarrow A(b, l), A \quad A(b, l) \Rightarrow A(b, l), A}{A(a, l) \Rightarrow A(b, l), A} \, Subst$$

Then the first premiss must be derivable. It is not an axiom, and unless $a = b$, it does not follow by *Irref*. *Split* only repeats the problem, leading to an infinite regress. This leaves only the second substitution rule, and we have

$$\frac{l \neq m \Rightarrow A(b, l), A \quad A(a, m) \Rightarrow A(b, l), A}{A(a, l) \Rightarrow A(b, l), A} \, Subst$$

As in the first case, rules for apartness relations will not lead to the first premiss. Otherwise it could be derived only by uniqueness of parallels, but that is not available. By theorem 6.3.2, derivability in the system of rules is equivalent to derivability with axioms, and we conclude the

Theorem 6.6.6: *The uniqueness axiom for parallel lines is independent of the other axioms of plane affine geometry.*

In case of theorems with quantifiers, assuming classical logic, a theorem to be proved is first converted into prenex form, then the propositional matrix into the variant of conjunctive normal form used above. Each conjunct corresponds to a regular sequent, without logical structure, and the overall structure of the derivation is as follows: First the regular sequents are derived by nonlogical rules only, then the conjuncts by $L\&$, $R\vee$, and $R\supset$. Now $R\&$ collects all these into the propositional matrix, and right quantifier rules lead into the theorem. The nonlogical rules typically contain function constants resulting from quantifier elimination. In the constructive case, these methods apply to formulas in the prenex fragment that admits a propositional part in regular normal form.

An example may illustrate the above structure of derivations: Consider the formula expressing that, for any two points, if they are distinct, there is a line on which the points are incident:

$$\forall x \forall y (x \neq y \supset \exists z(I(x, z)\&I(y, z)))$$

In prenex normal form, with the propositional matrix in the implicational variant of conjunctive normal form, this is equivalent to

$$\forall x \forall y \exists z((x \neq y \& A(x, z) \supset \perp)\&(x \neq y \& A(y, z) \supset \perp))$$

In a quantifier-free approach, we have instead the connecting line construction, with incidence properties expressed by rules in a quantifier-free form:

$$\overline{a \neq b, A(a, ln(a, b)), \Gamma \Rightarrow \Delta}^{\,Inc} \qquad \overline{a \neq b, A(b, ln(a, b)), \Gamma \Rightarrow \Delta}^{\,Inc}$$

We have the following derivation:

$$
\cfrac{
\cfrac{
\cfrac{
\cfrac{\overline{x \neq y, A(x, ln(x, y)) \Rightarrow \perp}^{\,Inc}}{x \neq y \, \& A(x, ln(x, y)) \Rightarrow \perp}{}^{L\&}
}{\Rightarrow x \neq y \, \& A(x, ln(x, y)) \supset \perp}{}^{R\supset}
\qquad
\cfrac{
\cfrac{\overline{x \neq y, A(y, ln(x, y)) \Rightarrow \perp}^{\,Inc}}{x \neq y \, \& A(y, ln(x, y)) \Rightarrow \perp}{}^{L\&}
}{\Rightarrow x \neq y \, \& A(y, ln(x, y)) \supset \perp}{}^{R\supset}
}{
\cfrac{
\cfrac{\Rightarrow (x \neq y \, \& A(x, ln(x, y)) \supset \perp) \,\&(x \neq y \, \& A(y, ln(x, y)) \supset \perp)}{\Rightarrow \exists z((x \neq y \, \& A(x, z) \supset \perp) \,\&(x \neq y \, \& A(y, z) \supset \perp))}{}^{R\exists}
}{\Rightarrow \forall x \forall y \exists z((x \neq y \, \& A(x, z) \supset \perp) \,\&(x \neq y \, \& A(y, z) \supset \perp))}{}^{R\forall, R\forall}
}{}^{R\&}
$$

Derivations with nonlogical rules and all but two of the logical rules of multi-succedent sequent calculi, $R\supset$ and $R\forall$, do not show whether a system is classical or constructive. The difference appears only if classical logic is needed in the conversion of axioms into rules.

NOTES TO CHAPTER 6

Most of the materials of this chapter come from Negri (1999) and Negri and von Plato (1998). The former work contains a single succedent approach to extension of contraction- and cut-free calculi with nonlogical rules. These calculi are used for a proof-theoretical analysis of derivations in theories of apartness and order, leading to conservativity results which have not been treated here. The latter work uses a multisuccedent approach. The examples in subsections (b) and (c) of Section 6.6 are treated in detail in Negri (1999). The proof-theoretic treatment of constructive linear order in subsection (c) is extended in Negri (1999a) to a theory of constructive ordered fields. The geometrical example in subsection (e) comes from von Plato (1998b).

Our proof of Theorem 6.6.1 was suggested by the proof in Buss (1998, sec. 2.5.1).

In Section 6.4, we mentioned some previous attempts at extending cut elimination to axiomatic systems. The work of Uesu (1984) contains the correct way of presenting atomic axioms as rules of inference. As to the use of conjunctive normal form in sequent calculus, we owe it to Ketonen's thesis of 1944, in which the invertible sequent calculus for classical propositional logic was discovered.

7

Intermediate Logical Systems

Intermediate logical systems, or "intermediate logics" as they are often called, are systems between intuitionistic and classical logic in deductive strength. Axiomatic versions of intermediate logical systems are obtained by the addition of different, classically valid axioms to intuitionistic logic. A drawback of this approach is that the proof-theoretic properties of axiomatic systems are weak.

In this chapter, we shall study intermediate logical systems by various methods: One is to translate well-known natural deduction rules into sequent calculus. Another is to add axioms in the style of the rule of excluded middle of Chapter 5 and the nonlogical rules of Chapter 6. We have seen that failure of the strict subformula property is no obstacle to structural proof analysis: It is sufficient to have some limit to the weight of formulas that can disappear in a derivation. A third approach to intermediate logical systems is to relax the right implication rule of multisuccedent intuitionistic sequent calculus by permitting formulas of certain types to appear in the succedent of its premiss, in addition to the single formula of the intuitionistic rule.

From a result of Gödel (1932) it follows that there is an infinity of nonequivalent intermediate logical systems. Some of these arise from natural axioms, such as the law of double negation, the weak law of excluded middle, etc.

There are approaches to intermediate logical systems, in which some property such as validity of an interpolation theorem or some property of algebraic models is assumed. The general open problem behind these researches concerns the structure of the **implicational lattice** of intuitionistic logic (in the first place, propositional logic). This is the problem of generating inductively all the classes of equivalent formulas between $\bot$ and $\bot \supset \bot$, ordered by implication. For formulas in one atom (and $\bot$, of course), this structure, the **free Heyting algebra with one generator**, is known, but above that only special cases have been mastered.

Our aim here is to present a few natural classes of intermediate logical systems and to study their proof-theoretical properties by elementary means. We shall study, in particular, the following logical systems:

1. Logic with the **weak law of excluded middle** $\sim A \vee \sim\sim A$.
2. **Stable logic**, characterized by the law of double-negation $\sim\sim A \supset A$.
3. **Dummett logic**, characterized by the law $(A \supset B) \vee (B \supset A)$.

We shall consider only the propositional parts of intermediate logical systems in what follows.

7.1. A SEQUENT CALCULUS FOR THE WEAK LAW OF EXCLUDED MIDDLE

We study the weak law of excluded middle by adding a rule to a single succedent calculus, analogous to the rule of excluded middle of Section 5.4.

We add to **G3ip** a rule of weak excluded middle for atomic formulas P:

$$\frac{\sim P, \Gamma \Rightarrow C \quad \sim\sim P, \Gamma \Rightarrow C}{\Gamma \Rightarrow C} \ {\scriptstyle Wem\text{-}at}$$

The weak law of excluded middle for atoms follows. In the other direction, that law in the form of an axiomatic sequent $\Rightarrow \sim P \vee \sim\sim P$, together with a cut on $\sim P \vee \sim\sim P$, leads to the rule of weak excluded middle.

In a proof-theoretical analysis of **G3ip**+*Wem-at*, we first have to establish inversion lemmas and, with their help, the admissibility of structural rules. Finally, we have to investigate the admissibility of the rule for arbitrary formulas. Since this method is by now familiar, we indicate only the main results.

We prove inversion lemmas by noting that application of the rule commutes with the inversions of the invertible rules of **G3ip**. Proofs of admissibility of weakening, contraction, and cut go through similarly to the corresponding proofs for the rule of excluded middle in Section 5.4. This is so because the rule has no principal formula: If an application of the rule is followed by weakening or contraction, we simply permute the order of application of the rules. With cut, we show the conversion for the case that the left premiss is derived by the rule,

$$\frac{\dfrac{\sim P, \Gamma \Rightarrow A \quad \sim\sim P, \Gamma \Rightarrow A}{\Gamma \Rightarrow A} \ {\scriptstyle Wem\text{-}at} \quad A, \Gamma \Rightarrow C}{\Gamma \Rightarrow C} \ {\scriptstyle Cut}$$

This is transformed into the derivation with lower cut-height,

$$\frac{\dfrac{\sim P, \Gamma \Rightarrow A \quad A, \Gamma \Rightarrow C}{\sim P, \Gamma, \Gamma \Rightarrow C} \ {\scriptstyle Cut} \quad \dfrac{\sim\sim P, \Gamma \Rightarrow A \quad A, \Gamma \Rightarrow C}{\sim\sim P, \Gamma, \Gamma \Rightarrow C} \ {\scriptstyle Cut}}{\Gamma, \Gamma \Rightarrow C} \ {\scriptstyle Wem\text{-}at}$$

and similarly if the right premiss has been derived by *Wem-at*.

Admissibility of the rule

$$\frac{\sim A, \Gamma \Rightarrow C \quad \sim\sim A, \Gamma \Rightarrow C}{\Gamma \Rightarrow C} \textit{Wem}$$

for arbitrary formulas A is proved by induction on weight of A. If $A = \bot$, the rule is derivable using the left premiss only:

$$\frac{\dfrac{\bot \Rightarrow \bot}{\Rightarrow \sim\bot} R\supset \quad \sim\bot, \Gamma \Rightarrow C}{\Gamma \Rightarrow C} \textit{Cut}$$

If $A = P$, we have the rule *Wem-at*. For the rest, it is easily shown that if the weak law of excluded middle holds for A and B, it holds for $A\&B$, $A \vee B$, and $A \supset B$ as well.

A logical system with the weak law of excluded middle *Wem-at* is well-behaved proof-theoretically. The subformula property needs to be adjusted into: All formulas in derivations are subformulas of the endsequent or of negations of negations of atoms.

7.2. A SEQUENT CALCULUS FOR STABLE LOGIC

We shall investigate the single succedent sequent calculus corresponding to the system of natural deduction with a principle of indirect proof for atomic formulas. Translation of this principle into sequent calculus gives the rule

$$\frac{\sim P, \Gamma \Rightarrow \bot}{\Gamma \Rightarrow P} \textit{Raa-at}$$

The calculus **G3ip**+*Raa-at* has the same strength as a calculus with the rule corresponding to stability for atoms:

$$\frac{\Gamma \Rightarrow \sim\sim P}{\Gamma \Rightarrow P}$$

Rule *Raa-at* is admissible in **G3ip**+*Gem-at*:

$$\frac{P, \Gamma \Rightarrow P \quad \dfrac{\sim P, \Gamma \Rightarrow \bot \quad \bot \Rightarrow P}{\sim P, \Gamma \Rightarrow P} \textit{Cut}}{\Gamma \Rightarrow P} \textit{Gem-at}$$

The other direction, from *Raa-at* to *Gem-at*, does not work:

Theorem 7.2.1: *The calculus* **G3ip**+Raa-at *is not complete for classical propositional logic.*

Proof: Assume there is a (cut-free) derivation of $\Rightarrow P \vee \sim P$. The last rule cannot be *Raa-at*; therefore it is $R\vee$, and $\Rightarrow P$ or $\Rightarrow \sim P$ is derivable. In the first case, $\Rightarrow P$ was derived by *Raa-at* but this is impossible because the premiss $\sim P \Rightarrow \bot$ would then have to be derivable in **G3ip**. In the second case, $\Rightarrow \sim P$ was derived by $R \supset$, but this is again impossible since $P \Rightarrow \bot$ is not derivable in **G3ip**. QED.

The sequent $\sim(P \vee \sim P) \Rightarrow \bot$ is easily derived in **G3ip**. Application of the rule of indirect proof to $P \vee \sim P$ would give a derivation of $\Rightarrow P \vee \sim P$, and we conclude that the rule of indirect proof for arbitrary formulas is not admissible in **G3ip**+*Raa-at*. This is already obvious from the fact that $A \vee B$ is not intuitionistically derivable from $\sim\sim(A \vee B)$, $\sim\sim A \supset A$ and $\sim\sim B \supset B$.

Theorem 7.2.2: *The structural rules are admissible in* **G3ip**+Raa-at.

Proof: Consider an instance of rule *Raa-at* in the derivation. Weakening and contraction can be permuted up since there is no principal formula in the antecedent of the conclusion. For cut, if the right premiss has been derived by rule *Raa-at*, it can be permuted with cut. If the left premiss has been derived by *Raa-at*, we have

$$\dfrac{\dfrac{\sim P, \Gamma \Rightarrow \bot}{\Gamma \Rightarrow P} \; \textit{Raa-at} \qquad P, \Delta \Rightarrow C}{\Gamma, \Delta \Rightarrow C} \; \textit{Cut}$$

Consider the right premiss. If it is an axiom, either C is an atom in Δ and the conclusion of cut also is an axiom, or $C = P$ and the conclusion of cut follows by weakening from the premiss $\Gamma \Rightarrow P$. If the right premiss has been concluded by $L\bot$, the conclusion of cut also follows by $L\bot$. If the right premiss has been concluded by a logical rule, cut is permuted up to its premisses, for P is an atom and cannot be principal in the right premiss. QED.

Theorem 7.2.3: *Rule* Raa *for arbitrary formulas is admissible in* **G3ip**+Raa-at *for the disjunction-free fragment of propositional logic.*

Proof: By adapting the conversions for & and $\supset$ in the proof of Theorem 5.4.6 to **G3ip**+Raa-at. QED.

If to **G3ip** we add rule *Raa* for arbitrary formulas, a rule that corresponds to Gentzen's original rules of natural deduction for classical propositional logic, we obtain a complete calculus: Application of the rule to the intuitionistically derivable premiss $\sim(P \vee \sim P) \Rightarrow \bot$ gives the conclusion $\Rightarrow P \vee \sim P$. Therefore this calculus is also closed with respect to cut, even if it does not permit a cut elimination procedure. To see the latter, consider the case that the left premiss of cut has been derived by *Raa*:

$$\dfrac{\dfrac{\sim A, \Gamma \Rightarrow \bot}{\Gamma \Rightarrow A} \; \textit{Raa} \qquad A, \Delta \Rightarrow C}{\Gamma, \Delta \Rightarrow C} \; \textit{Cut}$$

If A is principal in the right premiss, cut does not permute up. Now, if A is atomic, it is never principal in the left premiss, and we see why Prawitz had to restrict the rule of indirect proof to atomic formulas.

For the structure of derivations in **G3ip+Raa**, we obtain the following: Consider the first application of rule *Raa*, with premiss $\sim A, \Delta \Rightarrow \perp$. We conclude instead by $R\supset$ the sequent $\Delta \Rightarrow \sim\sim A$. Continuing in this way, the derivation of $\Gamma \Rightarrow C$ in **G3ip+Raa** is transformed into a derivation of $\Gamma^* \Rightarrow C^*$ in **G3ip**, where Γ^* and C^* are **partial double-negation translations** of Γ and C: Those parts of Γ, C that are principal in instances of *Raa* in the derivation are substituted by their double negations.

The first one to suggest a translation from classical to intuitionistic logic was Kolmogorov (1925) and related translations were found by Gödel, Gentzen, and Bernays in the early 1930s. In Kolmogorov's translation, each subformula of a given formula A is substituted with its double negation, with the result that the translated formula A^* is intuitionistically derivable if and only if A is classically derivable. Moreover, $A \supset\subset A^*$ is classically derivable. The Gödel–Gentzen translations, in turn, make disjunction and existence disappear with a result on translated formulas analogous to that of Kolmogorov.

The translation we have defined is not only a coding of classically derivable formulas into intuitionistically derivable ones, but is produced by the translation of a classical derivation into an intuitionistic one. Further, as long as we only consider propositional logic, the translation can be simplified: If $\Rightarrow C$ is classically derivable, then $\Rightarrow \sim\sim C$ also is, and by Theorem 5.4.9, $\Rightarrow \sim\sim C$ is derivable in **G3ip**. Here the last rule must be $R\supset$, so we have the intuitionistic derivation

$$\frac{\sim C \Rightarrow \perp}{\Rightarrow \sim\sim C}\ R\supset$$

If instead of $R\supset$ we apply rule *Raa*, we obtain the derivation

$$\frac{\sim C \Rightarrow \perp}{\Rightarrow C}\ Raa$$

The premiss is derivable in **G3ip**, so there is only one application of the classical rule, namely, the last.

Note that the laws of double-negation and weak excluded middle together are equivalent to the classical law of excluded middle.

7.3. Sequent Calculi for Dummett Logic

The classically valid propositional law $(A \supset B) \vee (B \supset A)$ first gained attention in Dummett's study of logical systems with a linearly ordered set of "truth values." This law is rather counterintuitive to most people: One instance is that Goldbach's conjecture implies Riemann's hypothesis or Riemann's hypothesis implies Goldbach's conjecture, but hardly anyone thinks these have much to do with each other. Any two propositions can be substituted for A and B, and Dummett's law is valid, not because of derivability of one from the other, but by classical two-valued semantics: If A is true, $B \supset A$ is true irrespective of B, and so is $(A \supset B) \vee (B \supset A)$. If A is false, truth of $(A \supset B) \vee (B \supset A)$ equally follows. A somewhat more intuitive formulation of Dummett's law is the equivalent disjunction property under hypotheses, $(A \supset B \vee C) \supset (A \supset B) \vee (A \supset C)$.

Underivability of the law $(P \supset Q) \vee (Q \supset P)$ for atoms P, Q in intuitionistic logic is easily shown. It is of interest to study the corresponding proof theory of what is usually called **Dummett logic**. We look at two approaches to this intermediate logical system:

(a) A left rule for Dummett logic: We shall first add to **G3ipm** a left rule called *Dmt-at*:

$$\frac{P \supset Q, \Gamma \Rightarrow C \quad Q \supset P, \Gamma \Rightarrow C}{\Gamma \Rightarrow C} \; Dmt\text{-}at$$

The corresponding rule for arbitrary A, B in place of P, Q will make the formula $(A \supset B) \vee (B \supset A)$ derivable. Inversion lemmas follow as for the rule *Wem-at*, and so does admissibility of all the structural rules as there is no principal formula. Since only atomic implications, i.e., implications in which both antecedent and consequent are atoms, disappear in derivations, the rule supports a weak subformula principle: All formulas in a derivation are subformulas of the endsequent or of atomic implications.

Admissibility of the left Dummett law for arbitrary formulas can be posed as the claim that the law for a formula follows intuitionistically from Dummett law for its components. If one of A and B, say, B, is equal to $\perp$, the Dummett law $(A \supset \perp) \vee (\perp \supset A)$ follows since $\perp \supset A$ is provable. If A is a conjunction or disjunction, the proofs go through, but if A is an implication $C \supset D$, we obtain $((C \supset D) \supset B) \vee (B \supset (C \supset D))$. Application of Dummett law to the components brings six cases, two of which, $C \supset B$, $D \supset B$, $D \supset C$ and $B \supset C$, $D \supset B$, $D \supset C$, do not imply the Dummett law. Rule *Dmt-at* is not sufficient for obtaining Dummett logic. A formulation as a left rule for arbitrary formulas A, B in place of the atoms of rule *Dmt-at* does not give any subformula property, and there is no satisfactory proof theory under this approach.

(b) Dummett logic through a right implication rule: We can obtain a sequent calculus for Dummett logic by relaxing the constraint on the succedent of the premiss of rule $R\supset$ of **G3ipm** by permitting any number of implications in the succedent of the premiss. Following Sonobe (1975), the right implication rule can introduce simultaneously n implications $A_1 \supset B_1, \ldots, A_n \supset B_n$ in the succedent of the conclusion. The right contexts Δ_i of the premises consist of all the implicational formulas of Δ except $A_i \supset B_i$, and in the succedent of the conclusion Δ can contain other formulas that are not implications:

$$\frac{A_1, \Gamma \Rightarrow \Delta_1, B_1 \quad \ldots \quad A_n, \Gamma \Rightarrow \Delta_n, B_n}{\Gamma \Rightarrow \Delta, A_1 \supset B_1, \ldots, A_n \supset B_n} \, SR\supset$$

The left implication rule of **G3ipm** has to be modified by allowing the succedent Δ of the conclusion to appear as a context also in the succedent of its left premiss:

$$\frac{A \supset B, \Gamma \Rightarrow \Delta, A \quad B, \Gamma \Rightarrow \Delta}{A \supset B, \Gamma \Rightarrow \Delta} \, L\supset$$

The calculus thus obtained will be called **G3LC**. Admissibility of all the structural rules for **G3LC** can now be proved by inductive means:

Lemma 7.3.1: *The rules of left and right weakening are admissible in* **G3LC**.

Proof: Admissibility of left weakening is routinely proved by induction on derivation height. For right weakening we use induction on formula length and height of derivation. If A is nonimplicational, we just apply the inductive hypothesis on the premises of the last rule (lower derivation height) and then the rule. We proceed similarly if A is implicational and the last step is not $R\supset$.

If A is an implicational formula $C \supset D$ and the last step is $SR\supset$ we obtain from the n premises $A_i, \Gamma \Rightarrow \Delta_i, B_i$ the stronger conclusion $\Gamma \Rightarrow A_1 \supset B_1, \ldots, A_n \supset B_n$. Using admissibility of left weakening and the inductive hypothesis on the lighter formulas C, D we obtain

$$C, \Gamma \Rightarrow D, A_1 \supset B_1, \ldots, A_n \supset B_n \tag{1}$$

By applying the inductive hypothesis with a lower derivation height we obtain from the n premises also the derivability of $A_i, \Gamma \Rightarrow \Delta_i, B_i, C \supset D$ for $1 \leqslant i \leqslant n$. This, together with (1) gives by $SR\supset$ the conclusion $\Gamma \Rightarrow \Delta, C \supset D$. QED.

We prove by induction on the length of A the

Lemma 7.3.2: *All sequents of the form $A \Rightarrow A$ are derivable in* **G3LC**.

Thus, by admissibility of left and right weakening, we obtain

Corollary 7.3.3: *All sequents of the form $A, \Gamma \Rightarrow \Delta, A$ are derivable in* **G3LC**.

Lemma 7.3.4: *The rule*

$$\frac{B \supset C, \Gamma \Rightarrow \Delta}{C, \Gamma \Rightarrow \Delta}$$

is admissible in **G3LC**.

Proof: By induction on derivation height. QED.

Lemma 7.3.5: *The rule*

$$\frac{\Gamma \Rightarrow \Delta, B \supset C}{B, \Gamma \Rightarrow \Delta, C}$$

is admissible in **G3LC**.

Proof: By induction on derivation height. If $B \supset C$ is not principal in the last step, use the inductive hypothesis and apply the rule. If it is principal, one of the premisses is $B, \Gamma \Rightarrow \Delta', C$ for some Δ' contained in Δ. The conclusion is then obtained by admissibility of right weakening. QED.

Proposition 7.3.6: *The rules of left and right contraction are admissible in* **G3LC**.

Proof: Admissibility for both rules is proved by induction on the length of A with subinduction on derivation height. We shall consider only those cases in which the proof differs from the proof already given for the system **G3im**.

For left contraction, assume that A is principal and not atomic, the last rule in the derivation being $L\supset$. Thus $A = B \supset C$, and the derivation ends with

$$\frac{B \supset C, B \supset C, \Gamma \Rightarrow \Delta, B \quad B \supset C, C, \Gamma \Rightarrow \Delta}{B \supset C, B \supset C, \Gamma \Rightarrow \Delta} L\supset$$

From the left premiss, we obtain by the inductive hypothesis a derivation of the sequent $B \supset C, \Gamma \Rightarrow \Delta, B$, and by Lemma 7.3.4 applied to the right premiss we get a derivation of $C, C, \Gamma \Rightarrow \Delta$, and hence, by length induction, a derivation of $C, \Gamma \Rightarrow \Delta$. The conclusion follows by applying $L\supset$.

For right contraction assume that the last step of the derivation is $R\supset$, i.e., A is $B \supset C$ and the derivation ends with

$$\frac{B, \Gamma \Rightarrow \Delta', C, B \supset C \quad B, \Gamma \Rightarrow \Delta', C, B \supset C \quad \{B_i, \Gamma \Rightarrow \Delta_i, B \supset C, B \supset C, C_i,\}_{i=1}^n}{\Gamma \Rightarrow \Delta, B \supset C, B \supset C} SR\supset$$

By Lemma 7.3.5 applied to $B, \Gamma \Rightarrow \Delta', C, B \supset C$, we obtain $B, B, \Gamma \Rightarrow \Delta', C, C$ and thus, by induction on formula length and left contraction a derivation of $B, \Gamma \Rightarrow \Delta', C$. Induction on height of derivation applied to all the n other premisses gives $\{B_i, \Gamma \Rightarrow \Delta_i, B \supset C, C_i\}_{i=1}^n$, and the conclusion follows by applying $SR\supset$ to these $n + 1$ premisses. QED.

Theorem 7.3.7: *The rule of cut is admissible in* **G3LC**.

Proof: The proof is by induction on length of the cut formula with subinduction on the height of cut. The only new case with respect to the proof detailed for the system **G3im** is when the cut formula A is an implication $B \supset C$ that is principal in both premisses. In this case the step

$$\dfrac{B, \Gamma \Rightarrow \Delta'', C \ \{B_i, \Gamma \Rightarrow \Delta_i, C_i, B \supset C\}_{i=1}^n}{\Gamma \Rightarrow \Delta, B \supset C} SR\supset \qquad \dfrac{B \supset C, \Gamma' \Rightarrow \Delta', B \quad C, \Gamma' \Rightarrow \Delta'}{B \supset C, \Gamma' \Rightarrow \Delta'} L\supset$$
$$\overline{\qquad\qquad\qquad\qquad \Gamma, \Gamma' \Rightarrow \Delta, \Delta' \qquad\qquad\qquad\qquad} Cut$$

is replaced by one cut of lower height and two cuts on shorter formulas:

$$\dfrac{\dfrac{\dfrac{\Gamma \Rightarrow \Delta, B \supset C \quad B \supset C, \Gamma' \Rightarrow \Delta', B}{\Gamma, \Gamma' \Rightarrow \Delta, \Delta', B} Cut \quad B, \Gamma \Rightarrow \Delta'', C}{\dfrac{\Gamma, \Gamma, \Gamma' \Rightarrow \Delta, \Delta', \Delta'', C}{} Cut \quad C, \Gamma' \Rightarrow \Delta'}{\dfrac{\Gamma, \Gamma, \Gamma', \Gamma' \Rightarrow \Delta, \Delta', \Delta', \Delta''}{} Cut} W*}{\dfrac{\Gamma, \Gamma, \Gamma', \Gamma' \Rightarrow \Delta, \Delta', \Delta', \Delta}{\Gamma, \Gamma' \Rightarrow \Delta, \Delta'} C*}$$

where W^* and C^* denote possibly repeated applications of left and right weakening and contraction. QED.

NOTES TO CHAPTER 7

For formulas in one atom and $\perp$, the structure of the implicational lattice of intuitionistic logic was determined by Rieger (1949) and Nishimura (1960). Further partial results can be found reported in the book by Balbes and Dwinger, *Distributive Lattices*, of 1974. It is somewhat odd for a logician to find studies of intuitionistic logic repeated there in an algebraic disguise.

We have studied only the propositional parts of intermediate logical systems in this chapter. An intermediate system characterized through a law for quantified formulas is the "logic of constant domains" (Görnemann 1971, van Dalen 1986).

Dummett logic was first studied by Dummett (1959), whose idea was to have a generalized linearly ordered set of truth values such as the unit interval [0, 1], instead of the two classical values 0 and 1. Linearity is expressed as a condition on valuations: For any valuation v and any two formulas A and B, either $v(A) \leqslant v(B)$ or $v(B) \leqslant v(A)$. In the former case, $v(A \supset B) = 1$, so also $v((A \supset B) \vee (B \supset A)) = 1$, and similarly in the latter case $v((A \supset B) \vee (B \supset A)) = 1$. Thus the Dummett law is validated in a linearly ordered set of truth values. Sometimes the name of Gödel is also mentioned in this connection. The reason is that in the proof of the result of Gödel (1932), the impossibility of interpreting intuitionistic logic as a many-valued logical system with a finite number of truth values, a denumerable sequence of formulas is constructed which, as observed by Dummett, determines as a limit Dummett logic.

The proof of admissibility of structural rules for the system **G3LC** presented here is due to Roy Dyckhoff. A terminating propositional system for Dummett logic, **G4LC**, based on the calculus **G4ip** is given in Dyckhoff (1999).

8

Back to Natural Deduction

The derivability relation of single succedent sequent calculus, written $\Gamma \Rightarrow C$, is closely related to the derivability relation of natural deduction, written $\Gamma \vdash C$ in Chapter 1. Usually the latter is intended as: There exists a natural deduction derivation tree finishing with C and with open assumptions **contained** in Γ. Thus the derivability relation is not a formal but a metamathematical one. As a consequence, weakening is "smuggled in": If C is derivable from Γ and if each formula of Γ is contained in Δ, then C is derivable from Δ. If the metamathematical derivability relation is used, it will be difficult to state in terms of natural deduction what weakening amounts to. We shall consider only a formal derivability relation for natural deduction, in which Γ is precisely the multiset of open assumptions in a natural deduction derivation.

One consequence of the use of a formal derivability relation is that not all sequent calculus derivations have a corresponding natural deduction derivation. For example, if the last step is a left weakening, it will have no correspondence in natural deduction and similarly if the last step is a contraction. However, such steps are artificial additions to a derivation. Equivalence of derivability in natural deduction and sequent calculus will obtain if no such "useless" weakenings or contractions are present.

We shall show in detail that weakening is, in terms of natural deduction, the same as the **vacuous** discharge of assumptions and that contraction is the same as **multiple** discharge. This explanation was already indicated in Section 1.3. In the other direction, a logical inference in natural deduction that at the same time discharges assumptions, vacuously or multiply, consists, in terms of sequent calculus, of two steps that have been purposely made independent: There is the logical step in which a formula is active, and there is a preceding weakening or contraction step in which the formula was principal.

The availability of weakening and contraction as independent steps of inference leads in sequent calculus to instances of the cut rule that do not have any correspondence in natural deduction. We shall call such instances **nonprincipal** cuts. Different ways of permuting up a nonprincipal cut can lead to different cut-free derivations, where a corresponding natural deduction derivation permits of just one conversion toward normal form.

In Section 1.3, we found a way from natural deduction to sequent calculus. It was essential that the elimination rules for conjunction and implication were formulated as general elimination rules analogous to disjunction elimination. In usual systems of natural deduction, only the special elimination rules for conjunction and implication are available. We shall show that it is these rules and the rule of universal elimination of predicate logic that are responsible for the lack of structural correspondence between derivations in natural deduction and in sequent calculus. With the general rules, the two ways of formalizing logical inferences are seen to be variants of one and the same thing.

Gentzen found the rules of natural deduction through an analysis of actual mathematical proofs, and they have been accepted ever since as "the rules" of natural deduction. How natural are the general elimination rules in comparison? In an informal proof, we would use an assumption of form $A \mathbin{\&} B$ by analyzing it into A and B and by deriving consequences directly from them, without the two intermediate logical steps of the usual conjunction elimination rules. Similarly, we use $A \supset B$ by decomposing it into A and B, then deriving consequences from B, and if at some stage A obtains, those consequences obtain. The same natural use of logic is found when $A \vee B$ is split into A and B in a proof by cases.

A further reason for the general elimination rules is that they follow from a uniform inversion principle, as in Section 1.2. Semantically, the change to general elimination rules is neutral as the meaning explanations for the connectives and quantifiers are given in terms of the introduction rules.

8.1. NATURAL DEDUCTION WITH GENERAL ELIMINATION RULES

In the formalist tradition originating with Hilbert, rules of inference operate on formulas to produce new formulas as conclusions. In Section 1.2, it was emphasized that rules of inference informally act on assertions. On a formal level, they act on derivations of the premises to yield a derivation of the conclusion.

Discharge of assumptions in natural deduction is indicated by the "little numbers" written next to the mnemonic sign for the rule of inference. The corresponding discharged assumptions are put in brackets and the number written on top of them. The way these little numbers are managed has the same importance as the rules of weakening and contraction in sequent calculus.

In natural deduction, the number of times an assumption has been made is well determined, and we shall consider open assumptions in derivations to form multisets with the same notational conventions as in the previous chapters on sequent calculi. For each instance of a rule that can discharge assumptions, it must be uniquely determined what assumptions are discharged, through a **label** written next to the sign of the inference rule and on top of the discharged, bracketed assumptions. We shall refer to these as **discharge labels** and **assumption labels**, respectively, and use the numbers 1, 2, 3, . . . as labels. Uniqueness of discharge is achieved by the following

Principle 8.1.1: Unique discharge of assumptions. No two instances of rules in a derivation can have a common discharge label.

We shall now give an inductive definition of the **derivation of a formula A from open assumptions** Γ. Derivability in natural deduction will then be a relation between a formula and a multiset. Whenever more than one derivation is assumed given in the definition, it is also assumed that these derivations do not have common discharge labels. Similarly, new labels must be chosen fresh.

Definition 8.1.2: *A derivation from open assumptions* in intuitionistic natural deduction *is defined by the following clauses:*

1. A is a derivation of A from the open assumption A.

2. Given derivations

$$
\begin{array}{cc}
\Gamma & \Delta \\
\vdots & \vdots \\
A & B
\end{array}
$$

of A from open assumptions Γ and of B from open assumptions Δ,

$$
\cfrac{
\begin{array}{cc}
\Gamma & \Delta \\
\vdots & \vdots \\
A & B
\end{array}
}{A \& B} \, {}^{\&I}
$$

is a derivation of $A\&B$ from open assumptions Γ, Δ.

3. Given derivations

$$
\begin{array}{cc}
\Gamma & \Delta \\
\vdots & \vdots \\
A & B
\end{array}
$$

with assumptions and conclusions as indicated,

$$
\cfrac{\begin{array}{c}\Gamma \\ \vdots \\ A\end{array}}{A \vee B}\,{}^{\vee I_1}
\qquad
\cfrac{\begin{array}{c}\Delta \\ \vdots \\ B\end{array}}{A \vee B}\,{}^{\vee I_2}
$$

are derivations of $A \vee B$ from open assumptions Γ and from Δ, respectively.

4. Given a derivation

$$A^m, \Gamma$$
$$\vdots$$
$$B$$

as indicated, with $m \geqslant 0$,

$$[\overset{1.}{A^m}], \Gamma$$
$$\vdots$$
$$\frac{B}{A \supset B} \supset I, 1.$$

is a derivation of $A \supset B$ from open assumptions Γ.

5. Given a derivation

$$\Gamma$$
$$\vdots$$
$$A(y/x)$$

of $A(y/x)$ from open assumptions Γ, if y does not occur free in Γ, $\forall x A$,

$$\Gamma$$
$$\vdots$$
$$\frac{A(y/x)}{\forall x A} \forall I$$

is a derivation of $\forall x A$ from open assumptions Γ.

6. Given a derivation

$$\Gamma$$
$$\vdots$$
$$A(t/x)$$

of $A(t/x)$ from open assumptions Γ,

$$\Gamma$$
$$\vdots$$
$$\frac{A(t/x)}{\exists x A} \exists I$$

is a derivation of $\exists x A$ from open assumptions Γ.

7. *Given derivations*

$$
\begin{array}{cc}
\Gamma & A^m, B^n, \Delta \\
\vdots & \vdots \\
A \& B & C
\end{array}
$$

as indicated, with $m, n \geqslant 0$,

$$
\cfrac{
\begin{array}{cc}
\Gamma & \overset{1.}{[A^m]}, \overset{2.}{[B^n]}, \Delta \\
\vdots & \vdots \\
A \& B & C
\end{array}
}{C} \; {\scriptstyle \&E,1.,2.}
$$

is a derivation of C from open assumptions Γ, Δ.

8. *Given derivations*

$$
\begin{array}{ccc}
\Gamma & A^m, \Delta & B^n, \Theta \\
\vdots & \vdots & \vdots \\
A \vee B & C & C
\end{array}
$$

as indicated, with $m, n \geqslant 0$,

$$
\cfrac{
\begin{array}{ccc}
\Gamma & \overset{1.}{[A^m]}, \Delta & \overset{2.}{[B^n]}, \Theta \\
\vdots & \vdots & \vdots \\
A \vee B & C & C
\end{array}
}{C} \; {\scriptstyle \vee E,1.,2.}
$$

is a derivation of C from open assumptions Γ, Δ, Θ.

9. *Given derivations*

$$
\begin{array}{ccc}
\Gamma & \Delta & B^n, \Theta \\
\vdots & \vdots & \vdots \\
A \supset B & A & C
\end{array}
$$

as indicated, with $n \geqslant 0$,

$$
\cfrac{
\begin{array}{ccc}
\Gamma & \Delta & \overset{1.}{[B^n]}, \Theta \\
\vdots & \vdots & \vdots \\
A \supset B & A & C
\end{array}
}{C} \; {\scriptstyle \supset E,1.}
$$

is a derivation of C from open assumptions Γ, Δ, Θ.

10. Given a derivation

$$\begin{array}{c} \Gamma \\ \vdots \\ \bot \end{array}$$

of ⊥ from open assumptions Γ,

$$\begin{array}{c} \Gamma \\ \vdots \\ \dfrac{\bot}{C} \;\bot E \end{array}$$

is a derivation of C from open assumptions Γ.

11. Given derivations

$$\begin{array}{cc} \Gamma & A(t/x)^m,\, \Delta \\ \vdots & \vdots \\ \forall x A & C \end{array}$$

of ∀x A from open assumptions Γ and of C from open assumptions $A(t/x)^m$, Δ,

$$\begin{array}{cc} & 1. \\ \Gamma & [A(t/x)^m],\, \Delta \\ \vdots & \vdots \\ \dfrac{\forall x A \qquad C}{C} & \forall E,1. \end{array}$$

is a derivation of C from open assumption Γ, Δ.

12. Given derivations

$$\begin{array}{cc} \Gamma & A(y/x)^m,\, \Delta \\ \vdots & \vdots \\ \exists x A & C \end{array}$$

of ∃x A from open assumptions Γ and of C from open assumptions $A(y/x)^m$, Δ, if y does not occur free in ∃x A, C, Δ,

$$\begin{array}{cc} & 1. \\ \Gamma & [A(y/x)^m],\, \Delta \\ \vdots & \vdots \\ \dfrac{\exists x A \qquad C}{C} & \exists E,1. \end{array}$$

is a derivation of C from open assumption Γ, Δ.

In 7 and 8 the labels must be chosen distinct. Note that formulas indicated as discharged from open assumptions can have other occurrences in the contexts.

The definition makes formal the observation in Section 1.2 that logical rules do not act on formulas or even assertions, but on derivations. **Derivability** of a formula A from open assumptions Γ naturally means that there is a derivation.

In the definition, the formula with the connective or quantifier in the elimination rules is the **major premiss** of the inference and the antecedent of implication in $\supset E$, the **minor premiss**. The discharged formulas in elimination rules are often referred to as "auxiliary" assumptions and the derivations in which they are made as "auxiliary" derivations. Conjunction and disjunction eliminations have special cases in which $A = B$.

Definition 8.1.3:
    (i) *The* height *of a derivation is the greatest number of consecutive rules of inference in it.*
   (ii) *A discharge is* vacuous *if in Definition 8.1.2* $m = 0$ *or* $n = 0$.
  (iii) *A discharge is* multiple *if in Definition 8.1.2* $m > 1$ *or* $n > 1$.

Theorem 8.1.4: Composition of derivations. *If*

$$
\begin{array}{cc}
\Gamma & A, \Delta \\
\vdots & \vdots \\
A & C
\end{array}
$$

are derivations of A from Γ *and of C from* A, Δ, *respectively, with disjoint discharge labels and with no clashes of free variables, then*

$$
\begin{array}{c}
\Gamma \\
\vdots \\
A, \Delta \\
\vdots \\
C
\end{array}
$$

is a derivation of C from Γ, Δ.

Proof: The proof is by induction on the height of the given derivation of C from A, Δ. If it is 0, then $C = A$ and Δ is empty, so the second derivation is A. The composition of derivations is the same as the first derivation. In the inductive case, the proof is according to the last rule used in deriving C from A, Γ, and there are 12 cases. For each case, the inductive hypothesis is applied to the derivations of the premisses of the last rule, and then the rule is applied.   QED.

In practice, labels and variables are renamed if the conditions regarding them are not met. The property of derivations stated by the theorem is often referred to as **closure under substitution**. When derivations in natural deduction are written in sequent calculus style, as in the examples of Section 1.2, composition of derivations can be expressed by the rule of substitution:

$$\frac{\Gamma \vdash A \quad A, \Delta \vdash C}{\Gamma, \Delta \vdash C} \; Subst$$

This rule resembles cut, but is different in nature: Closure under substitution just states that substitution through the putting together of derivations produces a correct derivation. This is seen clearly from the proof of admissibility of substitution. In natural deduction in sequent calculus style, there are no principal formulas in the antecedent, and therefore the substitution formula in the right premiss also appears in at least some premiss of the rule concluding the right premiss. Substitution is permuted up until the right premiss is an assumption. Elimination of substitution is very different from the elimination of cut.

By Theorem 8.1.4, the practice of pasting together derivations in natural deduction is justified. This is not perhaps clear a priori: Consider the reverse of cutting a derivation into two pieces at any formula in it. The two parts will not usually be formal derivations as defined in 8.1.2, because of the nonlocal character of natural deduction derivations. The contexts $\Gamma, \Delta, \ldots$ in the rules of natural deduction must be arbitrary for the compositionality of derivations to obtain. They must also be independent: With shared contexts, as in the **G3** sequent calculi, Theorem 8.1.4 would fail.

Substitution produces a non-normality whenever in $A, \Delta \vdash C$ the formula A is a major premiss of an elimination rule.

The multiplicity of open assumptions grows in general exponentially in the composition of derivations. This is exemplified by the composition of a derivation of A from Γ and of C from A^m, Δ, by the application of the composition of Theorem 8.1.4 m times:

$$
\begin{array}{cc}
\Gamma & \Gamma \\
\vdots & \vdots \\
A, & \overset{m\times}{\ldots}, A, \Delta \\
& \vdots \\
& C
\end{array}
$$

The composition gives a derivation of C from Γ^m, Δ.

8.2. TRANSLATION FROM SEQUENT CALCULUS TO NATURAL DEDUCTION

We shall give an inductive definition of a translation from cut-free derivations in the sequent calculus **G0i** of Section 5.1 to natural deduction derivations with general elimination rules. As mentioned, it is sometimes thought that natural deduction would not be able to express the rule of weakening and therefore derivability in natural deduction is defined as: C is derivable from Γ if there is a derivation with open assumptions contained in Γ. We shall instead consider the formal derivability relation of natural deduction, Definition 8.1.2, and only translate sequent calculus derivations in which all formulas principal in weakening or contraction are **used** in a logical rule:

Definition 8.2.1: *A formula in a sequent calculus derivation is* used *if it is active in an antecedent in a logical rule.*

Rules that use a formula make it disappear from an antecedent. In natural deduction, this corresponds to the discharge of assumptions, and a count of the assumption labels in the translated derivation will tell if there were weakenings or contractions in the sequent calculus derivation.

(a) The translation: The translation from cut-free sequent calculus derivations in **G0i** will be defined for derivations that contain no unused weakening or contraction formulas. The translation starts with the last step and works root-first step by step until it reaches axioms or instances of $L\bot$. The translation produces labels whenever formulas are used. We also add square brackets and treat labeled and bracketed formulas in the same way as other formulas when continuing the translation. The natural deduction derivation comes out from the translation all finished. To satisfy Principle 8.1.1, each rule that discharges assumptions must have fresh discharge labels. Below, in each case of translation, we write the result of the first step of translation with a rule in natural deduction notation and the premises from which the translation continues in sequent calculus notation, except that formulas in antecedents may appear with brackets and labels.

In a derivation with no unused weakenings or contractions, the last rule is a logical one, and we therefore begin with derivations that end with a logical rule:

Translation of logical rules:

$$\frac{A, B, \Gamma \overset{\vdots}{\Rightarrow} C}{A \& B, \Gamma \Rightarrow C} L\& \quad \leadsto \quad \frac{A \& B \quad \overset{1.}{[A]}, \overset{2.}{[B]}, \Gamma \overset{\vdots}{\Rightarrow} C}{C} \&E, 1., 2.$$

$$\frac{\Gamma \overset{\vdots}{\Rightarrow} A \quad \Delta \overset{\vdots}{\Rightarrow} B}{\Gamma, \Delta \Rightarrow A \& B} R\& \quad \leadsto \quad \frac{\Gamma \overset{\vdots}{\Rightarrow} A \quad \Delta \overset{\vdots}{\Rightarrow} B}{A \& B} \&I$$

$$\frac{A, \Gamma \overset{\vdots}{\Rightarrow} C \quad B, \Delta \overset{\vdots}{\Rightarrow} C}{A \vee B, \Gamma, \Delta \Rightarrow C} L\vee \quad \leadsto \quad \frac{A \vee B \quad \overset{1.}{[A]}, \Gamma \overset{\vdots}{\Rightarrow} C \quad \overset{2.}{[B]}, \Delta \overset{\vdots}{\Rightarrow} C}{C} \vee E, 1., 2.$$

$$\frac{\Gamma \overset{\vdots}{\Rightarrow} A}{\Gamma \Rightarrow A \vee B} R\vee_1 \quad \leadsto \quad \frac{\Gamma \overset{\vdots}{\Rightarrow} A}{A \vee B} \vee I_1 \qquad \frac{\Gamma \overset{\vdots}{\Rightarrow} B}{\Gamma \Rightarrow A \vee B} R\vee_2 \quad \leadsto \quad \frac{\Gamma \overset{\vdots}{\Rightarrow} B}{A \vee B} \vee I_2$$

$$\frac{\Gamma \overset{\vdots}{\Rightarrow} A \quad B, \Delta \overset{\vdots}{\Rightarrow} C}{A \supset B, \Gamma, \Delta \Rightarrow C} \, L\supset \quad \rightsquigarrow \quad \frac{A \supset B \quad \Gamma \overset{\vdots}{\Rightarrow} A \quad \overset{1.}{[B]}, \Delta \overset{\vdots}{\Rightarrow} C}{C} \, \supset E,1.$$

$$\frac{A, \Gamma \overset{\vdots}{\Rightarrow} B}{\Gamma \Rightarrow A \supset B} \, R\supset \quad \rightsquigarrow \quad \frac{\overset{1.}{[A]}, \Gamma \overset{\vdots}{\Rightarrow} B}{A \supset B} \, \supset I,1.$$

$$\frac{A(t/x), \Gamma \overset{\vdots}{\Rightarrow} C}{\forall x A, \Gamma \Rightarrow C} \, L\forall \quad \rightsquigarrow \quad \frac{\forall x A \quad \overset{1.}{[A(t/x)]}, \Gamma \overset{\vdots}{\Rightarrow} C}{C} \, \forall E,1.$$

$$\frac{\Gamma \overset{\vdots}{\Rightarrow} A(y/x)}{\Gamma \Rightarrow \forall x A} \, R\forall \quad \rightsquigarrow \quad \frac{\Gamma \overset{\vdots}{\Rightarrow} A(y/x)}{\forall x A} \, \forall I$$

$$\frac{A(y/x), \Gamma \overset{\vdots}{\Rightarrow} C}{\exists x A, \Gamma \Rightarrow C} \, L\exists \quad \rightsquigarrow \quad \frac{\exists x A \quad \overset{1.}{[A(y/x)]}, \Gamma \overset{\vdots}{\Rightarrow} C}{C} \, \exists E,1.$$

$$\frac{\Gamma \overset{\vdots}{\Rightarrow} A(t/x)}{\Gamma \Rightarrow \exists x A} \, R\exists \quad \rightsquigarrow \quad \frac{\Gamma \overset{\vdots}{\Rightarrow} A(t/x)}{\exists x A} \, \exists I$$

Translation of weakening:

$$\frac{\Gamma \overset{\vdots}{\Rightarrow} C}{\overset{n.}{[A]}, \Gamma \Rightarrow C} \, Wk \quad \rightsquigarrow \quad \Gamma \overset{\vdots}{\Rightarrow} C$$

Translation of contraction:

$$\frac{A, A, \Gamma \overset{\vdots}{\Rightarrow} C}{\overset{n.}{[A]}, \Gamma \Rightarrow C} \, Ctr \quad \rightsquigarrow \quad \overset{n.}{[A]}, \overset{n.}{[A]}, \Gamma \overset{\vdots}{\Rightarrow} C$$

Translation of axioms and $L\bot$:

$$A \Rightarrow A \quad \rightsquigarrow \quad A \qquad \frac{}{\bot \Rightarrow C} L\bot \quad \rightsquigarrow \quad \frac{\bot}{C} \bot E$$

By the assumption of no unused weakening or contraction formulas, the translation can reach only weakening or contraction formulas indicated as discharged by square brackets. The topsequents of derivations are axioms or instances of $L\bot$. If the translation arrives at these sequents and they do not have labels, their antecedents turn into **open** assumptions of the natural deduction derivation. When a formula is used, the translation produces formulas with labels and we can reach topsequents $\overset{n.}{[A]} \Rightarrow A$ and $\overset{n.}{[\bot]} \Rightarrow C$ with a label in the antecedent. These are translated into $[A]$ and $\frac{[\bot]}{C} \bot E$, with discharged assumptions. Note that if a labeled formula gets decomposed further up in the derivation, the labeled formula itself becomes a major premiss of an elimination rule that has been assumed. The components, instead, do not inherit that label but only those indicated in the above translations. The translation produces derivations in which the major premisses of elimination rules always are (open or discharged) assumptions:

Definition 8.2.2: *A derivation in natural deduction is in* full normal form *if all major premisses of E-rules are assumptions.*

We shall refer to such derivations briefly as normal. Note that $\bot$ in $\bot E$ is counted as a major premiss of an E-rule.

The translation from sequent calculus to natural deduction is an algorithm that works its way up from the endsequent in a local way, reflecting the local character of sequent calculus rules. It produces syntactically correct derivation trees with discharges fully formalized. The variable restrictions in rules $\forall I$ and $\exists E$ follow from those in rules $R\forall$ and $L\exists$. The translation of derivations with cuts will be treated in Section 8.4.

(b) The meaning of weakening and contraction: The translation of applications of the rule of weakening into natural deduction may seem somewhat surprising, but it will lead to a useful insight about the nature of this rule. Natural deduction rules permit the discharge of formulas that have not occurred in a derivation. Similarly, natural deduction rules permit the discharge of any number of occurrences of an assumption, not just the occurrence indicated in the schematic rule. Unfolding Definition 8.2.1, we have:

Observation 8.2.3: *Rule $\supset I$ and the elimination rules produce a* vacuous (multiple) *discharge whenever one of the following occurs:*

1. In $\supset I$ concluding $A \supset B$, no occurrence (more than one occurrence) of assumption A was discharged.

2. In $\&E$ and $\vee E$ with major premisses $A\&B$ and $A \vee B$, no occurrence of A or B (more than one occurrence of A or B, or more than two if $A = B$) was discharged.

3. In $\supset E$ with major premiss $A \supset B$, no (more than one) occurrence of B was discharged.

4. In $\forall E$ and $\exists E$ with major premiss $\forall x A$ or $\exists x A$, no (more than one) occurrence of $A(t/x)$ or $A(y/x)$ was discharged.

A **weakening formula** (respectively, **contraction formula**) is a formula A introduced by weakening (contraction) in a derivation. There can be applications of weakening and contraction that have no correspondence in natural deduction: Whenever we have a derivation with a weakening or contraction formula A that is not used, the endsequent is of the form $A, \Gamma \Rightarrow C$, where A is an **inactive** weakening or contraction formula throughout.

The condition of no inactive weakening or contraction formulas in a sequent calculus derivation permits a correspondence with the formal derivability relation of natural deduction:

Theorem 8.2.4: *Given a derivation of $\Gamma \Rightarrow C$ in **G0i** with no inactive weakening or contraction formulas, there is a natural deduction derivation of C from open assumptions Γ.*

Proof: The proof is by induction on the height of the given derivation and uses the translation from sequent calculus. If $\Gamma \Rightarrow C$ is an axiom or instance of $L\perp$, $\Gamma = C$ or $\Gamma = \perp$ and the translation gives the natural deduction derivations C and $\frac{\perp}{C}\perp E$ with open assumptions C and $\perp$, respectively. If the last rule is $L\&$, we have $\Gamma = A\&B, \Gamma'$, and the translation gives

$$\frac{A\&B \quad \overset{1.}{[A]}, \overset{2.}{[B]}, \Gamma' \overset{\vdots}{\Rightarrow} C}{C} \&E,1.,2.$$

If there are no inactive weakenings or contractions in the derivation of $A, B, \Gamma' \Rightarrow C$, there is by inductive hypothesis a natural deduction derivation of C from open assumptions A, B, Γ'. Now assume $A\&B$ and apply $\&E$ to obtain a derivation of C from $A\&B, \Gamma'$.

If there is an inactive weakening or contraction formula in the derivation of $A, B, \Gamma' \Rightarrow C$, it is by assumption not in Γ', so it is A or B or both. Deleting the weakenings and contractions with unused formulas, we obtain a derivation of $A^m, B^n, \Gamma' \Rightarrow C$, with $m, n \geqslant 0$ copies of A and B, respectively. By the inductive hypothesis, there is a corresponding natural deduction derivation with open assumptions A^m, B^n, Γ'. Application of $\&E$ now gives a derivation of C from $A\&B, \Gamma'$. All the other cases of logical rules are dealt with similarly.

The last step cannot be weakening or contraction by the assumption about no inactive weakening or contraction formulas. QED.

By the translation, the natural deduction derivation in Theorem 8.2.4 is normal. Later we show the converse of the theorem. Equivalence of derivability between sequent calculus and natural deduction applies only if unused weakenings and contractions are absent. The usual accounts of translation from sequent calculus to natural deduction pass silently over such problems, by use of a metamathematical derivability relation for natural deduction instead of the formal one.

Theorem 8.2.5: *Given a derivation of $\Gamma \Rightarrow C$ in **G0i** with no inactive weakening or contraction formulas, if A is a weakening (contraction) formula in the* *derivation, then A is vacuously (multiply) discharged in the translation to a natural deduction derivation.*

Proof: Formula A can be used in left rules and $R\supset$ only. In the translation to natural deduction, A becomes a labeled formula in the antecedent. It disappears when a weakening with A is reached and is multiplied when a contraction on A is reached.   QED.

If a derivation of $\Gamma \Rightarrow C$ contains unused weakenings or contractions, we can delete them to obtain a derivation of $\Gamma^* \Rightarrow C$ such that each formula in Γ^* also occurs in Γ. Then Γ^* is a multiset reduct of Γ as defined in 5.2.1. Now the translation to natural deduction can be applied to $\Gamma^* \Rightarrow C$.

Sometimes one sees systems of natural deduction with explicit weakening and contraction rules. They have the same effect as a metamathematical derivability relation, and we shall not use them.

Perhaps the simplest example of a derivation with weakening is, with the corresponding natural deduction obtained through translation at right,

$$
\dfrac{\dfrac{\dfrac{A \Rightarrow A}{A, B \Rightarrow A}\, Wk}{A\&B \Rightarrow A}\, L\&}{\Rightarrow A\&B \supset A}\, R\supset
\qquad \rightsquigarrow \qquad
\dfrac{\dfrac{\overset{1.}{[A\&B]} \quad \overset{2.}{[A]}}{A}\, \&E,2.,3.}{A\&B \supset A}\, \supset I,1.
$$

In the natural deduction derivation, B is vacuously discharged. The translation produces the "ghost" label 3 to which no open assumption corresponds. An intermediate stage of the translation just before the disappearance of the weakening formula is

$$
\dfrac{\overset{1.}{[A\&B]} \quad \dfrac{A \Rightarrow A}{\overset{2.}{[A]}, \overset{3.}{[B]} \Rightarrow A}\, Wk}{\dfrac{A}{A\&B \supset A}\, \supset I,1.}\, \&E,2.,3.
$$

In Gentzen's original sequent calculus there were two left rules for conjunction:

$$
\dfrac{A, \Gamma \Rightarrow C}{A\&B, \Gamma \Rightarrow C}\, L\&_1
\qquad\qquad
\dfrac{B, \Gamma \Rightarrow C}{A\&B, \Gamma \Rightarrow C}\, L\&_2
$$

These left rules correspond to the usual elimination rules for conjunction, and the derivation of $A\&B \supset A$ and its translation become

$$
\dfrac{\dfrac{A \Rightarrow A}{A\&B \Rightarrow A}\, L\&_1}{\Rightarrow A\&B \supset A}\, R\supset
\qquad \rightsquigarrow \qquad
\dfrac{\dfrac{\overset{1.}{[A\&B]}}{A}\, \&E_1}{A\&B \supset A}\, \supset I,1.
$$

Weakening is hidden in Gentzen's left conjunction rules and vacuous discharge in the special conjunction elimination rules. It is not possible to state fully the meaning of weakening in terms of natural deduction without using the general elimination rules.

The premiss of a contraction step in **G0i** can arise in three ways: First, the duplication A, A comes from a rule with two premisses, each having one occurrence of A. Second, A is the principal formula of a left rule and a premiss had A already in the antecedent. Third, weakening is applied to a premiss having A in the antecedent. Only the first two have a correspondence in natural deduction.

The simplest example of a multiple discharge should be the derivation of $A \supset A \& A$, given here both in **G0i** with a contraction and in a translation to natural deduction with a double discharge:

$$
\cfrac{\cfrac{\cfrac{A \Rightarrow A \quad A \Rightarrow A}{A, A \Rightarrow A\&A} \, R\&}{A \Rightarrow A\&A} \, Ctr}{\Rightarrow A \supset A\&A} \, R\supset
\qquad \rightsquigarrow \qquad
\cfrac{\cfrac{\overset{1.}{[A]} \quad \overset{1.}{[A]}}{A\&A} \, \&I}{A \supset A\&A} \, \supset I,1.
$$

In Definition 8.2.3, the clause about more than two occurrences of the discharged formula in $\&E$ and $\vee E$, in case of $A = B$, is exemplified by the derivation of $A \vee A \supset A$ in sequent calculus and its translation:

$$
\cfrac{\cfrac{A \Rightarrow A \quad A \Rightarrow A}{A \vee A \Rightarrow A} \, L\&}{\Rightarrow A \vee A \supset A} \, R\supset
\qquad \rightsquigarrow \qquad
\cfrac{\cfrac{\overset{1.}{[A \vee A]} \quad \overset{2.}{[A]} \quad \overset{3.}{[A]}}{A} \, \vee E,2.,3.}{A \vee A \supset A} \, \supset I,1.
$$

Here there is no contraction even if two occurrences of A are discharged at $\vee E$.

Often in the literature one sees translations of $L\&$ and $L\supset$ with the "dotted" inference below the inference line,

$$
\cfrac{A\&B}{A} \qquad \cfrac{\cfrac{A\&B}{B}}{\underset{C}{\vdots}} \qquad \cfrac{\cfrac{A \supset B \quad A}{B}}{\underset{C}{\vdots}}
$$

but these have the effect of confounding different sequent calculus derivations. When general elimination rules are used, the order of rules in a sequent calculus derivation is reflected in natural deduction. Consider, for example, the derivations

$$
\cfrac{\cfrac{\cfrac{A \Rightarrow A}{A, B \Rightarrow A} \, Wk}{A\&B \Rightarrow A} \, L\& \quad \cfrac{\cfrac{B \Rightarrow B}{A, B \Rightarrow B} \, Wk}{A\&B \Rightarrow B} \, L\&}{A\&B, A\&B \Rightarrow A\&B} \, R\&
\qquad
\cfrac{\cfrac{A \Rightarrow A \quad B \Rightarrow B}{A, B \Rightarrow A\&B} \, R\&}{A\&B \Rightarrow A\&B} \, L\&
$$

The usual translation does not distinguish between the two derivations, but gives for both the natural deduction derivation

$$\dfrac{\dfrac{A\&B}{A}\&E \quad \dfrac{A\&B}{B}\&E}{A\&B}\&I$$

We get instead the two translations

$$\dfrac{\dfrac{A\&B \quad [A]^{1.}}{A}\&E,1. \quad \dfrac{A\&B \quad [B]^{2.}}{B}\&E,2.}{A\&B}\&I \qquad \dfrac{A\&B \quad \dfrac{[A]^{1.} \quad [B]^{2.}}{A\&B}\&I}{A\&B}\&E,1.,2.$$

We have defined the translation root-first, rule after rule, and the order of logical rules in the natural deduction derivation is the same as that in the sequent calculus derivation.

(c) Translation from sequent calculus in natural deduction style: The above translation works on derivations in the calculus **G0i**. A translation from the sequent calculus in natural deduction style **GN** to natural deduction is simpler as there are no structural rules to be translated. The translation differs from the above only with rules that use assumptions. Rule $L\&$ is translated by

$$\dfrac{A^m, B^n, \Gamma \Rightarrow C}{A\&B, \Gamma \Rightarrow C}L\& \quad \rightsquigarrow \quad \dfrac{A\&B \quad \overset{\vdots}{[A^m]^{1.}, [B^n]^{2.}, \Gamma \Rightarrow C}}{C}\&E,1.,2.$$

where m and n occurrences of A and B, respectively, are turned into discharged assumptions. The other rules are translated in the same way.

A translation from **G3i** to natural deduction is obtained by use of the connection between **G3i** and **G0i** or **GN** of Section 5.2(c). Proof editor PESCA produces natural deduction derivations through proof-search in **G3i** and a translation to natural deduction, as in the example of Section C.2(a).

8.3. TRANSLATION FROM NATURAL DEDUCTION TO SEQUENT CALCULUS

We first define a translation from natural deduction to the calculus **G0i**, then indicate how derivations in **GN** are obtained through a simplification of the translation, and last consider the translation of the special elimination rules of natural deduction.

(a) The translation: Translation from fully normal natural deduction derivations with unique discharge to the calculus **G0i** is defined inductively according to the last rule used:

1. The last rule is &I:

$$
\frac{\begin{array}{cc} \Gamma & \Delta \\ \vdots & \vdots \\ A & B \end{array}}{A\&B} \,\&I \qquad \rightsquigarrow \qquad \frac{\begin{array}{cc} \Gamma & \Delta \\ \vdots & \vdots \\ A & B \end{array}}{\Gamma, \Delta \Rightarrow A\&B} \,R\&
$$

2. The last rule is &E: The natural deduction derivation is

$$
\frac{A\&B \qquad \begin{array}{c} \overset{1.}{[A^m]}, \overset{2.}{[B^n]}, \Gamma \\ \vdots \\ C \end{array}}{C} \,\&E,1.,2.
$$

The translation is by cases according to values of m and n:

$m = 0, n = 0$:

$$
\frac{\dfrac{\dfrac{\begin{array}{c} \Gamma \\ \vdots \\ C \end{array}}{A, \Gamma \Rightarrow C} \,Wk}{A, B, \Gamma \Rightarrow C} \,Wk}{A\&B, \Gamma \Rightarrow C} \,L\&
$$

$m = 1, \ n = 1$:

$$
\frac{\begin{array}{c} A, B, \Gamma \\ \vdots \\ C \end{array}}{A\&B, \Gamma \Rightarrow C} \,L\&
$$

Note that the closed assumptions have been opened anew by removal of the discharge labels and brackets. The cases of $m = 1, n = 0$ and $m = 0, n = 1$ have one weakening step before the $L\&$ inference.

$m > 1, n = 0$:

$$
\frac{\dfrac{\dfrac{\begin{array}{c} A^m, \Gamma \\ \vdots \\ C \end{array}}{A, \Gamma \Rightarrow C} \,Ctr^*}{A, B, \Gamma \Rightarrow C} \,Wk}{A\&B, \Gamma \Rightarrow C} \,L\&
$$

Here Ctr^* indicates an $m-1$ fold contraction, and m occurrences of the closed assumption A have been opened. The rest of the cases for $\&E$ are similar.

3. *The last rule is* $\vee I$:

$$\cfrac{\cfrac{\Gamma \\ \vdots \\ A}{A \vee B}}{}\vee I_1 \quad \rightsquigarrow \quad \cfrac{\cfrac{\Gamma \\ \vdots \\ A}{\Gamma \Rightarrow A \vee B}}{}R\vee_1 \qquad \cfrac{\cfrac{\Gamma \\ \vdots \\ B}{A \vee B}}{}\vee I_2 \quad \rightsquigarrow \quad \cfrac{\cfrac{\Gamma \\ \vdots \\ B}{\Gamma \Rightarrow A \vee B}}{}R\vee_2$$

4. *The last rule is* $\vee E$: The natural deduction derivation is

$$\cfrac{A \vee B \qquad \overset{1.}{[A^m]},\ \Gamma \quad \overset{2.}{[B^n]},\ \Delta \\ \qquad \begin{array}{cc} \vdots & \vdots \\ C & C \end{array}}{C}\vee E,1.,2.$$

and the translation is again by cases according to the values of m and n, as in 2. We shall indicate by Str the appropriate weakening and contraction steps. The case without any such steps is when $m = 1, n = 1$:

$$\cfrac{\begin{array}{cc} A, \Gamma & B, \Delta \\ \vdots & \vdots \\ C & C \end{array}}{A \vee B, \Gamma, \Delta \Rightarrow C}L\vee$$

The closed assumptions $[A]$ and $[B]$ have been opened. The general case is

$$\cfrac{\cfrac{\begin{array}{c} A^m, \Gamma \\ \vdots \\ C \end{array}}{A, \Gamma \Rightarrow C}Str \qquad \cfrac{\begin{array}{c} B^n, \Delta \\ \vdots \\ C \end{array}}{B, \Delta \Rightarrow C}Str}{A \vee B, \Gamma, \Delta \Rightarrow C}L\vee$$

5. *The last rule is* $\supset I$: The general case is translated by

$$\cfrac{\overset{1.}{[A^m]},\ \Gamma \\ \vdots \\ \cfrac{B}{A \supset B}}{}\supset I,1. \qquad \rightsquigarrow \qquad \cfrac{\cfrac{\begin{array}{c} A^m, \Gamma \\ \vdots \\ B \end{array}}{A, \Gamma \Rightarrow B}Str}{\Gamma \Rightarrow A \supset B}R\supset$$

Again closed assumptions have been opened. If $m = 1$, there is just the $R\supset$ rule.

6. *The last rule is $\supset E$:* The general case is translated as

$$
\dfrac{A \supset B \quad A \quad \begin{array}{c} \Gamma \quad \overset{1.}{[B^n]}, \Delta \\ \vdots \\ C \end{array}}{C} \supset E, 1. \quad \rightsquigarrow \quad \dfrac{\dfrac{\begin{array}{c}\Gamma \\ \vdots \\ A \end{array}}{\Gamma \Rightarrow A} \quad \dfrac{\dfrac{\begin{array}{c} B^n, \Delta \\ \vdots \\ C \end{array}}{B, \Delta \Rightarrow C} Str}{}}{A \supset B, \Gamma, \Delta \Rightarrow C} L\supset
$$

7. *The last rule is $\forall I$:*

$$
\dfrac{\begin{array}{c}\Gamma \\ \vdots \\ A(y/x) \end{array}}{\forall x A} \forall I \quad \rightsquigarrow \quad \dfrac{\begin{array}{c}\Gamma \\ \vdots \\ A(y/x) \end{array}}{\Gamma \Rightarrow \forall x A} R\forall
$$

8. *The last rule is $\forall E$:* The general case is translated as

$$
\dfrac{\forall x A \quad \begin{array}{c} \overset{1.}{[A(t/x)^m]}, \Gamma \\ \vdots \\ C \end{array}}{C} \forall E, 1. \quad \rightsquigarrow \quad \dfrac{\dfrac{\begin{array}{c} A(t/x)^m, \Gamma \\ \vdots \\ C \end{array}}{A(t/x), \Gamma \Rightarrow C} Str}{\forall x A, \Gamma \Rightarrow C} L\forall
$$

9. *The last rule is $\exists I$:*

$$
\dfrac{\begin{array}{c}\Gamma \\ \vdots \\ A(t/x) \end{array}}{\exists x A} \exists I \quad \rightsquigarrow \quad \dfrac{\begin{array}{c}\Gamma \\ \vdots \\ A(t/x) \end{array}}{\Gamma \Rightarrow \exists x A} R\exists
$$

10. *The last rule is $\exists E$:* The general case is translated as

$$
\dfrac{\exists x A \quad \begin{array}{c} \overset{1.}{[A(y/x)^m]}, \Gamma \\ \vdots \\ C \end{array}}{C} \forall E, 1. \quad \rightsquigarrow \quad \dfrac{\dfrac{\begin{array}{c} A(y/x)^m, \Gamma \\ \vdots \\ C \end{array}}{A(y/x), \Gamma \Rightarrow C} Str}{\exists x A, \Gamma \Rightarrow C} L\exists
$$

11. *The last rule is the rule of assumption:*

$$
A \quad \rightsquigarrow \quad A \Rightarrow A
$$

12. *The last rule is $\perp E$:*

$$
\dfrac{\perp}{C} \perp E \quad \rightsquigarrow \quad \dfrac{}{\perp \Rightarrow C} L\perp
$$

Note that in a fully normal derivation, the premiss of rule $\perp E$ is an assumption and nothing remains to be translated in step *12*. If in *11* or *12* there are discharges they are undone.

Theorem 8.3.1: *Given a fully normal natural deduction derivation of C from open assumptions* Γ, *there is a derivation of* $\Gamma \Rightarrow C$ *in* **G0i**.

Proof: By the translation defined. QED.

There are no unused weakenings or contractions in the derivation of $\Gamma \Rightarrow C$. By the translation, we obtain the converse of Theorem 8.2.5:

Theorem 8.3.2: *If A is vacuously (multiply) discharged in the derivation of C from open assumptions* Γ, *then A is a weakening (contraction) formula in the derivation of* $\Gamma \Rightarrow C$ *in* **G0i**.

The usual explanation of contraction runs something like this: "If you can derive a formula using assumption A twice, you can also derive it using A only once." But this is just a verbal statement of the rule of contraction. Logical rules of natural deduction that discharge assumptions vacuously or multiply are reproduced as weakenings or contractions plus a logical rule in sequent calculus. However, the weakening and contraction rules in themselves have no proof-theoretical meaning, as was pointed out by Gentzen (1936, pp. 513–14) already.

By the translation of a normal derivation in natural deduction to sequent calculus, each formula in the former appears in the latter. We therefore have, by the subformula property of **G0i**, a somewhat surprising proof of

Corollary 8.3.3: Subformula property. *In a normal derivation of C from open assumptions* Γ, *each formula in the derivation is a subformula of* Γ, C.

The translation of non-normal derivations will be given in Section 8.4.

(b) Isomorphic translation: The translations we have defined from natural deduction to the sequent calculus **G0i** and the other way around do not quite establish an isomorphism between the two: It is possible to permute weakenings and contractions on a formula A as long as A remains inactive so that isomorphism obtains modulo such permutations. This is a minor point that we could circumvent by adding to the requirement of no unused weakening or contraction the "last-minute" condition that that there must be no other logical rule between a weakening or contraction and the logical rules in which the weakening or contraction formula is used. Another way is to translate directly to the calculus **GN** that has no explicit weakening or contraction rules. This also dispenses with the cases on n and m, and vacuous and multiple discharges are turned into vacuous and multiple uses in perfect reverse to the translation from **GN** to natural deduction.

Normal natural deduction derivations and cut-free sequent calculus derivations differ only in notation.

A translation from natural deduction to **G3i** is obtained by the connection between **G3i** and **G0i** or **GN**.

(c) Translation of the special elimination rules: The translations of the special elimination rules of conjunction lead to the following sequent calculus rules:

$$\frac{}{A\&B, \Gamma \Rightarrow A}\,{}^{L\&S_1} \qquad \frac{}{A\&B, \Gamma \Rightarrow B}\,{}^{L\&S_2}$$

These zero-premiss rules are obtained from rule $L\&$ as special cases by setting $C = A$ and $C = B$, respectively, translating them, and deleting the premisses $A, B, \Gamma \Rightarrow A$ and $A, B, \Gamma \Rightarrow B$ that are derivable from $A \Rightarrow A$ and $B \Rightarrow B$ by weakening.

Consider the following derivation of $(A\&B)\&C \Rightarrow A$ with the special rules $L\&S$:

$$\frac{\dfrac{}{(A\&B)\&C \Rightarrow A\&B}\,{}^{L\&S_1} \quad \dfrac{}{A\&B \Rightarrow A}\,{}^{L\&S_1}}{(A\&B)\&C \Rightarrow A}\,{}^{Cut}$$

The conclusion is not an instance of rule $L\&S$, and therefore cut elimination fails in this case. We can rewrite the corresponding natural deduction derivation in terms of the general $\&E$ rule and then convert it into normal form, but the resulting derivation is not of the form of the special rules anymore.

The sequent calculus rule corresponding to modus ponens is

$$\frac{\Gamma \Rightarrow A}{A \supset B, \Gamma \Rightarrow B}\,{}^{L\supset S}$$

It can be obtained from rule $L\supset$ by setting $C = B$ and deleting the right premiss that is derivable. An example of failure of cut elimination when this special rule is used is given by

$$\frac{\dfrac{A \Rightarrow A}{A \supset (B \supset C), A \Rightarrow B \supset C}\,{}^{L\supset S} \quad \dfrac{B \Rightarrow B}{B \supset C, B \Rightarrow C}\,{}^{L\supset S}}{A \supset (B \supset C), A, B \Rightarrow C}\,{}^{Cut}$$

where the conclusion is not an instance of $L\supset S$ and cannot be obtained without cut. Thus we see that the use of special elimination rules in natural deduction involves "hidden cuts."

In Gentzen's original work, a translation of natural deduction derivations into sequent calculus is described (1934–35, sec. V. 4). Each formula C is first replaced by a sequent $\Gamma \Rightarrow C$, where Γ is a list of open assumptions C depends on, and then the rules are translated. Rules $\&I$ and $\vee I$ are translated in the obvious way.

Translations of $\supset I$ and $\vee E$ involve possible weakenings and contractions, corresponding to vacuous and multiple discharges. Whenever in the natural deduction there are instances of $\& E$ and $\supset E$, the first phase of the translation gives steps such as

$$\frac{\Gamma \Rightarrow A\&B}{\Gamma \Rightarrow A} \qquad \frac{\Gamma \Rightarrow A \supset B \quad \Delta \Rightarrow A}{\Gamma, \Delta \Rightarrow B}$$

These are turned into sequent calculus inferences by the following replacements, in which in the first derivation a left conjunction rule in Gentzen's original formulation occurs:

$$\frac{\Gamma \Rightarrow A\&B \quad \dfrac{A \Rightarrow A}{A\&B \Rightarrow A} {\scriptstyle L\&S_1}}{\Gamma \Rightarrow A} {\scriptstyle Cut} \qquad \frac{\Gamma \Rightarrow A \supset B \quad \dfrac{\Delta \Rightarrow A \quad B \Rightarrow B}{A \supset B, \Delta \Rightarrow B} {\scriptstyle L\supset}}{\Gamma, \Delta \Rightarrow B} {\scriptstyle Cut}$$

With the knowledge that the special elimination rules of natural deduction correspond to hidden cuts, it is to be expected that a normal derivation in the old sense translates into a sequent calculus derivation with cuts. In Gentzen's work, the "Hauptsatz" is proved in terms of sequent calculus, and the possibility of a formulation in terms of a normal form in intuitionistic natural deduction is only mentioned. No comment is made about the cuts that the translation of normal derivations to sequent calculus produces.

8.4. DERIVATIONS WITH CUTS AND NON-NORMAL DERIVATIONS

We first define a translation from sequent calculus with cuts of a suitable kind to natural deduction. Then a translation taking any non-normal derivation into a sequent calculus derivation with cuts is defined. The latter, in combination with cut elimination and translation back to a normal derivation, gives a normalization algorithm for natural deduction with general elimination rules.

(a) Derivations with cuts: We show that derivations with cuts can be translated into natural deduction if the cut formula is principal in both premisses or the right premiss. These **detour** cuts and **permutation** cuts are the **principal** cuts; the rest are **nonprincipal** cuts. Principal cuts correspond, in terms of natural deduction, to instances of rules of elimination in which the major premises are not assumptions. We shall call such premisses **conversion formulas**.

A sequent calculus derivation has an equivalent in natural deduction only if it has no unused weakening or contraction formulas. By this criterion, there is no correspondence in natural deduction for many of the nonprincipal cuts of sequent calculus. In particular, if the right premiss of cut has been derived by contraction, the contraction formula is not used in the derivation and there is no corresponding natural deduction derivation. This is precisely the problematic case that led Gentzen to use the rule of multicut. If cut and contraction are permuted, the right premiss of a cut becomes derived by another cut and there is likewise no translation.

In translating derivations with cuts, if the left premiss is an axiom, the cut is deleted. There are five detour cuts and another 25 permutation cuts with left premiss derived by a logical rule to be translated. We also translate principal cuts on $\bot$ as well as cases in which the left premiss has been derived by a structural rule, but derivations with other cases of cuts will not be translated. The translation of rules other than cut have been given in Section 8.2.

1. Detour cut on $A\&B$, and we have the derivation

$$\cfrac{\cfrac{\Gamma \overset{\vdots}{\Rightarrow} A \quad \Delta \overset{\vdots}{\Rightarrow} B}{\Gamma, \Delta \Rightarrow A\&B} \, {}_{R\&} \quad \cfrac{A, B, \Theta \overset{\vdots}{\Rightarrow} C}{A\&B, \Theta \Rightarrow C} \, {}_{L\&}}{\Gamma, \Delta, \Theta \Rightarrow C} \, {}_{Cut}$$

The translation is

$$\cfrac{\cfrac{\Gamma \overset{\vdots}{\Rightarrow} A \quad \Delta \overset{\vdots}{\Rightarrow} B}{A\&B} \, {}_{\&I} \quad \overset{1. \qquad 2.}{[A], [B], \Theta \overset{\vdots}{\Rightarrow} C}}{C} \, {}_{\&E,1.,2.}$$

Translation now continues from the premisses.

2–5. Detour cuts on $A \lor B$, $A \supset B$, $\forall x A$, and $\exists x A$. The translations are analogous to *1*, with the left and right rules translated as in Section 8.2.

6. Permutation cut on $C\&D$ with left premiss derived by $L\&$:

$$\cfrac{\cfrac{A, B, \Gamma \overset{\vdots}{\Rightarrow} C\&D}{A\&B, \Gamma \Rightarrow C\&D} \, {}_{L\&} \quad \cfrac{C, D, \Delta \overset{\vdots}{\Rightarrow} E}{C\&D, \Delta \Rightarrow E} \, {}_{L\&}}{A\&B, \Gamma, \Delta \Rightarrow E} \, {}_{Cut}$$

The translation is

$$\cfrac{\cfrac{A\&B \quad \overset{1. \quad 2.}{[A], [B], \Gamma \overset{\vdots}{\Rightarrow} C\&D}}{C\&D} \, {}_{\&E,1.,2.} \quad \overset{3. \quad 4.}{[C], [D], \Delta \overset{\vdots}{\Rightarrow} E}}{E} \, {}_{\&E,3.,4.}$$

Permutation cuts on $C\&D$ with left premiss derived by $L\lor$, $L\supset$, $L\forall$, and $L\exists$ are translated analogously, and the same when there are permutation cuts on $C \lor D$, $C \supset D$, $\forall xC$, and $\exists xC$.

7. We also have permutation cuts on $\bot E$ but no detour cuts since $\bot$ can never be principal in the left premiss. The derivation and its translation are, where L stands for a (one-premiss) left rule and E for an elimination,

$$\cfrac{\cfrac{\vdots}{\cfrac{\Gamma' \Rightarrow \bot}{\Gamma \Rightarrow \bot}\,L} \quad \overline{\bot \Rightarrow C}\,{}^{L\bot}}{\Gamma \Rightarrow C}\,Cut \qquad \rightsquigarrow \qquad \cfrac{\cfrac{\vdots}{\Gamma' \Rightarrow \bot}\,E}{\cfrac{\bot}{C}\,\bot E}$$

8. "Structural" cuts with left premiss derived by weakening, contraction, or cut. For weakening and contraction the translation reaches, by the condition of no unused weakening or contraction formulas, a conclusion of cut of the form $[\overset{n.}{A}], \Gamma, \Delta \Rightarrow C$. These are modified as follows and then the translations are continued:

$$\cfrac{\cfrac{\cfrac{\vdots}{\Gamma \overset{\centerdot}{\Rightarrow} B}\,Wk}{A, \Gamma \Rightarrow B} \quad \cfrac{\vdots}{B, \Delta \overset{\centerdot}{\Rightarrow} C}}{[\overset{n.}{A}], \Gamma, \Delta \Rightarrow C}\,Cut \qquad \rightsquigarrow \qquad \cfrac{\cfrac{\vdots}{\Gamma \overset{\centerdot}{\Rightarrow} B} \quad \cfrac{\vdots}{B, \Delta \overset{\centerdot}{\Rightarrow} C}}{\Gamma, \Delta \Rightarrow C}\,Cut$$

$$\cfrac{\cfrac{\cfrac{\vdots}{A, A, \Gamma \overset{\centerdot}{\Rightarrow} B}\,Ctr}{A, \Gamma \Rightarrow B} \quad \cfrac{\vdots}{B, \Delta \overset{\centerdot}{\Rightarrow} C}}{[\overset{n.}{A}], \Gamma, \Delta \Rightarrow C}\,Cut \qquad \rightsquigarrow \qquad \cfrac{\cfrac{\vdots}{[\overset{n.}{A}], [\overset{n.}{A}], \Gamma \overset{\centerdot}{\Rightarrow} B} \quad \cfrac{\vdots}{B, \Delta \overset{\centerdot}{\Rightarrow} C}}{[\overset{n.}{A}], \Gamma, \Delta \Rightarrow C}\,Cut$$

For left premiss of cut derived by another cut the translation is modular and the upper cut is handled as above.

(b) Non-normal derivations: In translating non-normal derivations into derivations in **G0i**, there are five cases of non-normality in which the major premiss of an elimination rule has been derived by the corresponding introduction rule:

1. The conversion formula has been derived by $\&I$ and the derivation is

$$\cfrac{\cfrac{\cfrac{\Gamma}{\vdots} \quad \cfrac{\Delta}{\vdots}}{\cfrac{A \quad B}{A\&B}\,\&I} \quad \cfrac{\overset{1.}{[A^m]}, \overset{2.}{[B^n]}, \Theta}{\cfrac{\vdots}{C}}}{C}\,\&E,1.,2.$$

The translation is by cases according to values of m and n. The general case is

$$
\cfrac{
  \cfrac{\Gamma \quad \Delta}{
    \cfrac{\begin{array}{cc}\vdots & \vdots\\ A & B\end{array}}{\Gamma, \Delta \Rightarrow A\&B}\,R\&
  }
  \qquad
  \cfrac{
    \cfrac{
      \cfrac{A^m, B^n, \Theta \;\; \vdots \;\; C}{A, B, \Theta \Rightarrow C}\,Str
    }{A\&B, \Theta \Rightarrow C}\,L\&
  }{}
}{\Gamma, \Delta, \Theta \Rightarrow C}\,Cut
$$

2–5. The conversion formula has been derived by $\vee I$, $\supset I$, $\forall I$, or $\exists I$, and the translation is analogous.

If the conversion formula has been derived by an elimination rule, we have again a number of cases:

6. If the rule is $\&E$, the derivation with conversion formula $C\&D$ is

$$
\cfrac{
  \cfrac{\dfrac{A\&B \qquad \overset{\overset{1. \quad\;\; 2.}{[A^m], [B^n], \Gamma}}{\underset{\vdots}{C\&D}}}{C\&D}\,\&E,1.,2.
  \qquad
  \overset{\overset{3. \quad\;\; 4.}{[C^k], [D^l], \Delta}}{\underset{\vdots}{E}}
}{E}\,\&E,3.,4.
$$

The translation is by cases according to values of m, n, k, l, with the general case

$$
\cfrac{
  \cfrac{
    \cfrac{\dfrac{A^m, B^n, \Gamma \;\; \vdots \;\; C\&D}{A, B, \Gamma \Rightarrow C\&D}\,Str}{A\&B, \Gamma \Rightarrow C\&D}\,L\&
  }{}
  \qquad
  \cfrac{
    \cfrac{\dfrac{C^k, D^l, \Delta \;\; \vdots \;\; E}{C, D, \Delta \Rightarrow E}\,Str}{C\&D, \Delta \Rightarrow E}\,L\&
  }{}
}{\Gamma, \Delta \Rightarrow E}\,Cut
$$

If $A\&B$ in turn is a conversion formula, another cut, on $A\&B$, is inserted after the $L\&$ rule that concludes the left premiss of the cut on $C\&D$.

There are altogether 25 cases of translations when the major premiss of an elimination rule has been derived by an elimination rule. All translations are analogous to the above.

Consider a typical principal cut, say, on $A\&B$:

$$
\cfrac{
  \Gamma \Rightarrow A\&B \qquad \cfrac{A, B, \Delta \overset{\vdots}{\Rightarrow} C}{A\&B, \Delta \Rightarrow C}\,L\&
}{\Gamma, \Delta \Rightarrow C}\,Cut
$$

We see that the cut is redundant, in the sense that its left premiss is an axiom, precisely when $A\&B$ is an assumption in the corresponding natural deduction derivation. In this case, the cut is not translated but deleted. We have, in general:

> A non-normal instance of a logical rule in natural deduction is represented in sequent calculus by the corresponding left rule and a cut.

Let us compare this explanation of cut to the presentation of cut as a combination of two lemmas $\Gamma \Rightarrow A$ and $A, \Delta \Rightarrow C$ into a theorem $\Gamma, \Delta \Rightarrow C$. Consider the derivation of C from assumptions A, Δ in natural deduction. Obviously A plays an essential role only if it is analyzed into components by an elimination rule; thus A is a major premiss of that elimination rule. If not, it acts just as a parameter in the derivation. Our explanation of cut makes more precise the idea of cut as a combination of lemmas: In terms of sequent calculus, the cut formula has to be principal in a left rule in the derivation of $A, \Delta \Rightarrow C$.

Given a non-normal derivation, translation to sequent calculus, followed by cut elimination and translation back to natural deduction, will produce a normal derivation:

Theorem 8.4.1: Normalization. *Given a natural deduction derivation of C from Γ, the derivation converts to a normal derivation of C from Γ^* where each formula in Γ^* is a formula in Γ.*

The normalization procedure will not produce a unique result since cut elimination has no unique result.

8.5. THE STRUCTURE OF NORMAL DERIVATIONS

We consider three different ways in which a natural deduction derivation with general elimination rules can fail to be normal, depending on how a major premiss of an elimination rule was derived. Then the subformula structure of normal derivations is detailed, with a direct proof of the subformula property. Last, we give a direct proof of normalization.

(a) Detour conversions: The usual definition of a normal derivation in natural deduction is that no conclusion of an introduction rule must be the major premiss of an elimination rule. Non-normal derivations are transformed into normal ones by **detour conversions** that delete each such pair of introduction and elimination rule instances, in the way shown in Section 1.2. To keep things simple, only the cases with no vacuous or multiple discharges were considered there. In a fully general form, a **detour convertibility** on the formula $A\&B$ obtains in a derivation whenever it has a part of the form

$$
\cfrac{\cfrac{\vdots\quad\vdots}{\cfrac{A\quad B}{A\&B}\&I}\quad\quad \cfrac{[\overset{1.}{A^m}],[\overset{2.}{B^n}]}{\vdots\atop C}}{C}\&E,1.,2.
$$
$$
\vdots
$$

Detour conversion on $A\&B$ gives, through simultaneous substitution, the modified derivation

$$
\cfrac{\vdots\quad\quad\vdots\quad\vdots\quad\quad\vdots}{A,\,\overset{m\times}{\vdots},\,A\quad B,\,\overset{n\times}{\vdots},\,B}
$$
$$
\vdots
$$
$$
C
$$
$$
\vdots
$$

A detour convertibility on $A \vee B$ is quite analogous. For implication, the situation is more complicated since a vacuous or multiple discharge is possible also in the introduction of the conversion formula:

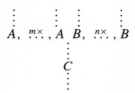

Detour conversion on $A \supset B$ gives the modified derivation

$$
\cfrac{\vdots\quad\quad\vdots\quad\quad\vdots\quad\quad\vdots}{A,\,\overset{m\times}{\vdots},\,A\quad A,\,\overset{m\times}{\vdots},\,A}
$$
$$
\cfrac{\vdots\quad\quad\quad\vdots}{B,\,\overset{n\times}{\vdots},\,B}
$$
$$
\vdots
$$
$$
C
$$
$$
\vdots
$$

Detour convertibilities on $\forall x\, A$ and $\exists x\, A$ are as follows:

$$
\cfrac{\cfrac{A(y/x)}{\forall x\, A}\ {\scriptstyle \forall I}\qquad \cfrac{\overset{\displaystyle \vdots}{\underset{[A(t/x)^m]}{\overset{1.}{}}}}{C}}{C}\ {\scriptstyle \forall E}
\qquad
\cfrac{\cfrac{A(t/x)}{\exists x\, A}\ {\scriptstyle \exists I}\qquad \cfrac{\overset{\displaystyle \vdots}{\underset{[A(y/x)^m]}{\overset{1.}{}}}}{C}}{C}\ {\scriptstyle \exists E}
$$

In the detour convertibility on $\forall x\, A$, the variable restriction on y permits the substitution of x by t in the derivation of $A(y/x)$. The resulting derivation of $A(t/x)$ is composed m times with the derivation of C from $A(t/x)^m$. In the detour convertibility on $\exists x\, A$, since in the auxiliary derivation C was derived from $A(y/x)$ for an arbitrary y, substitution of x by t produces a derivation of C from $A(t/x)^m$. The derivation of $A(t/x)$ is composed m times with it. Thus, detour convertibilities on $\forall x\, A$ and $\exists x\, A$ convert into one and the same derivation:

$$
\cfrac{A(t/x),\ \overset{m\times}{\ldots}\ ,\ A(t/x)}{C}
$$

In detour conversions, the open assumptions typically get multiplied into multiset reducts of the original assumptions, in the way shown in the cases of elimination of principal cuts in the calculus **GN** of Section 5.2. For example, the derivation

$$
\cfrac{\cfrac{A\quad B}{A\& B}\ {\scriptstyle \& I}\qquad \overset{1.}{[A]}}{A}\ {\scriptstyle \& E,1.}
$$

converts into the derivation A. The same result is obtained through translation to sequent calculus:

$$
\cfrac{\cfrac{A\Rightarrow A\quad B\Rightarrow B}{A,B\Rightarrow A\& B}\ {\scriptstyle R\&}\qquad \cfrac{\cfrac{A\Rightarrow A}{A,B\Rightarrow A}\ {\scriptstyle Wk}}{A\& B\Rightarrow A}\ {\scriptstyle L\&}}{A,B\Rightarrow A}\ {\scriptstyle Cut}
$$

Cut elimination produces the derivation

$$
\cfrac{A\Rightarrow A}{A,B\Rightarrow A}\ {\scriptstyle Wk}
$$

Deletion of the unused weakening gives the derivation $A \Rightarrow A$, corresponding to the result of the detour conversion.

(b) Permutation conversions for general elimination rules: Normal derivations with the usual natural deduction rules for conjunction and implication and without disjunctions have a pleasant property: In each step of inference, the formula below is an immediate subformula of a formula above, or the other way around. With disjunction elimination, this simple subformula structure along all branches of a normal derivation tree is lost. On the other hand, if the major premiss of an elimination step is concluded by disjunction elimination, the derivation can be transformed into a still more direct form through a **permutation conversion**. For example, if both steps are disjunction eliminations, we have

$$
\cfrac{
  A \vee B \quad
  \cfrac{
    \overset{\displaystyle \overset{1.}{[A]}}{\vdots} \quad
    \overset{\displaystyle \overset{2.}{[B]}}{\vdots}
  }{C \vee D \qquad C \vee D} {}_{\vee E,1.,2.}
  \qquad
  \cfrac{
    \overset{\displaystyle \overset{3.}{[C]}}{\vdots} \quad
    \overset{\displaystyle \overset{4.}{[D]}}{\vdots}
  }{E \qquad E}
}{E} {}_{\vee E,3.,4.}
$$

This derivation can be transformed into

$$
\cfrac{
  A \vee B \quad
  \cfrac{
    C \vee D \quad
    \overset{\displaystyle \overset{3.}{[C]}}{\vdots} \; E \quad
    \overset{\displaystyle \overset{4.}{[D]}}{\vdots} \; E
  }{E} {}_{\vee E,3.,4.}
  \quad
  \cfrac{
    C \vee D \quad
    \overset{\displaystyle \overset{5.}{[C]}}{\vdots} \; E \quad
    \overset{\displaystyle \overset{6.}{[D]}}{\vdots} \; E
  }{E} {}_{\vee E,5.,6.}
}{E} {}_{\vee E,1.,2.}
$$

by permutation of the second elimination up into the auxiliary derivations of the first elimination. Fresh discharge labels are introduced in accordance with the unique discharge principle. Consider now the possibility that $C \vee D$ in one of the auxiliary derivations of the unpermuted derivation is concluded by $\vee I$. After the permutation conversion, this occurrence of $C \vee D$ is both a major premiss of an elimination rule and a conclusion of an introduction rule. There obtains a "hidden" detour convertibility that becomes an actual one after the permutation conversion.

For predicate logic, there is a permutation conversion for existence elimination. Permutation conversions for disjunction and existence were found by Prawitz in 1965.

The above derivations with disjunction elimination are not fully normal in the sense of Definition 8.2.2. As was shown in the previous section, their translations to sequent calculus are

$$\dfrac{\dfrac{A \Rightarrow C \vee D \quad B \Rightarrow C \vee D}{A \vee B \Rightarrow C \vee D}\,{}_{L\vee} \quad \dfrac{C \Rightarrow E \quad D \Rightarrow E}{C \vee D \Rightarrow E}\,{}_{L\vee}}{A \vee B \Rightarrow E}\,{}_{Cut}$$

and

$$\dfrac{A \Rightarrow C \vee D \quad \dfrac{\dfrac{C \Rightarrow E \quad D \Rightarrow E}{C \vee D \Rightarrow E}\,{}_{L\vee}}{}}{A \Rightarrow E}\,{}_{Cut} \qquad \dfrac{B \Rightarrow C \vee D \quad \dfrac{\dfrac{C \Rightarrow E \quad D \Rightarrow E}{C \vee D \Rightarrow E}\,{}_{L\vee}}{}}{B \Rightarrow E}\,{}_{Cut}$$
$$\dfrac{}{A \vee B \Rightarrow E}\,{}_{L\vee}$$

Thus the conversion of the natural deduction derivation into a more direct form corresponds to a step of cut elimination, where a cut with cut formula principal in the right premiss only is permuted with $L\vee$ to move it upward in the derivation.

The general elimination rules for conjunction, implication, and universal quantification permit the permutation of eliminations up in the same way as with disjunction and existence elimination. Thus the structural properties of derivations with the three special elimination rules are quite different from those with the general elimination rules. To give an example, with the special rules we have the derivation

$$\dfrac{\dfrac{(A\&B)\&C}{A\&B}}{A}$$

With the general rule, this becomes the derivation

$$\dfrac{\dfrac{(A\&B)\&C \quad [A\&B]^{1.}}{A\&B}\,{}_{\&E,1.} \quad [A]^{2.}}{A}\,{}_{\&E,2.} \tag{1}$$

Here the major premiss of the second inference is a conclusion of $\&E$, and we have a permutation conversion into

$$\dfrac{(A\&B)\&C \quad \dfrac{[A\&B]^{1.} \quad [A]^{2.}}{A}\,{}_{\&E,2.}}{A}\,{}_{\&E,1.} \tag{2}$$

where the major premisses of both instances of $\&E$ are assumptions. With the special elimination rules, hidden convertibilities remain, in the form of major premisses of elimination rules that are not assumptions, as is made clear in the translation from non-normal derivations to sequent calculus derivations with cuts.

The above examples showed how permutation conversion works for disjunction elimination and general conjunction elimination. As an illustration of full normal form with general implication elimination, we solve a problem of normal form of Ekman (1998). Ekman found that a derivation of the formula $\sim(P \supset \subset \sim P)$, in which equivalence is implication in both directions, either is not normal or else has a subderivation of the form

$$
\cfrac{P \supset \sim P \quad \cfrac{\sim P \supset P \quad \sim P}{P}\,Mp}{\sim P}\,Mp
$$

The derivation has the redundancy, or is "indirect" in Ekman's terminology, in that the derivation of the conclusion could be replaced by the derivation of the first occurrence of $\sim P$. However, this will produce a non-normal derivation, for the top occurrence of $\sim P$ is the conclusion of $\supset I$ and the bottom occurrence is a major premiss of $\supset E$. This problem is solved by use of the general implication elimination rule:

$$
\cfrac{[(P\supset\sim P)\&(\sim P\supset P)]^{9} \quad \cfrac{[\sim P\supset P]^{8} \quad \cfrac{[P\supset\sim P]^{7} \quad [P]^{3} \quad \cfrac{\cfrac{[\sim P]^{2} \quad [P]^{3} \quad [\bot]^{1}}{\bot}\supset E,1}{\bot}\supset E,2}{\cfrac{\bot}{\sim P}\supset I,3}}{\cfrac{\bot}{\sim P}\supset I,3} \quad \cfrac{[P\supset\sim P]^{7} \quad [P]^{6} \quad \cfrac{[\sim P]^{5} \quad [P]^{6} \quad [\bot]^{4}}{\bot}\supset E,4}{\cfrac{\bot}{\bot}\supset E,5}\supset E,6}{\cfrac{\bot}{\sim((P\supset\sim P)\&(\sim P\supset P))}\supset I,9}\&E,7,8
$$

All major premisses of elimination rules in the derivation are assumptions, which is the characteristic property of normal derivations with general elimination rules. Further, the conclusion $\sim P$ by $\supset I$ is not a major premiss of $\supset E$.

The origin of the above problem is in the observation that Russell's paradox about "the set of all sets that are not members of themselves" can be derived intuitionistically, without the law of excluded middle. Deleting the last line of our derivation and reading P as "the set of all sets that are not members of themselves belongs to itself," we derive a contradiction from $(P \supset \sim P)\&(\sim P \supset P)$.

(c) Simplification conversions: Other reductions of natural deduction derivations exist besides detour and permutation conversions. In Prawitz (1971), a simplification of derivations in natural deduction is suggested, called properly **simplification conversion**. The convertibility arises from disjunction elimination when in at least one of the auxiliary derivations, say, the first one, a disjunct was not assumed:

$$
\cfrac{\begin{array}{c}\Gamma\\ \vdots\\ A\vee B\end{array} \quad \begin{array}{c}\Delta\\ \vdots\\ C\end{array} \quad \begin{array}{c}[B]^{1.}\!,\Theta\\ \vdots\\ C\end{array}}{C}\vee E,1.
$$

The elimination step is not needed, for C is already concluded in the first auxiliary derivation. With general elimination rules for conjunction and implication, we analogously have

$$\frac{\begin{array}{cc} \Gamma & \Delta \\ \vdots & \vdots \\ A\&B & C \end{array}}{C}\&E \qquad \frac{\begin{array}{ccc} \Gamma & \Delta & \Theta \\ \vdots & \vdots & \vdots \\ A \supset B & A & C \end{array}}{C}\supset E$$

In both inferences, C is already concluded without the elimination rule, and simplification conversion extends to all elimination rules, quantifier rules included. In terms of sequent calculus **GN**, Definition 5.2.7, there is in each of these inferences a (hereditarily) vacuous cut with cut formula concluded by a left rule in the right premiss. For example, translating the disjunction case to **GN** we have

$$\frac{\Gamma \Rightarrow A \vee B \quad \dfrac{\Delta \Rightarrow C \quad B^n, \Theta \Rightarrow C}{A \vee B, \Delta, \Theta \Rightarrow C}L\vee}{\Gamma, \Delta, \Theta \Rightarrow C}Cut$$

which converts to $\Delta \Rightarrow C$, and Δ is a multiset reduct of the antecedent of conclusion of the original cut. The other elimination rules lead to similar conversions. In the notion of vacuous cut, we find the systematic origin of simplification conversions, extending to all elimination rules. The notion is captured in terms of natural deduction by

Definition 8.5.1: *A simplification convertibility in a derivation is an instance of an E-rule with no discharged assumptions, or an instance of $\vee E$ with no discharges of at least one disjunct.*

A simplification convertibility can prevent the normalization of a derivation, as is shown by the following:

$$\frac{\dfrac{\dfrac{\overset{1.}{[A]}}{A \supset A}\supset I, 1. \quad \dfrac{\overset{2.}{[B]}}{B \supset B}\supset I, 2.}{(A \supset A)\&(B \supset B)}\&I \quad \dfrac{\overset{3.}{[C]}}{C \supset C}\supset I, 3.}{C \supset C}\&E$$

There is a detour convertibility but the pieces of derivation do not fit together in the right way to remove it. Instead, a simplification conversion into the derivation

$$\frac{\overset{3.}{[C]}}{C \supset C}\supset I, 3.$$

will remove the detour convertibility.

It is possible that in a simplification convertibility with $\vee E$, both auxiliary assumptions are vacuously discharged. In this case, there are two converted derivations of the conclusion.

(d) The subformula structure of general elimination rules: With special elimination rules in the $\vee$- and $\exists$-free fragment, there is a simple subformula structure along all **branches** of a normal derivation, from assumptions to a minor premiss of rule $\supset E$ or to the conclusion. In fully normal derivations with general elimination rules, branches are replaced by **threads** that jump from major premisses to their auxiliary assumptions. Contrary to first appearance, a greater uniformity in the structure of derivations, for the full language of predicate logic, is achieved.

The subformula property in natural deduction is more complicated than in sequent calculus because of the nonlocal character of the rules of inference. It is obtained through the notion of **thread** where for simplicity we assume that no simplification convertibilities obtain:

Definition 8.5.2: *A thread in a natural deduction derivation of C from open assumptions Γ without simplification convertibilities is a sequence of formulas $A_1, \ldots, A_n$ such that*

1. *A_n is either C or a minor premiss of $\supset E$.*
2. *A_{i-1} is either a major premiss with auxiliary assumption A_i in an E-rule, or a minor premiss with $A_{i-1} = A_i$ in an E-rule, or a premiss with conclusion A_i in an I-rule.*
3. *A_1 is a top formula not discharged by an E-rule.*

Threads typically run through a sequence of major premisses of E-rules until the conclusion of the innermost major premiss is built up by I-rules, and so on. If vacuous instances of elimination rules are admitted, there can be threads that stop at the major premiss.

Threads in a normal derivation, briefly, **normal threads**, have the following structure:

$$\overbrace{(\, A_1, \ldots, A}^{E\text{-part}} \underbrace{m, \ldots, A_n \,)}_{I\text{-part}}$$

In the E-part, the major premisses follow in succession and A_{i+1} is an immediate subformula of A_i. In the I-part, either A_{i+1} is equal to A_i or A_i is an immediate subformula of A_{i+1}.

We concluded in Corollary 8.3.3 the subformula property of normal derivations with general elimination rules by a corresponding result that is immediate for the sequent calculus **G0i**. A more direct proof in terms of natural deduction sheds some light on the structure of threads:

Direct proof of the subformula property: Each formula A is in at least one normal thread, and it is a subformula of the topformula or of the endformula of the thread. In the former case, the topformula is either an open assumption and the subformula property follows, else it is discharged by $\supset I$ and A is a subformula of the endformula of the thread. If the endformula is the conclusion of the whole derivation, the subformula property follows. If it is the endformula of a minor thread, it is also a subformula of the corresponding major premiss. The major premiss is either an open assumption and the subformula property follows. Else the major premiss is discharged by $\supset I$ and belongs to some normal thread with the endformula further down in the derivation. If this endformula is the conclusion of the derivation, the subformula property follows; if not, by repetition of the argument, the conclusion is reached. QED.

In sequent calculus, the rule of falsity elimination is represented by a sequent $\perp \Rightarrow C$ by which derivations can start. In standard natural deduction, instead, falsity elimination can apply at any stage of a derivation. This discrepancy is now explained as a hidden convertibility. In particular, if the conversion formula is $\perp$ derived by $\perp E$, we have a derivation with two non-normal instances of $\perp E$. Since $\perp E$ has only a major premiss, a permutation conversion just removes one of these instances:

$$
\begin{array}{c}
\Gamma \\
\vdots \\
\dfrac{\perp}{\dfrac{\perp}{C}}{\scriptstyle \perp E}{\scriptstyle \perp E}
\end{array}
\qquad \rightsquigarrow \qquad
\begin{array}{c}
\Gamma \\
\vdots \\
\dfrac{\perp}{C}{\scriptstyle \perp E}
\end{array}
$$

The first derivation has the translation to sequent calculus

$$
\dfrac{\dfrac{\overset{\vdots}{\Gamma \Rightarrow \perp} \quad \perp \Rightarrow \perp}{\Gamma \Rightarrow \perp}{\scriptstyle Cut} \quad \dfrac{}{\perp \Rightarrow C}{\scriptstyle L\perp}}{\Gamma \Rightarrow C}{\scriptstyle Cut}
$$

and the converted one

$$
\dfrac{\overset{\vdots}{\Gamma \Rightarrow \perp} \quad \dfrac{}{\perp \Rightarrow C}{\scriptstyle L\perp}}{\Gamma \Rightarrow C}{\scriptstyle Cut}
$$

Fully normal derivations do not have redundant iterations of $\perp E$. In Prawitz (1965, p. 20), the effect of the above permutation conversion is achieved by the ad hoc restriction that in $\perp E$ the conclusion be different from $\perp$.

In a typical application of $\perp E$ in natural deduction with the special elimination rules we have, using the modus ponens rule,

$$\frac{\dfrac{A \supset \perp \quad A}{\dfrac{\perp}{C}\,{}_{\perp E}} \, Mp}{}$$

With the more general implication elimination rule, the derivation and its permutation conversion are

$$\frac{A \supset \perp \quad A \quad \overset{1.}{[\perp]}}{\dfrac{\perp}{C}\,{}_{\perp E}}\, {}_{\supset E, 1.} \qquad \rightsquigarrow \qquad \frac{A \supset \perp \quad A \quad \dfrac{\overset{1.}{[\perp]}}{C}\,{}_{\perp E}}{C}\, {}_{\supset E, 1.}$$

Here the premiss $\perp$ is converted into a topformula of the derivation. The same applies in general and we thus obtain

Proposition 8.5.3: *A fully normal intuitionistic derivation begins with assumptions and instances of the intuitionistic rule $\perp E$, followed by a subderivation in minimal logic.*

This fact will give a natural translation of intuitionistic into minimal logic: Consider an intuitionistic derivation of C in full normal form. The conclusions of falsity elimination are derivable from falsity eliminations concluding atoms. By the subformula property, these are atoms of C, and let them be $P_1, \ldots, P_n$. Each step $\frac{\perp}{P_i}$ is replaced by an assumption $\perp \supset P_i$, and P_i is concluded from $\perp$ by $\supset E$ instead of $\perp E$. Collecting all the new assumptions, we obtain

Theorem 8.5.4: *Formula C is intuitionistically derivable if and only if*

$$(\perp \supset P_1) \& \ldots \& (\perp \supset P_n) \supset C$$

is derivable in minimal logic.

In normal derivations with special elimination rules in the $\vee$- and $\exists$-free fragment, there is a simple subformula structure along all branches from assumptions to a minor premiss of rule $\supset E$ or the conclusion. In normal derivations with general elimination rules, branches are replaced by threads that jump from major premisses to their auxiliary assumptions. Contrary to first appearance, a greater uniformity in the structure of derivations is achieved.

(e) Normalization: Theorem 8.4.1 gave a proof of normalization for intuitionistic natural deduction with general elimination rules through a translation to sequent calculus, cut elimination, and translation back to natural deduction. A direct proof of normalization is also possible. To simplify matters, we assume that no simplification convertibilities are met in normalization. The proof is presented in its main lines.

Direct proof of normalization: In order to prove normalization, we shall define an ordering on threads of a derivation depending on their conversion formulas and a secondary ordering depending on the position of major premisses of elimination rules in them.

With each thread is associated a **multiset of convertible formulas**, giving the number of convertible formulas of length 0, length 1, length 2, ..., of a maximum length l. These multisets are ordered as follows: Of two multisets, the one with the shorter formula of maximum length comes first. If both have maximum length l, the one with a lesser number formulas of that length comes first. If these numbers are equal, consider formulas of length $l - 1$, and so on. Multisets of convertible formulas in threads are ordered so that detour conversions reduce threads in the ordering.

The **height along a thread** of a major premiss A_i is measured as follows. Let h_1 be the number of steps from the topformula to a first major premiss in the thread and h_i the number of steps from the auxiliary assumption of major premiss A_{i-1} to major premiss A_i. The height of A_i in the thread is the sum $h_1 + \cdots + h_i$. Thus, if $A_1 \circ B_1, \ldots, A_m \circ B_m$ are the major premisses of a thread $(A, \ldots, C)$, height along the thread can be depicted as follows, where major premisses are separated by a semicolon from the auxiliary assumptions that follow them:

$$(A, \ldots, \underbrace{A_1 \circ B_1; A_1, \ldots}_{h_1}, \underbrace{A_2 \circ B_2; A_2, \ldots}_{h_2}; \ldots; \underbrace{A_{m-1}, \ldots}_{h_m}, A_m \circ B_m; A_m, \ldots, C)$$

From the construction of threads it is immediate that each formula in a derivation is in at least one thread. The height of each major premiss along normal threads is equal to zero. It is easily seen that the converse also holds. A permutation conversion on A_i has the effect of diminishing the height of A_i by one while maintaining the heights of major premisses coming before A_i along the thread. The threads are ordered lexicographically, according to the height of their first major premiss, second major premiss, and so on, with the effect that permutation conversions give threads that are reduced in the ordering.

Next we control the effect of conversions on threads. Given a non-normal derivation, its major premisses are the possible conversion formulas and no new possible conversion formulas are created under conversions:

1. Detour conversion on &: Assume the relevant part of the full derivation to be as in Section 8.5(a). The convertibility on the formula $A\&B$ in a thread such as

$$(\ldots, A, A\&B; A, \ldots, C, C, \ldots)$$

disappears, and there is a possible new convertibility on the formula A in the corresponding thread

$$(\ldots, A, \ldots, C, \ldots)$$

so that the multiset of convertible formulas is reduced.

2. Detour conversion on $\vee$: This case is identical to the above, save for changing & into $\vee$.

3. Detour conversion on $\supset$: Assume that the thread comes from a derivation such as the one in Section 8.5(a), and assume that the thread includes the discharged assumption A of $\supset I$. Before the conversion, we have a **major thread**

$$(A, \ldots, B, A \supset B; B, \ldots, C, C, \ldots)$$

In the conversion, the derivation of A is substituted for the assumption A, which creates a converted thread

$$(\ldots, A, \ldots, B, \ldots, C, \ldots)$$

The addition of a **minor thread** $(\ldots, A)$ leading to the minor premiss A in the beginning of the converted thread could add new convertible formulas longer than any in the original major thread. We instead do the following: In a detour conversion on implication, the discharged assumptions A in $\supset I$ are temporarily replaced by open assumptions A, with threads of type

$$(A, \ldots, B, \ldots, C, \ldots)$$

as a result, and a separate derivation of the minor premiss A. Thus the derivation is cut into parts, and we show that these parts and any subsequent parts that might turn up normalize, after which the assumptions A can be substituted by their normal derivations. If this creates new convertibilities, they are on strictly shorter formulas so that the process terminates.

We shall next show that permutation conversions reduce the heights of major premisses along threads without importing new convertible formulas. There are many cases, but we consider only one of them as the rest act in the same way on threads.

4. Permutation conversion on & with major premiss $C \& D$ derived by $\& E$, the latter rule having $A \& B$ as major premiss: Typical threads before and after conversion are

$$(\ldots, A \& B; A, \ldots, C \& D, C \& D; C, \ldots, E, E, \ldots)$$
$$\rightsquigarrow \quad (\ldots, A \& B; A, \ldots, C \& D; C, \ldots, E, E, E, \ldots)$$

Height along thread of major premiss $C\&D$ is reduced by one while heights of major premisses preceding it remain the same. The multiset of convertible formulas does not increase.

The effect of conversions on threads for the case of quantifiers is analogous to the above and will not be detailed.

The cutting into parts of the original derivation whenever a detour convertibility on implication is met can happen only a bounded number of times, since no new major premisses of elimination rules are created by any conversions. Each detour conversion reduces the multiset of convertible formulas and each permutation conversion does not increase it but reduces the height of a major premiss along threads. When no convertibilities remain, the assumptions in detour convertibilities on implication are substituted by the derivations of the minor premisses. Since these formulas are proper subformulas of the original convertible formulas, also this process terminates, and we have threads with no convertible formulas and zero heights along threads for major premisses of elimination rules. QED.

The proof of normalization almost establishes **strong normalization**, that is, the termination of conversions in any order whatsoever. The only restriction is that the detour conversions on implication are completed only after all other convertibilities have been exhausted. This restriction is not essential, as is shown by the proof of strong normalization and uniqueness of normal form for our system of natural deduction given by Joachimski and Matthes (2001). Their proof uses a system of term assignment.

It would be a redundancy in a normal derivation if it had major premisses of elimination rules that are derivable formulas:

Definition 8.5.5: *A major premiss of an elimination rule is a* proper assumption *if it is underivable.*

Theorem 8.5.6: *Given a derivation, there is a derivation in which all major premisses of elimination rules are proper assumptions.*

Proof: Consider a derivable major premiss A. In a normal derivation of A, the last rule must be an I-rule since an E-rule would leave an open assumption. A substitution of assumption A with a normal derivation creates a detour convertibility. From the conversion schemes, we observe that no conversion ever produces new major premisses of E-rules and that detour conversions produce shorter convertible formulas. Therefore the process of substituting derivable major premisses of E-rules with their derivations and subsequent normalization terminates in a derivation with proper assumptions. QED.

By the undecidability of predicate logic, the theorem does not give an effective proof transformation. A translation to sequent calculus gives

Corollary 8.5.7: *If the sequent* $\Gamma \Rightarrow C$ *is derivable in* **G0i***, it has a derivation in which all formulas principal in left rules are underivable.*

The eliminability of derivable principal formulas in left rules was discovered by Mints (1993). The formulation in terms of natural deduction makes it clear what the result means. The result can, of course, be extended to all assumptions.

8.6. CLASSICAL NATURAL DEDUCTION FOR PROPOSITIONAL LOGIC

We shall add to the sequent calculus **G0ip** the law of excluded middle in the form of a rule for atomic formulas similar to the one in Section 5.4, but with independent contexts. We then note that cut remains admissible, and that the rule itself is admissible for arbitrary propositional formulas. It follows that the calculus is complete for classical propositional logic. Then the translation from sequent calculus to natural deduction is extended by the translation of the rule of excluded middle.

(a) The rule of excluded middle: With P an atom and Γ, Δ, C arbitrary, the rule of excluded middle for atoms is

$$\frac{P, \Gamma \Rightarrow C \quad \sim P, \Delta \Rightarrow C}{\Gamma, \Delta \Rightarrow C} \ Gem0\text{-}at$$

The only difference with respect to the rule of excluded middle added to the calculus **G3ip** in Section 5.4 is that now we have independent contexts. Proofs of all essential results go through without difficulty.

Theorem 8.6.1: *The rule of cut is admissible in* **G0ip**+Gem0-at.

Proof: The proof is a continuation of the proof of cut elimination for the intuitionistic calculus **G0ip** by induction on the length of cut formula A with subinduction on the sum of heights of premisses of cut. Cut is permuted upward to cuts on the same formula but with lower cut-height, entirely analogously to the proof of Theorem 5.4.4. QED.

Theorem 8.6.2: *The rule of excluded middle for arbitrary formulas,*

$$\frac{A, \Gamma \Rightarrow C \quad \sim A, \Delta \Rightarrow C}{\Gamma, \Delta \Rightarrow C} \ Gem0$$

is admissible in **G0ip**+Gem0-at.

Proof: The proof is by induction on formula length. Cut is admissible and can be used in the proof.

For $A := \bot$, we derive the conclusion from the right premiss already by a cut with the derivable sequent $\Gamma \Rightarrow \bot \supset \bot$. For $A := P$, excluded middle is the rule *Gem0-at*.

For $A := A\&B$, we have the premisses $A\&B, \Gamma \Rightarrow C$, and $\sim(A\&B), \Delta \Rightarrow C$. Cut of the first by the derivable sequent $A, B \Rightarrow A\&B$ gives $A, B, \Gamma \Rightarrow C$, and cut of the second by the derivable sequent $\sim A \Rightarrow \sim(A\&B)$ gives $\sim A, \Delta \Rightarrow C$. By inductive hypothesis, the rule *Gem0* applied to A gives $B, \Gamma, \Delta \Rightarrow C$. Cut of the second premiss by the derivable sequent $\sim B \Rightarrow \sim(A\&B)$ gives $\sim B, \Delta \Rightarrow C$, and rule *Gem0* applied to B now gives $\Gamma, \Delta, \Delta \Rightarrow C$ that can be contracted to $\Gamma, \Delta \Rightarrow C$.

The cases of disjunction and implication are similar. QED.

Using the rule, we easily derive the sequent $\Rightarrow A \lor \sim A$. Since cut is admissible and the law of excluded middle derivable, the calculus is complete for classical propositional logic.

(b) Translation to natural deduction and back: The rule of excluded middle, in a notation not indicating possible vacuous or multiple discharges, is given by

$$\frac{\overset{1.}{[P]} \quad \overset{2.}{[\sim P]}}{\underset{\displaystyle C}{\quad \vdots \qquad \vdots \quad}} \\ \frac{C \qquad C}{C} \; \textit{Nem-at},1.,2.$$

where the assumptions P and $\sim P$ are discharged at the inference. The common natural deduction rule in classical logic, concluding an atom P if $\sim P$ leads to falsity, is a special case, as we shall see.

The translation from sequent calculus to classical natural deduction is obtained by the addition of the following to the previous translation in Section 8.2:

$$\frac{P, \Gamma \Rightarrow C \quad \sim P, \Delta \Rightarrow C}{\Gamma, \Delta \Rightarrow C} \; \textit{Gem0-at} \quad \rightsquigarrow \quad \frac{\overset{1.}{[P]}, \Gamma \overset{\vdots}{\Rightarrow} C \quad \overset{2.}{[\sim P]}, \Delta \overset{\vdots}{\Rightarrow} C}{C} \; \textit{Nem-at},1.,2.$$

The converse translation of an instance of the rule *Nem-at* requires the use of unique labeling for the m discharged assumptions P and n assumptions $\sim P$, where $\sim P^n$ denotes n copies of formula $\sim P$:

$$\frac{\overset{1.}{[P^m]}, \Gamma \quad \overset{2.}{[\sim P^n]}, \Delta}{\underset{\displaystyle C}{\quad \vdots \qquad \vdots \quad}} \\ \frac{C \qquad C}{C} \; \textit{Nem-at},1.,2.$$

The translation is by cases according to the values of m, n. The general case is

$$\cfrac{\cfrac{\begin{array}{c} P^m, \Gamma \\ \vdots \\ C \end{array}}{P, \Gamma \Rightarrow C}\,{}^{Str} \quad \cfrac{\begin{array}{c} \sim P^n, \Delta \\ \vdots \\ C \end{array}}{\sim P, \Delta \Rightarrow C}\,{}^{Str}}{\Gamma, \Delta \Rightarrow C}\,{}^{Gem0\text{-}at}$$

The closed assumptions have been opened. If $m = n = 1$, there is just rule *Gem0-at* and no weakening or contraction.

We obtain a sequent calculus closer to classical natural deduction by starting with the calculus **GN** and adding to it a rule of excluded middle with weakening and contraction with P and $\sim P$ built in:

$$\frac{P^m, \Gamma \Rightarrow C \quad \sim P^n, \Delta \Rightarrow C}{\Gamma, \Delta \Rightarrow C}$$

Translation to natural deduction and back is simplified in the same way as with the translation of logical rules in Section 8.2(c).

(c) Full normal form for classical propositional logic: We can now conclude the main results for normal derivations in classical natural deduction from the corresponding results in the single succedent sequent calculus formulation.

The usual system of classical natural deduction uses the rule of indirect proof for atoms, a special case of our rule. The rule of indirect proof is derivable from *Nem-at*: Assume that there is a derivation of $\perp$ from $\sim P$; then use rule $\perp E$ to get the derivation

$$\frac{\begin{array}{c} 1. \\ {[P]} \end{array} \quad \cfrac{\begin{array}{c} 2. \\ {[\sim P]} \\ \vdots \\ \perp \end{array}}{P}\,{}^{\perp E}}{P}\,{}^{Nem\text{-}at,\, 1.,2.}$$

Contrary to the rule of indirect proof, the premiss $\sim P$ is not discharged after $\perp$, but one step later. It is possible to convert indirect inferences on disjunctions into their components only if the disjunctions are also major premisses in a $\vee E$-rule. Thus full normal form fails for this rule (see also Prawitz 1965, p. 39, and Stålmarck 1991), but this is repaired in natural deduction for classical propositional logic with general elimination rules and the rule of excluded middle: Translations into natural deduction of the derivations in the proof of admissibility of excluded middle for arbitrary formulas, Theorem 5.4.6, give a uniform method for converting instances of the natural deduction rule of excluded middle for arbitrary formulas A

$$\frac{\overset{1.}{[A^m]} \quad \overset{2.}{[\sim A^n]}}{\underset{\displaystyle C}{\overset{\displaystyle C \qquad C}{}}} \textit{Nem},1.,2.$$

into ones with the rule *Nem-at*.

Lemma 8.6.3: *Application of the rule of natural deduction excluded middle converts to applications of natural deduction excluded middle to atoms.*

Proof: Consider the case in which indirect proof is insufficient, that of a disjunction $A \lor B$: We assume given the two derivations

$$\begin{array}{cc} A \lor B & \sim(A \lor B) \\ \vdots & \vdots \\ C & C \end{array}$$

Derivation of C by *Nem* applied to $A \lor B$,

$$\frac{[A \lor B] \quad [\sim(A \lor B)]}{\underset{\displaystyle C}{\overset{\displaystyle C \qquad\qquad C}{}}} \textit{Nem}$$

is converted into the derivation with *Nem* applied to A and B,

The other cases of conversions of *Nem-at* are obtained by translating to natural deduction the rest of the transformations in the proof of Theorem 5.4.6. QED.

Definition 8.6.4: *A derivation in intuitionistic natural deduction* +Nem-at *is in* full normal form *if no instance of* Nem-at *is followed by a logical rule and all subderivations up to instances of* Nem-at *are fully normal intuitionistic derivations.*

Theorem 8.6.5: *If a formula C is derivable from open assumptions* Γ *in intuitionistic natural deduction* +Nem-at *there is a derivation of C from open assumptions* Γ^* *in full normal form, where* Γ^* *is a multiset reduct of* Γ.

Proof: A routine verification shows that rule *Nem-at* commutes with the logical rules, modulo possible multiplications of open assumptions.   QED.

Thus, if formula C is classically derivable, the corresponding natural deduction derivation has by the theorem a neat separation of minimal, intuitionistic and classical parts: It begins with assumptions and instances of the intuitionistic $\frac{1}{C}\bot E$ rule, followed by a minimal subderivation, and ends with purely classical applications of the rule of excluded middle. As in sequent calculus, these can further be restricted to atoms of C:

Theorem 8.6.6: *In a derivation of C from open assumptions* Γ *in intuitionistic natural deduction* +Nem-at, *instances of* Nem-at *can be restricted to atoms of C.*

Proof: Permute down applications of *Nem-at* so that those on atoms not in C come right after the intuitionistic subderivation. Let the first of these be on an atom P:

$$\dfrac{\overset{\displaystyle \overset{1.}{[P]},\Gamma'}{\underset{\displaystyle C}{\vdots}} \quad \overset{\displaystyle \overset{2.}{[\sim P]},\Gamma''}{\underset{\displaystyle C}{\vdots}}}{C}\ \textit{Nem}, 1., 2.$$

The first subderivation of C, from P, Γ', is transformed into a derivation of $\sim P$ from $\sim C$, Γ' which is then substituted for the assumption of $\sim P$ in the second subderivation, followed by an application of *Nem* to C:

$$\dfrac{\overset{2.}{[C]} \qquad \dfrac{\dfrac{\overset{3.}{[\sim C]} \quad \overset{\displaystyle \overset{1.}{[P]},\Gamma'}{\underset{\displaystyle C}{\vdots}}}{\dfrac{\bot}{\sim P}\ \supset I, 1.}\ \supset E \qquad \overset{\displaystyle \Gamma''}{\underset{\displaystyle C}{\vdots}}}{C}}{C}\ \textit{Nem}, 2., 3.$$

By Lemma 8.6.3, the application of *Nem* to C converts to atoms of C. The proof transformation is repeated for the remaining atoms that are not atoms in C.   QED.

Full normal form gives a translation from classical to intuitionistic and minimal propositional logic:

Proposition 8.6.7: *Let $P_1, \ldots, P_n$ be the atoms of C. Then C is classically derivable if and only if*

$$(P_1 \vee \sim P_1) \& \ldots \& (P_n \vee \sim P_n) \supset C$$

is intuitionistically derivable. Further, C is classically derivable if and only if

$$(\perp \supset P_1) \& \ldots \& (\perp \supset P_n) \& (P_1 \vee \sim P_1) \& \ldots \& (P_n \vee \sim P_n) \supset C$$

is derivable in minimal logic.

Proof: In a given classical derivation of C, transform each instance of *Nem-at*, at the bottom of the intuitionistic subderivation, into an intuitionistic inference concluding C by $\vee E$ through the assumptions $P_i \vee \sim P_i$ for all atoms P_i of C. Now collect all these assumptions together and use the translation from intuitionistic to minimal logic, Theorem 8.5.4. QED.

Given a classical derivation with rule *Nem*, normalization will give a derivation with instances of *Nem-at*. The above translations can be optimized by leaving out those assumptions $P_i \vee \sim P_i$ for which there is no corresponding instance of *Nem-at* in the derivation.

Next consider the translation of a classical derivation of C into an intuitionistic derivation of $(P_1 \vee \sim P_1) \& \ldots \& (P_n \vee \sim P_n) \supset C$. If C is already intuitionistically derivable, the antecedent is empty and we can identify C as the set of its formal proofs through the Curry–Howard isomorphism. If not, the proof of Proposition 8.6.7 shows how, in terms of natural deduction, the proof of C reduces to the intuitionistic subderivations. The usual way of applying the formulas-as-types principle to classical theories is to assume the law of excluded middle for arbitrary formulas. In the notation of constructive type theory, this can be done by the type declaration $em : A \vee \sim A$. No one, of course, can in general tell how such a function em should be evaluated. However, now we can look on the fully normal classical derivation of C as an instruction on how to construct the function that converts decisions on atoms $P_1, \ldots, P_n$ into a proof of C. For example, if we have a set S with a decidable equality $Eq : (S)(S)Prop$, the declaration $deq : (a, b : S)Or(Eq(a, b), \sim Eq(a, b))$ will implement a classical logic for the language of this equality. Sometimes, as with equality for natural numbers, we can actually define such a function deq.

By adding proof terms to the rule of excluded middle, a satisfactory formulation of the Curry–Howard isomorphism for classical propositional logic is obtained, but we shall not go into the details.

208   STRUCTURAL PROOF THEORY

NOTES TO CHAPTER 8

In Prawitz' book of 1965, a system of sequent calculus is given with shared contexts treated as sets and axioms of the form $A, \Gamma \Rightarrow A$, thus with weakening built into the system and no cut rule. A proof of closure with respect to cut is sketched, through completeness of natural deduction and translation of normal derivations into cut-free sequent calculus derivations (p. 91). Prawitz uses a sequent calculus with Gentzen's original $L\&$ rules, and therefore the special conjunction elimination rules do not produce cuts. Uses of modus ponens can also be turned into a cut-free sequent calculus derivation, for a detailed study shows that a cut elimination procedure is infact contained in its translation.

In Zucker's paper (1974) on translation from sequent calculus to natural deduction, it is stated that "one or both of these systems must be modified to some extent," but the change on natural deduction just concerns discharge of assumptions. In a related paper by Pottinger (1977) an example demonstrates the failure of isomorphism between sequent calculus and natural deduction with the usual $\&E$ rules (p. 350, see also Zucker, p. 2). No one seems to have followed the idea that it is this system of natural deduction, not sequent calculus, that lies at the back of the failure of isomorphism between derivations in the two calculi. In Herbelin (1995) (see also Dyckhoff and Pinto 1998), a sequent calculus is given that does not distinguish between derivations obtainable in standard calculi from each other through certain permutations, and a unique correspondence with natural deduction derivations with special elimination rules is achieved.

There is another way of arriving at the general elimination rule for conjunction than the inversion principle we have used, namely, constructive type theory. The general rule comes straight out by suppressing the proof objects in the typed rule, as in Martin-Löf (1984, p. 44). In the other direction, typing our general implication elimination rule will result in a new selector, **generalized application**:

$$\frac{c : A \supset B \quad a : A \quad \overset{[x \,:\, B]}{\underset{\vdots}{d : C}}}{gap(c, a, (x)d) : C}$$

A full type-theoretical rule uses the function type $(A)B$ that has no correspondence in first-order logic. The usual first-order selector ap that corresponds to modus ponens is defined, for $B = C$, by $ap(c, a) = gap(c, a, (x)x) : B$. Normality means that each selector term has a variable as first argument. A direct proof of strong normalization for natural deduction with general elimination rules was found by Joachimski and Matthes (2001), through a term assignment system. (They also suggested the term generalized application for general implication elimination typed.)

The solution of Ekman's problem comes from von Plato (2000).

The natural deduction formulation of the rule of excluded middle was studied already in Tennant (1978) but has remained relatively unknown even if its first appearance is due to Gentzen (1936). The reason should be that no subformula property had been proved within a natural deduction approach. However, Tennant proves a natural deduction version of our Lemma 8.6.3 and the important result that applications of the rule reduce to applications on atoms.

In Gordeev (1987), a sequent calculus rule corresponding to Peirce's law is given, concluding $\Gamma \Rightarrow A$ from $A \supset B, \Gamma \Rightarrow A$. Admissibility of cut and the subformula property are proved. This calculus is complete for classical propositional logic, but cannot be extended with nonlogical rules of inference while maintaining cut elimination, contrary to our classical calculus.

Extensions of the sequent calculus **G3i** by nonlogical rules translate into normal natural deductions, in which the nonlogical natural deduction rules are obtained from the translation of the single succedent rule-scheme for sequents:

$$\frac{P_1 \ldots P_m \quad \overset{[Q_1]}{\overset{\vdots}{C}} \ldots \overset{[Q_n]}{\overset{\vdots}{C}}}{C}$$

This is the natural deduction scheme for **nonlogical elimination rules**. An early work that uses rules of this form, for equality and apartness, is Van Dalen and Statman (1979).

In Hallnäs and Schroeder-Heister (1990), regular sequents $P_1, \ldots, P_m \Rightarrow Q$ are translated, for the purposes of logic programming, into natural deduction rules of the form

$$\frac{P_1 \ldots P_m}{Q}$$

The two kinds of nonlogical natural deduction rules are interderivable. The relation of these two ways of extending natural deduction is analogous to the situation in sequent calculus: Extension of sequent systems with regular sequents does not in general permit cut-free derivations, whereas extension with nonlogical rules does. This is seen clearly in the example of predicate logic with equality.

CONCLUSION

Diversity and Unity in Structural Proof Theory

COMPARING SEQUENT CALCULUS AND NATURAL DEDUCTION

Structural proof theory was born in two forms, natural deduction and sequent calculus. The former has been the more accessible way to proof theory, used in teaching. The latter, instead, has yielded better to structural proof analysis. For example, the underivability results for intuitionistic predicate logic in Section 4.3 were obtained for sequent calculus in the early 1950s.

Even if natural deduction gives the easier access, in the end proofs are easier to find in sequent calculus. It formalizes the analysis into subgoals of the theorem to be proved, whereas in natural deduction this has to be done intuitively. Furthermore, the sequent calculi we studied in Chapters 2–4, with their shared contexts in two-premiss rules, support root-first proof search.

With independent contexts, we found sequent calculi that come very close to natural deduction, especially if in the latter general elimination rules are used. One essential difference, the presence in sequent calculus of explicit rules of weakening and contraction, was overcome by a suitable change of the logical rules of sequent calculus to permit implicit weakening and contraction similarly to natural deduction. Then cut-free proofs in sequent calculus and normal proofs in natural deduction became mere notational variants of one and the same proof. Isomorphic translation turned the sequent calculus derivation with its locally applied rules into a standard nonlocal natural deduction derivation. One difference between the two types of calculi remained: Where the logical rules of natural deduction admit of non-normal instances, sequent calculus uses a logical rule and a cut. It is a cut with the cut formula principal in at least the right premiss. Because of the presence of an independent rule of cut, transformation of a sequent calculus derivation into cut-free form is profoundly different from the conversion of a natural deduction derivation into normal form. There are cuts with nonprincipal formulas that have no interpretation in terms of natural deduction. Even here the gap between the two formulations of structural proof theory was narrowed: first,

211

by the restriction of cut elimination to the hereditarily principal cuts of Section 5.2, and secondly, by the implicit treatment of weakening and contraction that does away with a number of cuts and permutations of cuts to which nothing corresponds in natural deduction.

Gentzen's doctoral thesis gave the rules of sequent calculus in two groups, the structural rules as the first group, and the logical rules as the second. A few years later he called the rules of weakening, contraction, exchange and change of bound variable "Strukturänderungen," structural modifications (1936, p. 513). (Gentzen treated contexts as lists in which the exchange of order was needed, contrary to multisets.) All of the structural modifications except weakening "do not change the meaning of a sequent . . . all these possibilities of modification are of a purely formal nature. It is only because of special features of the formalism that these rules must be expressly given." (ibid., pp. 513–14). Weakening can be justified by admitting that if a proposition is correct under given assumptions, it should remain correct if arbitrary additional assumptions are made. Despite these words of Gentzen, a lot of work has been done in structural proof theory in the search of some ultimate meaning of weakening and contraction in themselves, independent of a formalism of logical rules. We have shown that the change to general elimination rules permits the interpretation of weakening and contraction in terms of natural deduction, as vacuous and multiple discharge of assumptions, respectively. Here again, as with the rule of cut, the formulation of weakening and contraction as independent rules brings cases that have no correspondence in natural deduction.

There is more work to be done in relating cut elimination to normalization. The rule of cut has been usually left as something one should not touch; but going back to the historical origins in Gentzen (in particular, his first paper that appeared in 1932), we find that he considered various forms of rules, one of which is the cut rule. It descends from work of Paul Hertz in the 1920s: In Hertz (1929, p. 462), a rule of "syllogism" is suggested that we can write as

$$\frac{\Gamma_1 \Rightarrow A_1 \quad \ldots \quad \Gamma_n \Rightarrow A_n \quad A_1, \ldots, A_n, \Delta \Rightarrow B}{\Gamma_1, \ldots, \Gamma_n, \Delta \Rightarrow B} Syl$$

The n formulas $A_1, \ldots, A_n$ in the rightmost premiss are cut in one step of inference, so we can call Hertz' rule **simultaneous cut**. Gentzen's rule of cut is the special case of $n = 1$. The rule of multicut of Section 5.1 is a somewhat different generalization of cut, with just one premiss $\Gamma \Rightarrow A$ on the left and all of the A_i identical to A, thus with n copies of A in the right premiss deleted in one step. Gentzen's **mix** rule in (1934–35, III.3.1) is a cut in which n copies of the cut formula in the left premiss and m copies in the right one are deleted.

Another rule relating to cut is the "chain rule" ("Kettenschluss," Gentzen 1936, p. 543) that we can write as

$$\frac{\Gamma_1 \Rightarrow C_1 \quad C_1, \Gamma_2 \Rightarrow C_2 \quad \cdots \quad C_{n-1}, \Gamma_n \Rightarrow C_n}{\Gamma_1, \ldots, \Gamma_n \Rightarrow C_n} Chn$$

Looking at the indices, one sees that the idea of the rule comes from the way recursion on natural numbers works.

As Gentzen (1932, p. 332) remarks, the rule of simultaneous cut follows by repeated applications of the rule of cut. However, we should note that once a series of cuts reproduces the conclusion of a simultaneous cut, these cuts can be eliminated in any order, typically leading to different cut-free proofs. The rule of simultaneous cut with all of the premises $\Gamma_i \Rightarrow A_i$ identical corresponds exactly to the simultaneous substitution created by a detour conversion, as in Section 8.5.

It seems plausible that one can define a cut elimination procedure that uses some generalization of cut in intermediate stages, with results analogous to those of strong normalization and confluence (uniqueness of normal form) in natural deduction. Moreover, if this can be done for a single succedent intuitionistic calculus, the calculus **GN** in the first place, it should be possible for a classical multisuccedent calculus such as **GM** as well.

When sequent calculus and natural deduction are compared, an outstanding difference is the elegant treatment of classical predicate logic in multisuccedent sequent calculus. Natural deduction has a satisfactory proof theory of classical propositional logic, as shown in Section 8.6, but equally satisfactory classical rules of natural deduction for the full language of predicate logic have not been found.

A UNIFORM LOGICAL CALCULUS

Modifications in sequent calculus can bring it closer to natural deduction, but one would also like to relate the two approaches to structural proof theory in a more direct manner. There are some more general ways of viewing proof theory based on semantical considerations. Constructive type theory formalizes the computational semantics of intuitionistic logic and goes beyond the division of proof theory into systems of natural deduction and sequent calculus.

In this last section, we suggest a syntactic approach to unifying proof theory through a logical calculus, the rules of which contain as particular instances the rules of sequent calculus and natural deduction. The rules of the logical calculus to be presented can be described as the obvious formulation of a "multiple conclusion natural deduction calculus with general logical rules, written in sequent calculus style." By general logical rules is meant a formulation with general elimination rules and their dual **general introduction rules**. As indicated, the rules of natural deduction in sequent calculus style as well as the rules of sequent calculus itself, in single succedent and multisuccedent versions, will come out as instances. Further, the inverses of the sequent calculus rules will be instances.

In the rules of the uniform calculus, denoted by **MG**, contexts will be treated as multisets. Gentzen's original single-arrow notation for the formal derivability relation is used to distinguish it from the turnstile we have occasionally used when reasoning about derivations in natural deduction and from the double arrow of sequent calculus. In the logical rules, the major premiss is the sequent with the connective. The other premisses are minor premisses. Rules that display multiple occurrences A^m, B^n of formulas have instances for any $m, n \geqslant 0$. We show only the propositional rules.

MG

Rule of assumption:

$A \to A$

Logical rules:

$$\frac{A \& B, \Gamma \to \Delta \quad \Gamma' \to \Delta', A^m \quad \Gamma'' \to \Delta'', B^n}{\Gamma, \Gamma', \Gamma'' \to \Delta, \Delta', \Delta''} \, \&I$$

$$\frac{\Gamma \to \Delta, A \& B \quad A^m, B^n, \Gamma' \to \Delta'}{\Gamma, \Gamma' \to \Delta, \Delta'} \, \&E$$

$$\frac{A \vee B, \Gamma \to \Delta \quad \Gamma' \to \Delta', A^m, B^n}{\Gamma, \Gamma' \to \Delta, \Delta'} \, \vee I$$

$$\frac{\Gamma \to \Delta, A \vee B \quad A^m, \Gamma' \to \Delta' \quad B^n, \Gamma'' \to \Delta''}{\Gamma, \Gamma', \Gamma'' \to \Delta, \Delta', \Delta''} \, \vee E$$

$$\frac{A \supset B, \Gamma \to \Delta \quad A^m, \Gamma' \to \Delta', B^n}{\Gamma, \Gamma' \to \Delta, \Delta'} \, \supset I$$

$$\frac{\Gamma \to \Delta, A \supset B \quad \Gamma' \to \Delta', A^m \quad B^n, \Gamma'' \to \Delta''}{\Gamma, \Gamma', \Gamma'' \to \Delta, \Delta', \Delta''} \, \supset E$$

$$\frac{\Gamma \to \Delta, \bot}{\Gamma \to \Delta, \Delta'} \, \bot E$$

General introduction rules are formulated in perfect symmetry to the general elimination rules. From the calculus it is clear that the rule for falsity is a logical elimination rule, a fact confounded in sequent calculi when $\bot \Rightarrow C$ is treated as an axiom on a par with $A \Rightarrow A$.

The previous definition of normality for natural deduction with general elimination rules is extended to introduction rules also, by requiring that the major premisses of all rules in a derivation are assumptions. Normalization for the uniform calculus **MG** is obtained through a translation to the sequent calculus **GM**+*Cut*: General introduction rules are translated as right rules followed by a cut and general elimination rules dually as left rules followed by a cut. For example, rule $\&I$ of the above table is translated as

$$\frac{\dfrac{\Gamma' \Rightarrow \Delta', A^m \quad \Gamma'' \Rightarrow \Delta'', B^n}{\Gamma', \Gamma'' \Rightarrow \Delta', \Delta'', A\&B} R\& \qquad A\&B, \Gamma \Rightarrow \Delta}{\Gamma, \Gamma', \Gamma'' \Rightarrow \Delta, \Delta', \Delta''} Cut$$

Other introduction rules have analogous translations. For elimination rules, consider as an example $\supset E$. It has the translation

$$\frac{\Gamma \Rightarrow \Delta, A \supset B \quad \dfrac{\Gamma' \Rightarrow \Delta', A^m \quad B^n, \Gamma'' \Rightarrow \Delta''}{A \supset B, \Gamma', \Gamma'' \Rightarrow \Delta', \Delta''} L\supset}{\Gamma, \Gamma', \Gamma'' \Rightarrow \Delta, \Delta', \Delta''} Cut$$

Given a derivation in **MG**, a translation to **GM**, cut elimination for **GM** (Corollary 5.2.14), and translation back to **MG** produces a derivation in which all major premisses are assumptions. In the translation back, rule $R\vee$, for example, is translated as follows:

$$\frac{\Gamma \Rightarrow \Delta, A^m, B^n}{\Gamma \Rightarrow \Delta, A \vee B} R\vee \quad \rightsquigarrow \quad \frac{A \vee B \to A \vee B \quad \Gamma \to \Delta, A^m, B^n}{\Gamma \to \Delta, A \vee B} \vee I$$

Other rules are translated similarly, with major premisses in **MG** rules always becoming assumptions. We therefore have the

Theorem: *A derivation of* $\Gamma \to \Delta$ *in* **MG** *can be transformed into a normal derivation of* $\Gamma^* \to \Delta^*$ *where* Γ^* *and* Δ^* *are multiset reducts of* Γ.

The uniform calculus is not strongly normalizing, for there are rules that can be permuted with each other with no end.

We shall show how to obtain the rules of various logical calculi from the uniform calculus.

(a) The rules of multisuccedent sequent calculus: To recover the sequent calculus rules, we do the following two things:

1. Find the substitution that makes the major premiss of each rule an assumption. Thus, for the introduction rules, we set $\Gamma = \emptyset$ and Δ equal to the principal formula and the other way around for the elimination rules.

2. Delete the major premiss that has become an assumption, and change the single arrow to a double one and the introduction rule symbols to right rule symbols and elimination rule symbols to left rule symbols.

The result of the above is the classical multisuccedent sequent calculus **GM** of Section 5.2. The standard logical rules of multisuccedent sequent calculus with independent contexts, the calculus **G0c** of Section 5.1, are obtained as special cases, with $m, n = 1$ in the above rules. Weakening and contraction must be added as primitive rules.

(b) Single succedent sequent calculus: By restricting in the rules of the uniform calculus the succedent in each premiss and conclusion to be one formula and otherwise proceeding as in points *1* and *2* above, we obtain the intuitionistic sequent calculus **GN** of Section 5.2. The two right disjunction rules arise quite naturally from the requirement of a single succedent formula.

As above, with $m, n = 1$ and weakening and contraction added, the intuitionistic single succedent calculus with independent contexts **G0i** of Section 5.1 is obtained.

(c) Inverses of sequent calculus rules: The rules of the uniform calculus give as instances the inverses of the rules of **GM**, by the following:

1. Find the substitution that makes the minor premisses of each rule assumptions.

2. Delete the minor premisses that have become assumptions, change the single arrow to a double one, etc.

Introduction rules of **MG** produce inverses of left rules of **GM** and elimination rules inverses of right rules. For obtaining these inverses, the axiom has to be formulated with just one formula and no context. The true reason for our formulation of the axiom is that it is needed for having uniquely determined first occurrences of certain formulas in cut elimination procedures, as in Section 5.2. The rule of falsity elimination has no minor premiss so there is nothing to invert.

Inverses of **GN** are obtained by setting the succedent to be empty or just one formula in the inverses of **GM**, whichever gives a single succedent rule. Inversions are produced for all but rule $R\vee$ and the first premiss of $L\supset$.

Remarkably, the inverses obtained as instances of the uniform calculus are all inverses of shared context rules.

(d) Natural deduction: We obtain systems of natural deduction from the uniform calculus by restricting the succedent in the rules for **MG** to one formula. Doing just this will give a system with general introduction and elimination rules, denoted by **NG**. Further restrictions lead to systems with general elimination rules only, and to the usual system in which only disjunction elimination is of the form of a general rule. The rules of natural deduction with general introduction and elimination rules are as follows:

NG

Rule of assumption:

$A \to A$

Logical rules:

$$\frac{A\&B, \Gamma \to C \quad \Gamma' \to A \quad \Gamma'' \to B}{\Gamma, \Gamma', \Gamma'' \to C}\&I \qquad \frac{\Gamma \to A\&B \quad A^m, B^n, \Gamma' \to C}{\Gamma, \Gamma' \to C}\&E$$

$$\frac{A \vee B, \Gamma \to C \quad \Gamma' \to A}{\Gamma, \Gamma' \to C}\vee I_1 \qquad \frac{A \vee B, \Gamma \to C \quad \Gamma' \to B}{\Gamma, \Gamma' \to C}\vee I_2$$

$$\frac{\Gamma \to A \vee B \quad A^m, \Gamma' \to C \quad B^n, \Gamma'' \to C}{\Gamma, \Gamma', \Gamma'' \to C}\vee E$$

$$\frac{A \supset B, \Gamma \to C \quad A^m, \Gamma' \to B}{\Gamma, \Gamma' \to C}\supset I \qquad \frac{\Gamma \to A \supset B \quad \Gamma' \to A \quad B^n, \Gamma'' \to C}{\Gamma, \Gamma', \Gamma'' \to C}\supset E$$

$$\frac{\Gamma \to \perp}{\Gamma \to C}\perp E$$

The general introduction rules have a more striking look if written in natural deduction style:

$$\frac{\begin{matrix}[A\&B]\\ \vdots \\ C\end{matrix} \quad A \quad B}{C}\&I \qquad \frac{\begin{matrix}[A \vee B]\\ \vdots \\ C\end{matrix} \quad A}{C}\vee I_1 \qquad \frac{\begin{matrix}[A \vee B]\\ \vdots \\ C\end{matrix} \quad B}{C}\vee I_2 \qquad \frac{\begin{matrix}[A \supset B] \quad [A]\\ \vdots \\ C\end{matrix} \quad B}{C}\supset I$$

A comparison with the general elimination rules, as in Section 1.2, displays the perfect symmetry of general introduction and elimination rules. We can also express it in words:

> *General introduction rules state that if a formula C follows from a formula A, then it already follows from the immediate grounds for A; general elimination rules state that if C follows from the immediate grounds for A, then it already follows from A.*

When the major premisses of introduction rules are assumptions and are left unwritten, the usual introduction rules of natural deduction and the general elimination rules remain. When also the minor premisses of $\&E$ and $\supset E$ are assumptions and left unwritten, Gentzen's original rules of natural deduction in sequent calculus style, a notational variant of the calculus that started structural proof theory, are obtained.

Simple Type Theory and Categorial Grammar

In this appendix we describe a general framework for functions that are used in categorial grammars. It is known as **simple type theory**. Then the grammars for the languages of propositional and predicate logic are given.

A.1. SIMPLE TYPE THEORY

We shall use the term **type** for domains and ranges of functions. Each object is **typed**, meaning that it always belongs to some type. There will be some **basic types**, upon which the functional hierarchy of simple type theory is built. Arbitrary types will be denoted by $\alpha, \beta, \gamma, \ldots$. Given two types α and β, we can form the **type of functions** from α to β, denoted $(\alpha)\beta$. The statement that a is an **object** in type α is written as $a : \alpha$ ("declaration of an object a of type α"). To get started with the formation of types of functions, we declare some basic types. Given a type of functions $(\alpha)\beta$, we can apply a function that is an object of that type, say $f : (\alpha)\beta$, to obtain as value an object of type β:

$$\frac{f : (\alpha)\beta \quad a : \alpha}{f(a) : \beta} \tag{1}$$

This is the scheme of **functional application**. (The first premiss looks more familiar if written in the usual notation in mathematics, $f : \alpha \to \beta$.) In the other direction, we have a scheme for **functional abstraction**. It is a way of forming a function from an expression containing a variable:

$$\frac{\begin{matrix}[x : \alpha]\\ \vdots\\ b : \beta\end{matrix}}{(x)b : (\alpha)\beta} \tag{2}$$

Assuming an arbitrary object x of type α given, if we are able to construct an object b of type β, then $(x)b$ is the functional abstract, an object of type $(\alpha)\beta$, where the parenthesis notation in $(x)b$ indicates the variable over which abstraction is taken. The square brackets $[x : \alpha]$ are used to indicate that the assumption $x : \alpha$ is **discharged** when the functional abstract is formed. A functional abstract is always applied through **substitution**: If the value a is given to x, the value of $(x)b$ applied to a is given by the expression $b(a/x)$, where the notation a/x indicates substitution of x by a. Formally, application through substitution is written as a rule that defines the value of the application of a functional abstract. A notation expressing **definitional equality** is needed for this. In general, the judgment that two objects a and b of a type α are equal is written as $a = b : \alpha$ ("a and b are equal in α"). In β-**conversion**, we conclude such an equality from two premisses:

$$\frac{\begin{array}{c}[x : \alpha]\\ \vdots\\ b : \beta \quad a : \alpha\end{array}}{((x)b)(a) = b(a/x) : \beta} \tag{3}$$

Extra parentheses are used to indicate the functional abstract $(x)b$ uniquely; if merely $(x)b(a)$ is written, it could also be the functional abstract, over x, of b applied to a.

Repeated functional application leads to expressions of the form $f(a)\ldots(c)$ that we write as $f(a, \ldots, c)$. Thus functions of several arguments are formally functions of one argument that have as values other functions.

Schemes (1)–(3) of functional abstraction, application, and β-conversion are the three principles of simple type theory.

Simple type theory is expressive enough to work as a categorial grammar of **predicate logic**. There we have a ground category of **individual objects**, the category of **propositions**, and **properties** over the category of individual objects, represented as **propositional functions**. These take individual objects as arguments and return propositions as values. The category of propositions is denoted by *Prop*. "Category" here is a synonym for type.

Before showing how logical languages are represented through categorial grammars, we look at propositions that do not have logical structure, namely those that are **atomic propositions** from a logical point of view.

In any discourse, **domains of individuals** are introduced. In arithmetic, we have the domain N of natural numbers. Individual objects are introduced by declarations of the form $n : N$; for example, $0 : N$ introduces the natural number zero. Next we have propositional functions over N, for example, a function we can call *Even*, the category of which is $(N)Prop$. Thus functional application gives us

$$\frac{Even : (N)Prop \quad 12 : N}{Even(12) : Prop}$$

In geometry, we have the two domains *Point* and *Line*, and a two-place propositional function

$$Incident : (Point)(Line)Prop$$

that gives, by successive application to $a : Point$ and $l : Line$, the proposition $Incident(a, l)$ as value. The usual way of expressing this, "point a is incident with line l," leaves implicit the functional form of the proposition.

A.2. CATEGORIAL GRAMMAR FOR LOGICAL LANGUAGES

In pure logic, the interest is in logical structure, not in the structure of the basic building blocks, the atomic propositions. In **propositional logic**, no structure at all is given to atomic propositions, but these are introduced just as pure parameters $P, Q, R, \ldots$, with the categorizations

$$P : Prop, \quad Q : Prop, \quad R : Prop, \quad \ldots$$

Connectives are functions for forming new propositions out of given ones. We have the constant function $\bot$, called **falsity**, for which we simply write $\bot : Prop$. Next we have **negation**, with the categorization

$$Not : (Prop)Prop$$

The two-place connectives *And*, *Or*, and *Implies* are categorized by

$$And : (Prop)(Prop)Prop$$
$$Or : (Prop)(Prop)Prop$$
$$Implies : (Prop)(Prop)Prop$$

The use of symbols used to be considered an essential characteristic of logical languages. We shall need symbols for expressing generality: First we have the atomic propositions that are denoted symbolically by $P, Q, R. \ldots$ Next we have **arbitrary** propositions, denoted by $A, B, C, \ldots$. For the rest, the only thing that matters is the categorization, and symbols serve only to make formulas shorter. They will be introduced through the following definitional equalities:

$$\sim\, = Not : (Prop)Prop$$
$$\&\, = And : (Prop)(Prop)Prop$$
$$\vee\, = Or : (Prop)(Prop)Prop$$
$$\supset\, = Implies : (Prop)(Prop)Prop$$

Further, the functional structure is hidden by an **infix** notation and by the dropping of parentheses, $\sim P$ for $\sim(P)$, $A\&B$ for $\&(A, B)$, and so on. This will create an ambiguity not present in the purely functional notation, such as $A \& B \supset C$ that could be both $\&(A, \supset (B, C))$ and $\supset (\&(A, B), C)$. We follow the usual convention of writing $A \& (B \supset C)$ for the former and $A \& B \supset C$ for the latter, and in general, having conjunction and disjunction bind more strongly than implication.

Equivalence is a defined notion:

$$A \supset\subset B = (A \supset B) \& (B \supset A) : Prop$$

The definition of negation through implication and falsity is given by

$$\sim A = A \supset \bot : Prop$$

These are somewhat abbreviated notations ("pattern-matching equations"). More formally, if an arbitrary proposition is given by the declaration $A : Prop$, functional application of $\supset$ gives, successively, $\supset (A) : (Prop)Prop$ and $\supset (A, \bot) : Prop$. Negation is the one-place defined connective $\sim = (A) \supset (A, \bot) : (Prop)Prop$, where the first A in parentheses indicates functional abstraction over A in $\supset (A, \bot)$. Then, by β-conversion, we get for the negation of a proposition B

$$((A) \supset (A, \bot))(B) = \supset (B, \bot) : Prop$$

Given an arbitrary proposition A, it is either the constant proposition $\bot$, an atomic proposition, or (the value of) conjunction, disjunction, or implication. The notation often used in categorial grammar is

$$A := \bot \mid P \mid A \& B \mid A \vee B \mid A \supset B$$

By the method of functional abstraction, propositions can be presented as values of constant functions from any type to the type $Prop$. For example, given a type α with $x : \alpha$, we can abstract vacuously over x in $\bot$ to obtain $(x)\bot : (\alpha)Prop$. The rule of β-conversion gives trivially the value $\bot$ to applications of the constant function $(x)\bot$; we always have $((x)\bot)(a) = \bot : Prop$, as there is no place in $\bot$ to substitute a value of x.

We can put **predicate logic** into the framework of simple type theory if we assume for simplicity that we deal with just one domain $\mathcal{D}$. The objects, individual **constants**, are denoted by $a, b, c, \ldots$. Instead of the propositional constants $P, Q, R, \ldots$, atomic propositions can be values of propositional functions over $\mathcal{D}$, thus categorized as $P : (\mathcal{D})Prop$, $Q : (\mathcal{D})(\mathcal{D})Prop$, and so on. Next we have individual **variables** $x, y, z, \ldots$ taking values in $\mathcal{D}$.[1] Following the usual custom, we permit free variables in propositions. Propositions with free variables

---

[1] In traditional terminology, we can think of the constants as the "given" objects, thought of as fixed in value, and of the variables as the "sought," as when a and b are assumed given, and a value is sought for x such that the condition $ax + b = 0$ is fulfilled.

are understood as propositions under assumptions, say, if $A : (\mathcal{D})Prop$, then $A(x) : Prop$ under the assumption $x : \mathcal{D}$. **Terms** $t, u, v, \ldots$ are either individual parameters or variables. The language of predicate logic is obtained by adding to propositional logic the **quantifiers** *Every* and *Some*, with the categorizations

$$Every : ((\mathcal{D})Prop)Prop$$
$$Some : ((\mathcal{D})Prop)Prop$$

These are relative to a given domain $\mathcal{D}$. Thus, for each domain $\mathcal{D}$, a quantifier over that domain is a function that takes as argument a one-place propositional function $A : (\mathcal{D})Prop$ and gives as value a proposition, here either $Every(A)$ or $Some(A)$. The symbolic notation for quantifiers is given in the definitions

$$\forall = Every : ((\mathcal{D})Prop)Prop$$
$$\exists = Some : ((\mathcal{D})Prop)Prop$$

The usual way of writing quantified propositions is either $\forall x\, A$ and $\exists x\, A$ or $\forall x\, A(x)$ and $\exists x\, A(x)$. In the latter, the expression $A(x)$ does not stand for the application of A to x, but just mentions the quantified variable.

For greater generality, we can consider the domain a parameter that can be varied. Each domain $\mathcal{D}$ is a **set**, or belongs to the type of sets, formally $\mathcal{D} : Set$. Simple type theory is not expressive enough for the categorization of **bounded** quantifiers that take as first argument a set $\mathcal{D}$ that acts as the domain, then a propositional function $A : (\mathcal{D})Prop$ depending on that set, and give as value a proposition, either $\forall(\mathcal{D}, A)$ or $\exists(\mathcal{D}, A)$. These propositions have convenient variable-free readings, in the style of the Aristotelian syllogisms: all $\mathcal{D}$'s are A, some $\mathcal{D}$ is A. With bounded quantifiers, we have in use any number of domains of individuals. The set over which a quantifier ranges can be indicated by a notation such as $(\forall x : \mathcal{D})A(x)$. We can now write propositions such as $(\forall x : Line)(\exists y : Point)Incident(y, x)$.

When logical languages are defined through categorial grammars, quantifiers always apply to one-place propositional functions, but not to arbitrary formulas. If we have, say, $B : (\mathcal{D})(\mathcal{D})Prop$ and $a : \mathcal{D}$, then $B(a) : (\mathcal{D})Prop$, and we get in the usual notation $\forall x\, B(a)$ which looks uncommon, instead of the functional notation $\forall(\mathcal{D}, B(a))$. A writing with free variables displayed would be $B(x, y) : Prop$, and functional abstraction over y gives $(y)B(x, y) : (\mathcal{D})Prop$, so we have $\forall x(y)B(x, y)$. Since we permit free variables in propositions, we can write $\forall x\, B(x, y)$, provided that we have made the assumption $y : \mathcal{D}$.

In general, from the purely functional notation $\forall(A)$, where $A : (\mathcal{D})Prop$, we see that two quantified propositions must be set equal if they differ only in the symbol used for the quantified variable. This can be effected in two ways: The first is to have an explicit rule of α-**conversion** that permits renaming variables bound by a quantifier. The second way is to build α-conversion into a logical system by a suitable formulation of the rules of inference for quantifiers, as in Section 4.1.

NOTES TO APPENDIX A

Functional abstraction, found by Alonzo Church in the 1930s, formalizes the common notion of a function as an expression with a variable. The application of a function consists of a substitution, and the computation of values of a function consists of steps of β-conversion, until no such conversion applies. None of these concepts, abstraction, application, and computation of values of a function, is recognized by the set-theoretical notion of a function. Thus the appreciation of Church's work as one of the most important contributions to foundations of mathematics has been slow in coming.

The early work of Church is found in his (1940) and in the monograph *The Calculi of Lambda-Conversion*, 1941. The impact of Church's λ-calculus in logic and computer science is described in Barendregt (1997). The definition of the language of predicate logic through type theory and categorial grammar is treated in detail in Ranta (1994).

Proof Theory and Constructive Type Theory

In this second appendix, we shall first introduce **lower-level** type theory, and then show how it can be used for the semantical justification of the rules of natural deduction. Next we introduce **higher-level** type theory, then show by an example how mathematical theories can be represented formally as **type systems**.

B.1. LOWER-LEVEL TYPE THEORY

We start with a type of propositions, designated *Prop*. Propositions are thought of as their sets of formal proofs, in accordance with the **propositions-as-sets** principle. For each kind of proposition, there will be rules of **formation**, **introduction**, **elimination**, and **computation**. The rules operate on assertions or, as one often says, **judgments**, of which there are four kinds. *Prop* and *Set* are considered synonyms, and we use whatever terminology is appropriate in a situation, logical or set-theoretical:

$A : Prop, A : Set,$ A is a proposition, A is a set,

$a : A,$ a is a proof of proposition A, a is an element of set A,

$a = b : A,$ a and b are equal elements of set A,

$A = B : Set,$ A and B are equal sets.

To emphasize the formal character of a proof a of a proposition A, it is often called a **proof-object** or also a **proof term**. We also call them simply **objects**, a word that can equally well mean the element of a set. The equality of two objects in the judgment $a = b : A$ is **definitional**. There will be two general rules for the definitional equality of objects:

$$a = a : A \qquad \frac{a = c : A \quad b = c : A}{a = b : A}$$

The first, premissless rule expresses the reflexivity and the second the transitivity of the definitional equality of two objects. In these rules, the premises include that $A : Set, a : A, b : A$, etc., but we do not write these out. Let us derive symmetry of definitional equality from the above rules: Assume $b = a : A$. By reflexivity, $a = a : A$, and therefore, by transitivity, $a = b : A$. Thus definitional equality of objects in a given set is an equivalence relation in that set.

Similarly to the equality of objects, there will be general rules for the equality of sets:

$$A = A : Set \qquad \frac{A = C : Set \quad B = C : Set}{A = B : Set}$$

Symmetry follows as in the case of equality of objects.

Dependent types are families of sets indexed by a set: If $B(x)$ is a set whenever $x : A$, we can form the **product type** $(x : A)B$. The notation is a generalization of that of simple type theory. In the latter, with B a constant type, we write $(A)B$ for the function type. Thus we can also call $(x : A)B$ a dependent function type. Connected to dependent types, we have **hypothetical** judgments, or judgments **in a context**, of all the four forms of judgment of type theory. Contexts are progressive lists of variable declarations. We now stipulate that A is a set in the context $x_1 : A_1, x_2 : A_2, \ldots, x_n : A_n$ if $A(a_1/x_1, \ldots, a_n/x_n)$ is a set whenever $a_1 : A_1, a_2 : A_2(a_1/x_1), \ldots, a_n : A_n(a_1/x_1, \ldots, a_{n-1}/x_{n-1})$. Similarly, $x : A$ in the context $x_1 : A_1, x_2 : A_2, \ldots, x_n : A_n$ if $x(a_1, \ldots, a_n) : A(a_1/x_1, \ldots, a_n/x_n)$ whenever $a_1 : A_1, a_2 : A_2(a_1/x_1), \ldots, a_n : A_n(a_1/x_1, \ldots, a_{n-1}/x_{n-1})$.

In type theory, it is usual to give the rules for forming propositions as explicit syntactical rules. The rules for propositional logic are

$$\frac{A : Prop \quad B : Prop}{A \& B : Prop} \qquad \frac{A : Prop \quad B : Prop}{A \vee B : Prop} \qquad \frac{A : Prop \quad B : Prop}{A \supset B : Prop}$$

The rules for quantified propositions are

$$\frac{\begin{matrix}[x : A]\\ \vdots\\ A : Prop \quad B : Prop\end{matrix}}{(\forall x : A)B : Prop} \qquad \frac{\begin{matrix}[x : A]\\ \vdots\\ A : Prop \quad B : Prop\end{matrix}}{(\exists x : A)B : Prop}$$

To these we add the zero-premiss rule $\bot : Prop$.

For each form of proposition, we have a function constant that gives as values objects of that set, called the **constructor**. Then we have **selectors**, functions that operate on the set. The constructor for conjunction introduction is a function, to be called *pair*, operating on proofs a of A and b of B, to give as value *pair*(a, b) a proof of $A \& B$, with the rule notation

$$\frac{a : A \quad b : B}{pair(a, b) : A \& B} \, \&I$$

The constructors for disjunction introduction are

$$\frac{a : A}{i(a) : A \vee B} \vee I_1 \qquad \frac{b : B}{j(b) : A \vee B} \vee I_2$$

The constructors i and j are the "canonical injections" into the "disjoint union" $A \vee B$.

The constructor for implication introduction is

$$\frac{\begin{array}{c}[x : A]\\ \vdots \\ b : B\end{array}}{(\lambda x)b : A \supset B} \supset I$$

For conjunction elimination, we have the two rules with selectors

$$\frac{c : A \& B}{p(c) : A} \& E_1 \qquad \frac{c : A \& B}{q(c) : B} \& E_2$$

The functions p and q are the **projections** of the "product" $A \& B$.

For disjunction, we have the elimination rule, where $(x)d$ and $(y)e$ are obtained by functional abstraction as in Appendix A,

$$\frac{c : A \vee B \quad \begin{array}{c}[x : A]\\ \vdots \\ d : C\end{array} \quad \begin{array}{c}[y : B]\\ \vdots \\ e : C\end{array}}{\vee E(c, (x)d, (y)e) : C} \vee E$$

Thus the selector $\vee E$ is a function that takes as arguments an arbitrary proof c of $A \vee B$, a function $(x)d$ that converts arbitrary proofs of A into proofs of C, and a function $(y)e$ that converts arbitrary proofs of B into proofs of C. The value of $\vee E$ is a proof of C.

For implication, we have

$$\frac{c : A \supset B \quad a : A}{ap(c, a) : B} \supset E$$

The quantifier introduction rules with constructors are

$$\frac{\begin{array}{c}[x : A]\\ \vdots \\ b : B\end{array}}{(\lambda x)b : (\forall x : A)B} \forall I \qquad \frac{a : A \quad b : B(a/x)}{pair(a, b) : (\exists x : A)B} \exists I$$

The corresponding elimination rule for universal quantification is

$$\frac{c : (\forall x : A)B \quad a : A}{ap(c, a) : B(a/x)} \ \forall E$$

For existential quantification, there are two rules:

$$\frac{c : (\exists x : A)B}{p(c) : A} \ \exists E_1 \qquad \frac{c : (\exists x : A)B}{q(c) : B(p(c)/x)} \ \exists E_2$$

To the introduction and elimination rules we add a rule corresponding to falsity elimination:

$$\frac{c : \bot}{efq(c) : C(c/x)} \ \bot E$$

The rule is more general than the corresponding rule in propositional logic because of the possible dependence of C on the proof-object of the premiss. The quantifier rules are almost the same as the rules for implication and conjunction: The only difference is that, in the latter two, B is a constant proposition. Indeed, the rules for implication are special cases of the rules for universal quantification and similarly for conjunction and existence. Thus it is sufficient to have the quantifier rules and the rules for disjunction and falsity.

If in the above rules we hide all the proof-objects, we are back to the usual rules of natural deduction.

The rules of type theory have judgments or assertions as premisses and as conclusion. In Chapter 1, we used a turnstile notation $\vdash A$ for emphasizing that rules of inference act on assertions, not propositions. The assertion $\vdash A$ is what remains from $a : A$ when the proof-object is deleted. Lower-level type theory can be seen as a formalization of the computational semantics of intuitionistic logic, and we see that the type-theoretical rules make the rules of natural deduction valid under this semantics.

In Dummett's constructive semantics, meaning is explained in terms of proof. A proof of an implication $A \supset B$ is a function that converts an arbitrary proof of A into some proof of B. Thus, to explain what a proof of $A \supset B$ is, we would have to accept the notion of an arbitrary proof. To avoid the circularity of this explanation, Dummett (1975) distinguished direct or canonical and indirect or noncanonical proofs and required that the semantical explanations for the latter reduce to those for the former. In type theory, a canonical proof-object is one that is of the form of a constructor, and a noncanonical is one of the form of a selector. The requirement is, in these terms, that noncanonical objects must always convert to canonical ones. The conversions are made explicit in **computation rules**, also called **equality rules**, that is, rules that prescribe how the values of selectors are computed into canonical form:

The computation rules for conjunction are

$$\frac{a : A \quad b : B}{p(pair(a, b)) = a : A}\ \&eq \qquad \frac{a : A \quad b : B}{q(pair(a, b)) = b : B}\ \&eq$$

The computation rules for disjunction are

$$\frac{\begin{array}{ccc}[x:A] & [y:B]\\ \vdots & \vdots\\ a:A \quad d:C & e:C\end{array}}{\vee E(i(a), (x)d, (y)e) = d(a/x) : C}\ \vee eq \qquad \frac{\begin{array}{ccc}[x:A] & [y:B]\\ \vdots & \vdots\\ b:B \quad d:C & e:C\end{array}}{\vee E(j(b), (x)d, (y)e) = e(b/y) : C}\ \vee eq$$

Finally, the computation rule for implication is

$$\frac{\begin{array}{c}[x:A]\\ \vdots\\ b:B\end{array}}{ap((\lambda x)b, a) = b(a/x) : B}\ \supset eq$$

The computation rules for the quantifiers are just like the rules for implication and conjunction, with a dependent type in place of the constant type *B*. The computation of a noncanonical expression proceeds outside–in and corresponds exactly to the detour conversion of natural deduction derivations as in Section 8.5; for example, the last computation rule above corresponds to converting an introduction of implication followed by elimination.

In order to explain the notion of an arbitrary proof, it is essential that the conversion of a noncanonical expression terminate in a finite number of steps in a unique canonical form. This was shown by Martin-Löf (1975).

Similarly to natural deduction with general elimination rules, we can give general elimination rules for conjunction and implication in type theory:

$$\frac{\begin{array}{cc}[x:A], [y:B]\\ \vdots\\ c:A\&B \qquad d:C\end{array}}{\&E(c, (x)(y)d) : C} \qquad \frac{\begin{array}{cc}[y:B]\\ \vdots\\ c:A \supset B \quad a:A \quad d:C\end{array}}{gap(c, a, (y)d) : C}$$

The selectors &*E* and *gap* are computed by the equalities

$$\&E(pair(a, b), (x)(y)d) = d(a/x, b/y) : C$$
$$gap((\lambda x)b, a, (y)d) = d(b(a/x)/y) : C$$

The selectors in the special elimination rules for conjunction can be defined:

$$p(c) = \&E(c, (x)(y)x) : A \qquad q(c) = \&E(c, (x)(y)y) : B$$

Similarly, the selector corresponding to modus ponens has the definition

$$ap(c, a) = gap(c, a, (y)y) : B$$

The rule of universal elimination has analogous general and special rules in type theory.

B.2. HIGHER-LEVEL TYPE THEORY

Simple type theory was briefly presented in Section A.1. Its generalization by dependent types results in higher-level type theory. A dependent type is a family of types parametrized by another type. The objects of a dependent type are functions over the parameter type such that the range of the function depends on the argument. More formally, β is a dependent type over α if α is a type and β is a type whenever an object $x : \alpha$ is given. The family of dependent types is written as $(x : \alpha)\beta$. In the case of dependent typing, the notation for typings of functions has to display the dependency, here the variable x acting as argument, and this is achieved by generalizing the scheme of functional abstraction of simple type theory into

$$\frac{\begin{array}{c} [x : \alpha] \\ \vdots \\ b : \beta \end{array}}{(x)b : (x : \alpha)\beta} \tag{1}$$

We write $\beta(x)$ for a **type in the context** $x : \alpha$. We can consider the notation $(\alpha)\beta$ of simple type theory as an abbreviation of $(x : \alpha)\beta$ when β is a constant type.

The scheme of functional application becomes

$$\frac{f : (x : \alpha)\beta \quad a : \alpha}{f(a) : \beta(a/x)} \tag{2}$$

Functions formed by abstraction are applied by the rule of β-conversion:

$$\frac{\begin{array}{c} [x : \alpha] \\ \vdots \\ b : \beta \quad a : \alpha \end{array}}{((x)b)(a) = b(a/x) : \beta(a/x)} \tag{3}$$

In lower-level type theory, we added proof-objects to predicate logic and gave rules of formation, introduction, elimination, and computation separately for each logical operation. With higher-level type theory, these rules come out as instances of the general schemes of functional application and β-conversion.

Higher-level type theory works as a general framework for categorial grammars. In the case of logical languages, the connectives are categorized as in simple type theory, but the categorization of bounded quantifiers requires dependent typing:

$$\forall : (A : Set)(B : (A)Prop)Prop, \quad \exists : (A : Set)(B : (A)Prop)Prop$$

The quantifiers are functions that take as arguments a set A, a propositional function B over A (that is, a property of objects of A), and give as value a proposition $\forall(A, B)$ or $\exists(A, B)$. Proof-objects for the introduction of universally quantified propositions have the categorization

$$\lambda : (A : Set)(B : (A)Prop)((x : A)B(x))\forall(A, B)$$

Here the proof-object for the third argument, of type $(x : A)B(x)$, is left out by the convention about constant types. Also, since B is a propositional function over A, it must be written out that it is applied to $x : A$ with the proposition $B(x)$ as value. By the typing, λ is a function that takes as arguments, in turn, a set A, a propositional function B over A, and a function that transforms any proof $a : A$ into a proof of $B(a)$, and gives as value a proof of $\forall(A, B)$. For the special case of B constant over A, we have

$$\lambda : (A : Prop)(B : Prop)((A)B)Implies(A, B)$$

Written as a rule, λ-abstraction looks very much like the rule for functional abstraction. The difference is that $Implies(A, B)$ is a set with a constructor λ, whereas $(\alpha)\beta$ is a type that does not need to have a constructor. This is also reflected in the order of the rules. For λ-abstraction, the rule for the constructor is given first, and then the selector rule is justified by the conversion rule. In general functional abstraction, instead, the application rule comes first in conceptual order. (See Ranta 1994, p. 166, for detailed explanation.)

For the introduction of the existential quantifier, we have the categorization

$$pair : (A : Set)(B : (A)Prop)(x : A)(B(x))\exists(A, B)$$

It is usual not to write out the type arguments A and B of $pair$, but to take it as a function in a context $A : Set$, $B : (A)Prop$. Thus the value is written as $pair(a, b)$ instead of $pair(A, B, a, b)$. Since all objects have to be typed, the first two arguments in the latter can be read off from a and b.

The selectors have the categorizations

$$ap : (A : Set)(B : (A)Prop)(\forall(A, B))(a : A)B(a)$$

$$p : (A : Set)(B : (A)Prop)(\exists(A, B))A$$

$$q : (A : Set)(B : (A)Prop)(\exists(A, B))B(p(A, B, c))$$

The third argument of p is $c : \exists(A, B)$. The categorizations for implication and conjunction are special cases. For disjunction introduction, we have

$$i : (A : Prop)(B : Prop)(A)Or(A, B)$$

$$j : (A : Prop)(B : Prop)(B)Or(A, B)$$

The categorization of the elimination rule has to be written on two lines:

$\vee E : (A : Prop)(B : Prop)(C : (Or(A, B))Prop)$
$\quad (c : Or(A, B))((a : A)C(i(A, B, a)))((b : B)C(j(A, B, b)))C(c)$

The general elimination rules for universal and existential quantification are typed analogously.

The computation rules are definitional equalities showing how selectors act on constructors. We show only the selectors corresponding to the special elimination rules. The computation rule for universal quantification concludes the equality

$$ap(A, B, \lambda(A, B, c), a) = c(a) : B(a)$$

The computation rules for existential quantification give

$$p(A, B, pair(A, B, a, b)) = a : A$$

$$q(A, B, pair(A, B, a, b)) = b : B$$

For disjunction, we have two computation rules with the equalities

$$\vee E(A, B, C, i(A, B, a), d, e) = d(a) : C(i(A, B, a))$$

$$\vee E(A, B, C, j(A, B, b), d, e) = e(b) : C(j(A, B, b))$$

B.3. TYPE SYSTEMS

Type theory has a reading as a constructive set theory, with the basic form of assertion $a : A$ read as: a is an element of the set A. Conjunction corresponds to the intersection and disjunction to the (disjoint) union of two sets, and universal quantification to the Cartesian product of a family of sets, existential quantification to the direct sum of a family of sets.

Another reading of type theory is that types $A, B, C, \ldots$ express problems to be solved, in particular, specifications of programming problems. Then objects can be interpreted as programs and the basic form of assertion $a : A$ as the statement that program a meets the specification A. In traditional programming languages, there is no formal way of expressing program specifications. In traditional logical languages, there is no formal way of expressing proofs, but type theory unites these two. The effect on programming methodology is that the **correctness** of a program becomes a formally well-defined property of correct typing.

Higher-level type theory offers a general framework for defining **type systems**, not limited to predicate logic. As a simple example, let us consider a type system for elementary geometry (compare also the example from Section 6.6(e)). We declare the basic sets of points and lines by $Pt : Set, Ln : Set$. Next we declare the basic relations of distinct points, distinct lines, and apartness of a point from a line, with some obvious abbreviations to make the declarations fit a line. Also, since *Prop* is a constant type, the arguments in $x : Pt, \ldots$ are left out:

$$DiPt : (Pt)(Pt)Prop, \quad DiLn : (Ln)(Ln)Prop, \quad Apt : (Pt)(Ln)Prop$$

The reason for using distinct instead of equal points and lines as basic relations is that the construction of **connecting** lines and **intersection** points requires conditions expressed by dependent typing. These two geometrical constructions are introduced by the declarations

$$ln : (a : Pt)(b : Pt)(DiPt(a, b))Ln, \quad pt : (l : Ln)(m : Ln)(DiLn(l, m))Pt$$

The usual geometrical axioms are: There is a unique connecting line for any two distinct points, and the points are incident with the line. There is a unique intersection point for any two distinct lines, and the point is incident with both lines, where incidence is defined as the negation of apartness. The incidences are expressed by the propositions $Inc(a, ln(a, b))$, $Inc(b, ln(a, b))$, $Inc(pt(l, m), l)$, and $Inc(pt(l, m), m)$ but these are well-formed only if the conditions in the constructions are verified. Thus we have a simple case of dependent typing going beyond the expressive means of usual predicate logic. Looking at the typings of the two constructions, we notice that they both have three arguments, the connecting line construction two points a and b and a proof, say w, of $DiPt(a, b)$. Thus functional application gives us $ln(a, b, w) : Ln$ in the context $a : Pt, b : Pt, w : DiPt(a, b)$.

In type theory, the incidence properties of constructed objects are implemented by declaring function constants that prove these properties:

$$inc\text{-}ln1 : (a : Pt)(b : Pt)(w : DiPt(a, b))Inc(a, ln(a, b, w))$$

$$inc\text{-}ln2 : (a : Pt)(b : Pt)(w : DiPt(a, b))Inc(b, ln(a, b, w))$$

The uniqueness of connecting lines can be formalized by

$$uni\text{-}ln : (a : Pt)(b : Pt)(w : DiPt(a, b))(l : Ln)(Inc(a, l))(Inc(b, l))EqLn(l, ln(a, b, w))$$

where equality of lines is defined as negation of *DiLn*. For a complete formalization, the axiomatic properties of the basic relations have to be given, as well as principles that permit the substitution of equal objects in the basic relations, as in Section 6.6.

When a theory is formalized as a type system, computer implementations of type theory, known as **proof editors**, can be used for the formal development of proofs. Such proof editors are interactive systems of proof and program development, in which each step is checked for correctness through the **type-checker**, which is the heart of the computer implementation. A formally verified proof can be seen as a program that converts whatever is needed to verify the assumptions of a theorem into what is claimed in the theorem. In terms of programming problems, a formally verified proof converts the data of a problem into its solution. This is particularly clear in geometrical problems: Each problem has some "given" objects with assumed properties, and the solution-program transforms these into the "sought" objects with required properties. The effect of constructivity is that such programs are provably terminating.

NOTES TO APPENDIX B

Type theory as understood here was developed by Martin-Löf on the basis of ideas such as dependent typing and proof-objects (also found in de Bruijn 1970) and the propositions-as-sets principle or "Curry–Howard isomorphism" in Howard (1969). An early exposition is Martin-Löf (1975). The lower-level theory is explained in the booklet Martin-Löf (1984). The rules he gives for propositional equality permit the conclusion of a definitional equality from propositional equality which turned out to be erroneous. This is corrected in later expositions of type theory, such as Nordström, Petersson, and Smith (1990) and Ranta (1994).

The use of type theory as a programming system with the possibility of program verification was explained in Martin-Löf (1982). Actual computer implementation benefitted from the introduction of the efficient higher-level notation developed by Martin-Löf since 1985 on the basis of the calculus of constructions of Coquand and Huet (1988). A concise exposition, with applications to logic and linguistics, is given in Ranta (1994). The example of formalization of elementary geometry as a type system comes from von Plato (1995).

PESCA – A Proof Editor for Sequent Calculus

by

AARNE RANTA

aarne@cs.chalmers.se

PESCA is a program that helps in the construction of proofs in sequent calculus. It works both as a proof editor and as an automatic theorem prover. Proofs constructed in PESCA can both be seen on the terminal and printed into LaTeX files. The user of PESCA can choose among different versions of classical and intuitionistic propositional and predicate calculi and extend them by systems of nonlogical axioms. The aim of this appendix is to show what PESCA can be used for, as well as to give an outline of its implementation, which is written in the functional programming language Haskell. PESCA is a simple and small program, and extending it by implementing various calculi and algorithms of this book can provide instructive student projects on a level more advanced than the mere use of the editor.

C.1. INTRODUCTION

It was already realized by Gentzen that sequent calculus is not very natural for humans actually to write proofs. It carries around a lot of information that humans tend to keep in their heads rather than to put on paper. Although greatly improving the performance of machines operating on proofs, this information easily obscures the human inspection of them, and actually writing sequent calculus proofs in full detail is tedious and error prone. Thus it is obviously a task for which a machine can be helpful.

The domain of sequent calculi allows for indefinitely many variations, which are not due to disagreements on what should be provable but to different decisions on the fine structure of proofs. In terms of provability, it is usually enough to tell whether a calculus is intuitionistic or classical. In the properties of proof structure, there are many more choices. The very implementation of PESCA precludes most of them, but it still leaves room for different calculi, only some of which are included in the basic distribution. These calculi can be characterized as having:

Shared multiset contexts, no structural rules.

However, calculi can have a single formula as well as several formulas in the succedents of their sequents.

The fundamental common property of the calculi treated by PESCA is **top-down determinacy**:

Given a conclusion and a rule, the premises are determined.

This property is essential for our method of **top-down proof search**. The user of PESCA starts with a conclusion and then tries to **refine** it by suggesting a rule. If the rule is applicable, the proof of the conclusion is completed by the derived premises, and the proof search can continue from them. A branch in a proof is successfully terminated when a rule gives an empty list of premises. A branch fails if no rule applies to it. This simple procedure, which has been adopted from the proof editor ALF (Magnusson 1994) for the much richer formalism of constructive type theory, is not applicable to calculi that are not top-down determinate.

The term top-down runs counter to the standard typographical layout of proof trees, in which premises appear above conclusions. The term is frequent in computer science, where it may come from the standard layout of syntax trees, in which the root is above the trees. In proof theory, the confusion is usually avoided by saying "root-first" instead. There will be slight notational differences as compared to earlier chapters.

C.2. TWO EXAMPLE SESSIONS

(a) A proof in propositional calculus: Let us first construct a proof of the law of commutativity for disjunction, in the sequent form,

$$A \lor B \Rightarrow B \lor A.$$

Having started PESCA, we see its prompt $|$ –. We then enter a **new goal** by the command n followed by the sequent written in ASCII notation:

$|$ – n A v B => B v A

The reply of PESCA is a new **proof tree**, consisting of the conclusion alone, which is the only **subgoal** of the tree. Subgoals are identified by sequences of digits starting from the root, as in the following example:

$$\frac{\frac{111 \quad 112}{11} \quad \frac{121}{12} \quad 13}{1}$$

The command s shows the current subgoals, of which we still have just one, numbered 1. More interestingly, the command a shows the **applicable rules** for a given subgoal. In the situation where we are in our proof, we have

```
|- a 1

r 1 A1 S1 Lv   -- A v B => B v A
r 1 A1 S1 Rv1 -- A v B => B v A
r 1 A1 S1 Rv2 -- A v B => B v A
```

Any of the displayed r commands (line segments preceding --) can be cut and pasted to a command line, and it gives rise to a **refinement** of the subgoal, an extension of the tree that is determined by the chosen rule. For instance, choosing the first alternative takes us to a proof by the left disjunction rule from two premisses,

```
|- r 1 A1 S1 Lv

A => B v A   B => B v A
----------------------
A v B => B v A
```

As will be explained below, the part A1 S1 specifies the active formulas in the antecedent and the succedent; when the first formulas are chosen, as here, this part could be omitted, as in the next refinement,

```
|- r 11 Rv2

A => A
----------
A => B v A   B => B v A
----------------------
A v B => B v A
```

The left branch 111 can now be refined by rule ax, which does not generate any more subgoals:

```
|- r 111 ax
```

The right branch 12 can be refined analogously with 11, by Rv1 followed by ax. When the proof is ready, its ASCII appearance is usually not of very good quality. Now, at last, it is rewarding to use the command

```
|- 1
```

to write the current proof into a LATEX document, which looks like

$$\dfrac{\dfrac{A \Rightarrow A^{ax}}{A \Rightarrow B \vee A}\,RV_2 \quad \dfrac{B \Rightarrow B^{ax}}{B \Rightarrow B \vee A}\,RV_1}{A \vee B \Rightarrow B \vee A}\,LV$$

when processed.
   Finally,

```
|- d
```

shows the proof translated into natural deduction:

$$\underset{A \vee B}{\text{ass.}} \quad \dfrac{\overset{\overset{1}{A}}{B \vee A}}{\quad}\,VI_2 \quad \dfrac{\overset{\overset{1}{B}}{B \vee A}}{B \vee A}\,VI_1 \atop \rule{4cm}{0.4pt}}{B \vee A}\,VE,1,1$$

(b) A proof in predicate calculus: In predicate calculus, one more command is usually needed than in propositional calculus: the command i for **instantiating** parameters. Logically, a parameter such as t in the existential introduction rule $R\exists$

$$\dfrac{\Gamma \Rightarrow A(t/x)}{\Gamma \Rightarrow (\exists x)A}$$

is just like another premiss, which calls for a construction to be complete. This logic is made explicit in the Σ introduction rule of Martin-Löf's type theory and simplifies greatly the implementation of inference rules. Here, staying faithful to the syntax of predicate calculus, we have to treat such parameters as hidden premisses in the rules. Thus a proof that has uninstantiated parameters is incomplete in the same way as a proof that has open subgoals.
   Let us prove the quantifier switch law;

$$(\exists y)(\forall x)C(x, y) \Rightarrow (\forall x)(\exists y)C(x, y).$$

After introducing the goal, we make a couple of ordinary refinements:

```
|- n (/Ey)(/Ax)C(x,y) => (/Ax)(/Ey)C(x,y)
|- r 1 A1 S1 R/A
|- r 11 L/E
|- r 111 A1 S1 R/E
```

The last refinement introduces a parameter t, which we instantiate by y:

```
|- i t y
```

Continuing by L/A, x, and ax, we obtain a complete proof, in which we have afterward marked next to the rule symbols the two instantiations made:

$$\frac{\dfrac{\dfrac{\dfrac{\dfrac{C(x, y), (\forall x)C(x, y) \Rightarrow C(x, y)^{ax}}{(\forall x)C(x, y) \Rightarrow C(x, y)} L\forall x}{(\forall x)C(x, y) \Rightarrow (\exists y)C(x, y)} R\exists y}{(\exists y)(\forall x)C(x, y) \Rightarrow (\exists y)C(x, y)} L\exists}{(\exists y)(\forall x)C(x, y) \Rightarrow (\forall x)(\exists y)C(x, y)} R\forall}$$

C.3. SOME COMMANDS

Some PESCA commands were already exemplified in the two sessions of the previous section. This section will give a synopsis of them, as well as explain a couple of other commands. For the full set of commands, we refer to the electronically distributed manual.

Each command consists of one character followed by zero or more arguments, some of which may be optional. In the following, as in the on-line help file of PESCA, the arguments are denoted by words indicating their types. Optional arguments are enclosed in brackets. Typewriter font is used for terminal symbols.

To understand the commands fully, one should know that a PESCA session takes place in an **environment** which changes as a function of the commands. The environment consists of a **current calculus** and a **current proof**. In the beginning, the calculus is the intuitionistic predicate calculus **G3i**, and the proof is the one consisting of the impossible empty sequent ⇒.

r goal [A int] [S int] rule (refine)

replaces the goal by an application of the rule, if applicable, and leaves the current proof unchanged otherwise. The goal is denoted by a sequence of digits, as explained in the previous section. The options [A int] [S int] reset the active formulas in the antecedent and the succedent of the goal – by default, the active formula is number 1, which in the antecedent is the first and in the succedent the last printed formula. If the number exceeds the length, resetting the active formula has no effect.

i parameter term (instantiate)

replaces all occurrences of the parameter in the current proof by the term.

t goal int (try to refine)

replaces the goal by the first proof that it finds by recursively trying to apply all rules maximally int times. This is the **automatic proof search method** of PESCA, based on brute force but always terminating. With certain calculi, such as **G4ip**, this method always finds a proof after a predictable number of steps. With predicate calculus rules that require instantiations, the method usually fails.

n sequent (new)

replaces the current proof by a proof consisting of the given sequent which thereby becomes its open subgoal number 1.

u subtree (undo)

replaces the subtree by a goal consisting of the conclusion of the subtree. Subtrees are identified by sequences of digits in the same way as subgoals: Subtree n is the tree, the root node of which is the node n.

s (show subgoals)

shows all open subgoals. If the system responds by showing nothing, the current proof is complete.

a goal (applicable rules)

shows all refinement commands applicable to the goal.

c calculus (change calculus)

changes the current calculus. The help command ? shows available calculi. As a calculus is just a set of rules, calculi can be unioned by the operation +. Thus the command c G3c + Geq selects classical predicate calculus with equality.

x file (read axioms)

reads a file with nonlogical axioms, parses it into rules, adds the rules into the current calculus, and writes the rules into a file.

l [file] (print proof in a LaTeX file)

prints the current proof in LaTeX format in the indicated file. To process the LaTeX file, the style file proof.sty (Tatsuta 1990) is needed.

d [file] (print proof in natural deduction in a LaTeX file)

prints the current proof in LaTeX format in the indicated file. It works for **G3i** and **G3ip** only.

All LaTeX-producing commands also call the system to run LaTeX and then create the xdvi image on the background.

C.4. AXIOM FILES

Certain kinds of formulas can be interpreted as sequent calculus rules that preserve
the possibility of cut elimination and are hence favorable for proof search. These
formulas are implications with conjunctions of atoms on their left-hand sides and
disjunctions of atoms on their right-hand sides. Either side can be empty. An empty
left-hand side is represented by the omission of the implication sign. An empty
right-hand side is represented by the absurdity $\perp$. Alternatively, the negation of a
left-hand side is interpreted as its implication of absurdity. All those variables that
occur on the right-hand side but not on the left-hand side are treated as parameters.

As an example, consider a set of axioms for the theory of lattices. What the
user of PESCA types into a file looks as follows:

```
-- starts header. Text above header is sent to latex
as such.
Mtl    (a \wedge b) \leq a
Mtr    (a \wedge b) \leq b
Jnl    a \leq (a \vee b)
Jnr    b \leq (a \vee b)
Unimt  c \leq a & c \leq b -> c \leq (a \wedge b)
Unijn  a \leq c & b \leq c -> (a \vee b) \leq c
Ref    a \leq a
Trans  a \leq b & b \leq c -> a \leq c
```

The command x makes PESCA read the file and construct a set of sequent calculus
rules. These rule are also printed in LaTeX:

$$\frac{a \wedge b \leq a, \Gamma \Rightarrow \Delta}{\Gamma \Rightarrow \Delta} \; Mtl$$

$$\frac{a \wedge b \leq b, \Gamma \Rightarrow \Delta}{\Gamma \Rightarrow \Delta} \; Mtr$$

$$\frac{a \leq a \vee b, \Gamma \Rightarrow \Delta}{\Gamma \Rightarrow \Delta} \; Jnl$$

$$\frac{b \leq a \vee b, \Gamma \Rightarrow \Delta}{\Gamma \Rightarrow \Delta} \; Jnr$$

$$\frac{c \leq a \wedge b, c \leq a, c \leq b, \Gamma \Rightarrow \Delta}{c \leq a, c \leq b, \Gamma \Rightarrow \Delta} \; Unimt$$

$$\frac{a \vee b \leq c, a \leq c, b \leq c, \Gamma \Rightarrow \Delta}{a \leq c, b \leq c, \Gamma \Rightarrow \Delta} \; Unijn$$

$$\frac{a \leq a, \Gamma \Rightarrow \Delta}{\Gamma \Rightarrow \Delta} \; Ref$$

$$\frac{a \leq c, a \leq b, b \leq c, \Gamma \Rightarrow \Delta}{a \leq b, b \leq c, \Gamma \Rightarrow \Delta} \; Trans$$

C.5. ON THE IMPLEMENTATION

The full source code of PESCA is approximately 1400 lines of Haskell code, divided into nine modules.

(a) Abstract syntax. The central module is the one defining an **abstract syntax** of sequents, formulas, singular terms, proofs, etc. On the level of abstract syntax, all these expressions are Haskell data objects, which could also be called **syntax trees**. With the exception of the communication with the user, PESCA always operates on syntax trees. There are lots of operations defined by pattern matching on the basic types of syntax trees: Sequent, Formula, Term, Proof, AbsRule.

The definition that expresses the top-down determinacy of PESCA (see Section C.1) is the definition of the type of abstract inference rules:

```
type AbsRule = Sequent -> Maybe [Either Sequent Ident]
```

A rule is a function that takes a conclusion sequent as its argument and returns either a list of premisses or a failure. Returning an empty list of premisses means that the proof of the conclusion is complete. The premisses can be either sequents or parameter identifiers. (It was already argued, in Section C.2 above, that parameters are really premisses.)

Sequents are treated as pairs of multisets of formulas. The definition of the type of sequents is as pairs of lists rather than multisets: The multiset aspect is implicit in various functions that consider variants of these lists obtained when one formula in turn is made the active formula. In most cases, this is enough, and it is not necessary to consider all permutations.

Some calculi restrict the succedents of sequents to be single formulas. PESCA does not use a distinct type for these calculi: It simply is the property of certain calculi, such as **G3ip**, that the rules never require or produce multisuccedent sequents.

(b) Parsing and printing. In user input and PESCA output, syntax trees are represented by **strings** of characters. The relation between syntax trees and strings is defined by the **parsing** and **printing**. Parsing follows the top-down combinator method explained in Wadler (1985). The printing functions produce terminal output for PESCA sessions as well as LaTeX to be written into files.

(c) Predefined calculi. Some intuitionistic and classical calculi are defined directly as sets of abstract rules. Because abstract rules are functions, they cannot be read from separate files at runtime, but must be compiled into PESCA. Also considered was a special syntax that could be used for reading calculi from files, but finally this was restricted to nonlogical axioms: There is, after all, so much variation and irregularity in the rules, that the syntax and the operations on it become complicated and are not guaranteed to cope with "arbitrary sequent calculus rules." However, to experiment with new calculi, it is enough to edit one file.

(d) Interaction – refinement and proof search. The central proof search functions are refinement, instantiation, undoing, and automatic search. All of these operations are based on the underlying operation of replacing a subtree by another tree,

```
replace :: Proof -> [Int] -> Proof -> Proof
```

where lists of integers denote subtrees. Another underlying operation is to list all those rules of a calculus that apply to a given sequent,

```
applicableRules :: AbsCalculus -> Sequent ->
                   [((Ident,AbsRule), [Either Sequent Ident])]
```

Automatic proving calls this function in performing top-down proof search, following the methods of Wadler (1985), just like parsing.

(e) Natural deduction. This translates proofs in calculi **G3i** and **G3ip** into natural deduction. This would easily extend to **G4i** and **G4ip**, but this and the more demanding other calculi are left as an exercise.

(f) Dialogue and command language. The dialogue is based on monadic input and output. While commands are executed in an environment of a current calculus and a current proof, they produce output to the screen and files and change the environment.

NOTES TO APPENDIX C

PESCA is an experimental system still under development. Contributions and bug reports are thus welcome.

ELECTRONIC REFERENCES

Haskell home page, http://www.haskell.org/
PESCA home page, http://www.cs.chalmers.se/~aarne/pesca/
M. Tatsuta (1990) proof.sty (Proof Figure Macros), LaTeX style file.

Home page of this book, http://prooftheory.helsinki.fi

Bibliography

Artin, E. (1957) *Geometric Algebra*, Wiley, New York.

Balbes, R. and P. Dwinger (1974) *Distributive Lattices*, University of Missouri Press, Columbia, Missouri.

Barendregt, H. (1997) The impact of the lambda calculus in logic and in computer science, *The Bulletin of Symbolic Logic*, vol. 2, pp. 181–215.

Bernays, P. (1945) Review of Ketonen (1944), *The Journal of Symbolic Logic*, vol. 10, pp. 127–130.

Beth, E. (1959) *The Foundations of Mathematics*, North-Holland, Amsterdam.

Bishop, E. and D. Bridges (1985) *Constructive Analysis*, Springer, Berlin.

de Bruijn, N. (1970) The mathematical language AUTOMATH, its usage and some of its extensions, as reprinted in R. Nederpelt et al., eds, *Selected Papers on Automath*, pp. 73–100, North-Holland, Amsterdam.

Buss, S. (1998) Introduction to proof theory, in S. Buss, ed, *Handbook of Proof Theory*, pp. 1–78, North-Holland, Amsterdam.

Church, A. (1940) A formulation of the simple theory of types, *The Journal of Symbolic Logic*, vol. 5, pp. 56–68.

Church, A. (1941) *The Calculi of Lambda-Conversion*, Princeton University Press.

Coquand, T. and G. Huet (1988) The calculus of constructions, *Information and Computation*, vol. 76, pp. 95–120.

Curry, H. (1963) *Foundations of Mathematical Logic*, as republished by Dover, New York, 1977.

van Dalen, D. (1986) Intuitionistic logic, in D. Gabbay and F. Guenthner, eds, *Handbook of Philosophical Logic*, vol. 3, pp. 225–339, Reidel, Dordrecht.

van Dalen, D. (1994) *Logic and Structure*, Springer, Berlin.

van Dalen, D. and R. Statman (1979) Equality in the presence of apartness, in J. Hintikka et al., eds, *Essays in Mathematical and Philosophical Logic*, pp. 95–116, Reidel, Dordrecht.

Dragalin, A. (1988) *Mathematical Intuitionism: Introduction to Proof Theory*, American Mathematical Society, Providence, Rhode Island.

Dummett, M. (1959) A propositional calculus with denumerable matrix, *The Journal of Symbolic Logic*, vol. 24, pp. 96–107.

Dummett, M. (1975) The philosophical basis of intuitionistic logic, as reprinted in M. Dummett, *Truth & Other Enigmas*, pp. 215–247, Duckworth, London, 1978.

Dummett, M. (1977) *Elements of Intuitionism*, Oxford University Press.

Dyckhoff, R. (1992) Contraction-free sequent calculi for intuitionistic logic, *The Journal of Symbolic Logic*, vol. 57, pp. 795–807.

Dyckhoff, R. (1997) Dragalin's proof of cut-admissibility for the intuitionistic sequent calculi **G3i** and **G3i′**, Research Report CS/97/8, Computer Science Division, St Andrews University.

Dyckhoff, R. (1999) A deterministic terminating sequent calculus for Gödel-Dummett logic, *Logic Journal of the IGPL*, vol. 7, pp. 319–326.

Dyckhoff, R. and S. Negri (2000) Admissibility of structural rules in contraction-free sequent calculi, *The Journal of Symbolic Logic*, vol. 65, pp. 1499–1518.

Dyckhoff, R. and S. Negri (2001) Admissibility of structural rules for extensions of contraction-free sequent calculi, *Logic Journal of the IGPL*, in press.

Dyckhoff, R. and L. Pinto (1998) Cut-elimination and a permutation-free sequent calculus for intuitionistic logic, *Studia Logica*, vol. 60, pp. 107–118.

Ekman, J. (1998) Propositions in propositional logic provable only by indirect proofs, *Mathematical Logic Quarterly*, vol. 44, pp. 69–91.

Feferman, S. (2000) Highlights in proof theory, in V. Hendricks et al., eds, *Proof Theory: History and Philosophical Significance*, pp. 11–31, Kluwer, Dordrecht.

Gentzen, G. (1932) Ueber die Existenz unabhängiger Axiomensysteme zu unendlichen Satzsystemen, *Mathematische Annalen*, vol. 107, pp. 329–350.

Gentzen, G. (1934–35) Untersuchungen über das logische Schliessen, *Mathematische Zeitschrift*, vol. 39, pp. 176–210 and 405–431.

Gentzen, G. (1936) Die Widerspruchsfreiheit der reinen Zahlentheorie, *Mathematische Annalen*, vol. 112, pp. 493–565.

Gentzen, G. (1938) Neue Fassung des Widerspruchsfreiheitsbeweises für die reine Zahlentheorie, *Forschungen zur Logik und zur Grundlegung der exakten Wissenschaften*, vol. 4, pp. 19–44.

Gentzen, G. (1969) *The Collected Papers of Gerhard Gentzen*, M. Szabo, ed, North-Holland, Amsterdam.

Girard, J.-Y. (1987) *Proof Theory and Logical Complexity*, Bibliopolis, Naples.

Gödel, K. (1931) On formally undecidable propositions of *Principia mathematica* and related systems I (English translation of German original), in van Heijenoort (1967), pp. 596–617.

Gödel, K. (1932) Zum intuitionistischen Aussagenkalkül, as reprinted in Gödel's *Collected Works*, vol. 1, pp. 222–225, Oxford University Press 1986.

Gödel, K. (1941) In what sense is intuitionistic logic constructive?, a lecture first published in Gödel's *Collected Works*, vol. 3, pp. 189–200, Oxford University Press 1995.

Gordeev, L. (1987) On cut elimination in the presence of Peirce rule, *Archiv für mathematische Logik*, vol. 26, pp. 147–164.

Görnemann, S. (1971) A logic stronger than intuitionism, *The Journal of Symbolic Logic*, vol. 36, pp. 249–261.

Hallnäs, L. and P. Schroeder-Heister (1990) A proof-theoretic approach to logic programming. I. Clauses as rules, *Journal of Logic and Computation*, vol. 1, pp. 261–283.

Harrop, R. (1960) Concerning formulas of the type $A \to B \vee C$, $A \to (Ex)B(x)$ in intuitionistic formal systems, *The Journal of Symbolic Logic*, vol. 25, pp. 27–32.

van Heijenoort, J., ed, (1967) *From Frege to Gödel, A Source Book in Mathematical Logic, 1879–1931*, Harvard University Press.

Herbelin, H. (1995) A λ-calculus structure isomorphic to Gentzen-style sequent calculus structure, *Lecture Notes in Computer Science*, vol. 933, pp. 61–75.

Hertz, P. (1929) Ueber Axiomensysteme für Beliebige Satzsysteme, *Mathematische Annalen*, vol. 101, pp. 457–514.

Hilbert, D. (1904) On the foundations of logic and arithmetic (English translation of German original), in van Heijenoort (1967), pp. 129–138.

Howard, W. (1969) The formulae-as-types notion of construction, published in 1980 in J. Seldin and J. Hindley, eds, *To H. B. Curry: Essays on Combinatory Logic, Lambda Calculus and Formalism*, pp. 480–490, Academic Press, New York.

Hudelmaier, J. (1992) Bounds for cut elimination in intuitionistic propositional logic, *Archive for Mathematical Logic*, vol. 31, pp. 331–354.

Joachimski, F. and R. Matthes (2001) Short proofs of normalization for the simply typed λ-calculus, permutative conversions, and Gödel's T, *Archive for Mathematical Logic*, vol. 40, in press.

Ketonen, O. (1941) Predikaattikalkyylin täydellisyydestä (On the completeness of predicate calculus), *Ajatus*, vol. 10, pp. 77–92.

Ketonen, O. (1943) "Luonnollisen päättelyn" kalkyylista (On the calculus of "natural deduction"), *Ajatus*, vol. 12, pp. 128–140.

Ketonen, O. (1944) *Untersuchungen zum Prädikatenkalkül* (Annales Acad. Sci. Fenn. Ser. A.I. 23), Helsinki.

Kleene, S. (1952), *Introduction to Metamathematics*, North-Holland, Amsterdam.

Kleene, S. (1952a) Permutability of inferences in Gentzen's calculi LK and LJ, *Memoirs of the American Mathematical Society*, No. 10, pp. 1–26.

Kleene, S. (1967), *Mathematical Logic*, Wiley, New York.

Kolmogorov, A. (1925) On the principle of excluded middle (English translation of Russian original), in van Heijenoort (1967), pp. 416–437.

Magnusson, L. (1994) *The Implementation of ALF – a Proof Editor Based on Martin-Löf's Monomorphic Type Theory With Explicit Substitution*, PhD thesis, Department of Computing Science, Chalmers University of Technology, Gothenburg.

Martin-Löf, P. (1975) An intuitionistic theory of types: predicative part, in H. Rose and J. Shepherson, eds, *Logic Colloquium '73*, pp. 73–118, North-Holland, Amsterdam.

Martin-Löf, P. (1982) Constructive mathematics and computer programming, in L. Cohen et al., eds, *Logic, Methodology and Philosophy of Science IV*, pp. 153–175, North-Holland, Amsterdam.

Martin-Löf, P. (1984) *Intuitionistic Type Theory*, Bibliopolis, Naples.

Martin-Löf, P. (1985) On the meanings of the logical constants and the justifications of the logical laws, in *Atti degli Incontri di Logica Matematica*, vol. 2, pp. 203–281,

Dipartimento di Matematica, Università di Siena. Republished in *Nordic Journal of Philosophical Logic*, vol. 1 (1996), pp. 11–60.

Menzler-Trott, E. (2001) *Gentzens Problem: Mathematische Logik im nationasozialistischen Deutschland*, Birkhauser, Basel.

Mints, G. (1993) A normal form for logical derivation implying one for arithmetic derivations, *Annals of Pure and Applied Logic*, vol. 62, pp. 65–79.

Mostowski, A. (1965) *Thirty Years of Foundational Studies*, Societas Philosophica Fennica, Helsinki. Also in Mostowski's collected works, vol. 1.

Negri, S. (1999) Sequent calculus proof theory of intuitionistic apartness and order relations, *Archive for Mathematical Logic*, vol. 38, pp. 521–547.

Negri, S. (1999a) A sequent calculus for constructive ordered fields, in U. Berger et al., eds, *Reuniting the Antipodes – Constructive and Nonstandard Views of the Continuum*, Kluwer, Dordrecht, in press.

Negri, S. (2000) Natural deduction and normal form for intuitionistic linear logic, to appear.

Negri, S. and J. von Plato (1998) Cut elimination in the presence of axioms, *The Bulletin of Symbolic Logic*, vol. 4, pp. 418–435.

Negri, S. and J. von Plato (1998a) From Kripke models to algebraic countervaluations, in H. de Swart, ed, *Automated Reasoning with Analytic Tableaux and Related Methods*, pp. 247–261 (LNAI, vol. 1397), Springer, Berlin.

Negri, S. and J. von Plato (2001) Sequent calculus in natural deduction style, *The Journal of Symbolic Logic*, vol. 66, in press.

Nishimura, I. (1960) On formulas of one variable in intuitionistic propositional calculus, *The Journal of Symbolic Logic*, vol. 25, pp. 327–331.

Nordström, B., K. Petersson and J. Smith (1990) *Programming in Martin-Löf's Type Theory: An Introduction*, Oxford University Press.

von Plato, J. (1995) The axioms of constructive geometry, *Annals of Pure and Applied Logic*, vol. 76, pp. 169–200.

von Plato, J. (1998) Natural deduction with general elimination rules, *Archive for Mathematical Logic*, in press.

von Plato, J. (1998a) Proof theory of full classical propositional logic, ms.

von Plato, J. (1998b) A structural proof theory of geometry, ms.

von Plato, J. (2000) A problem of normal form in natural deduction, *Mathematical Logic Quarterly*, vol. 46, pp. 121–124.

von Plato, J. (2001) A proof of Gentzen's *Hauptsatz* without multicut, *Archive for Mathematical Logic*, vol. 40, pp. 9–18.

Pottinger, G. (1977) Normalization as a homomorphic image of cut-elimination, *Annals of Mathematical Logic*, vol. 12, pp. 323–357.

Prawitz, D. (1965) *Natural Deduction: A Proof-Theoretical Study*. Almqvist & Wicksell, Stockholm.

Prawitz, D. (1971) Ideas and results in proof theory, in J. Fenstad, ed, *Proceedings of the Second Scandinavian Logic Symposium*, pp. 235–308, North-Holland.

Ranta, A. (1994) *Type-Theoretical Grammar*, Oxford University Press.

Rieger, L. (1949) On the lattice theory of Brouwerian propositional logic, *Acta Facultatis Rerum Nat. Univ. Carol.*, vol. 189, pp. 1–40.

Schroeder-Heister, P. (1984) A natural extension of natural deduction, *The Journal of Symbolic Logic*, vol. 49, pp. 1284–1300.

Schütte, K. (1950) Schlussweisen-Kalküle der Prädikatenlogik, *Mathematische Annalen*, vol. 122, pp. 47–65.

Schütte, K. (1956) Ein System des verknüpfenden Schliessens, *Archiv für mathematische Logik und Grundlagenforschung*, vol. 2, pp. 55–67.

Sonobe, O. (1975) A Gentzen-type formulation of some intermediate propositional logics, *Journal of the Tsuda College*, vol. 7, pp. 7–13.

Stålmarck, G. (1991) Normalization theorems for full first order classical natural deduction, *The Journal of Symbolic Logic*, vol. 56, pp. 129–149.

Takeuti, G. (1987) *Proof Theory*, 2nd ed, North-Holland, Amsterdam.

Tennant, N. (1978) *Natural Logic*, Edinburgh University Press.

Troelstra, A. and D. van Dalen (1988) *Constructivism in Mathematics*, 2 vols., North-Holland, Amsterdam.

Troelstra, A. and H. Schwichtenberg (1996) *Basic Proof Theory*, Cambridge University Press.

Uesu, T. (1984) An axiomatization of the apartness fragment of the theory DLO$^+$ of dense linear order, in *Logic Colloquium '84*, pp. 453–475 (Lecture Notes in Mathematics, vol. 1104), Springer, Berlin.

Ungar, A. (1992) *Normalization, Cut Elimination, and the Theory of Proofs*, CSLI Lecture Notes No. 28.

Vorob'ev, N. (1970) A new algorithm for derivability in the constructive propositional calculus, *American Mathematical Society Translations*, vol. 94, pp. 37–71.

Wadler, P. (1985) How to replace failure by a list of successes, *Proceedings of Conference on Functional Programming Languages and Computer Architecture*, pp. 113–128 (Lecture Notes in Computer Science, vol. 201), Springer, Berlin.

Zucker, J. (1974) Cut-elimination and normalization, *Annals of Mathematical Logic*, vol. 7, pp. 1–112.

Author Index

Subject Index

Pages containing definitions of concepts are sometimes indicated by *italics*.

I

impredicative, 28
incompleteness, *xi*, 22
inconsistent, 20, *137*
indirect proof, 12, 24, 26, 48, 66, 115, 158,
    160, 204
infix, 3, 221
intermediate logic, 22, 156–164
intuitionism, 25, 46
intuitionistic logic, *xiv*, 12, 19, 21, 25–46, 66,
    67, 76, 78, 156, 160, 164, 198, 206, 207
inversion lemma, 32, 46, 71, 75, 109,
    115, 121
inversion principle, *xiv*, *xv*, *xvi*, 6, 24, 65,
    86, 166
invertible rules, 19, 33, 34, 49, 60, 90,
    96, 216

K

König's lemma, 82

L

label, 166
lattice theory, 147–150, 241
leaf 127
linear logic, 125
lists, 14
logical axiom, 16
logical languages, *xiii*, 1–3
logical systems, 1–5
loop, 43, 121

M

major premiss, 6, 171
meaning, *xiv*, 5, 23, 26, 64, 166,
    212, 228
midsequent, 80
minimal logic, 12, 198, 206, 207
minor premiss, 8, 171
mix rule, *xvi*, 212
models, 22, 42, 43, 156
modus ponens, 9, 41, 184, 198,
    208, 229
multicut, *xvi*, 88, 186, 212
multiple conclusion, 47, 213
multiset, 15, 18
multiset reduct, 99, 100, 177,
    191, 206

N

negation, 2, 29, 52, 221, 222
noncanonical, 13, 228
nonconstructive, 82
nonlogical axiom, 133
normal form, *xvii*, 9, *175*, 183, 192, 197,
    198, 204, 205, 214
normal thread, 196
normalization, 9, 189, 195, 198–201,
    212, 214
    strong, 201, 208, 213

O

open assumption, 9, 47
open case, 47
operational, 5, 47, 48, 87, 115
order
    linear, 145, 164
    nondegenerate, 147
    partial, 146
ordinal proof theory, *xi*

P

Peirce's law, 121, 208
permutation, 34, 98, 100, 102, 165,
    183, 208
permutation conversion, 192–194
permutation cut, 185
PESCA, 179, 235–243
predicative, 27, 46
prenex form, 78
program correctness, 232, 234
program specification, *xiii*, 232
programming languages, *xiii*, 1, 3, 232
proof editor, *xiii*, *xv*, 179, 233, 235
proof search, *xv*, 15, 16, 17, 19, 20, 43, 50,
    60, 121, 211, 236, 239
proof term, 13, 201, 208, 225
proof-object, 13, 208, 225–234
proper assumption, 201
proposition, 2, 3, 13
propositional function, 220
propositions-as-sets, 13, 28, 207, 225

Q

quantifiers, 64–67, 86, 121, 191, 223, 226,
    227, 228, 230, 231
    second-order, 28

Index of Logical Systems

Nomenclature

The letter **G** stands for **sequent systems** (Gentzen systems):

G0-systems have **independent contexts** in two-premiss rules.
G3-systems have **shared contexts** in two-premiss rules.
p stands for the **propositional** part of a system.
i stands for an **intuitionistic** system.
c stands for a **classical** system.
m stands for a **multisuccedent** system.
G4ip is obtained from **G3ip** by changing the left implication rules.
Extension of a system by *Rule* is indicated through writing +*Rule*.
A superscript * denotes an arbitrary extension.

The letter **N** indicates **natural deduction** style:

GN is a sequent system in natural deduction style.
GM is the corresponding multisuccedent system.
NG is natural deduction in sequent style.
MG is the corresponding multisuccedent system.

Tables of Rules

www.ingramcontent.com/pod-product-compliance
Ingram Content Group UK Ltd.
Pitfield, Milton Keynes, MK11 3LW, UK
UKHW040704180125
453697UK00010B/395